Epigenetics in Ecology and Evolution

SCIENCES

Ecosystems and Environment
Field Directors – Françoise Gaill and Dominique Joly

Epigenetics, Subject Head – Dominique Joly

Epigenetics in Ecology and Evolution

Coordinated by
Christoph Grunau
Stéphane Maury

iSTE

WILEY

First published 2025 in Great Britain and the United States by ISTE Ltd and John Wiley & Sons, Inc.

ISTE Ltd
27-37 St George's Road
London SW19 4EU
UK

www.iste.co.uk

John Wiley & Sons, Inc.
111 River Street
Hoboken, NJ 07030
USA

www.wiley.com

Library of Congress Control Number: 2025931324

British Library Cataloguing-in-Publication Data
A CIP record for this book is available from the British Library
ISBN 978-1-78945-216-7

ERC code:
LS2 Integrative Biology: from Genes and Genomes to Systems
LS2_3 Epigenetics
LS8 Environmental Biology, Ecology and Evolution
LS8_6 Evolutionary ecology
LS8_7 Evolutionary genetics

Contents

Preface

Christoph GRUNAU[1] and Stéphane MAURY[2]
[1] *IHPE, Université de Perpignan Via Domitia, France*
[2] *P2E, Université d'Orléans, France*

Genetics and its mysterious little sister, epigenetics, have deep roots in the history of biology. After embryology's revival in the modern era, in which preformationist visions of development and epigenesis clashed, the latter triumphed in the 19th century, with the heredity of acquired traits being accepted. In the latter part of the 19th century, the notion of stable inheritance of characters through generations emerged, a notion inherited by the "classic" concept of the gene in the early 20th century. Genetics now focuses on the transmission of traits between generations, putting aside generation itself (in the sense of the genesis of a living being). Epigenetics has grown in this "all-genetic" context. First defined as the study of the links between genotype and phenotype by Conrad Waddington in the 1940s, epigenetics was redefined a few years later by David L. Nanney, during the emergence of molecular biology, as the set of factors that modulate gene function in a heritable way without changing the DNA sequence.

Today, the concept of epigenetics is important in a wide variety of fields ranging from molecular biology to disease diagnosis and treatment, as well as covering the improvement of animal and plant species for agricultural use, as well as the relationships between species, biodiversity and evolution. Business models based on the use of epigenetic markers and concepts are beginning to emerge, raising questions of public awareness and perception. Epigenetics remains a complex field, which is sometimes still enigmatic or misunderstood in certain situations, and not only by the general public.

This book attempts to provide answers to as wide an audience as possible, enabling readers to build or consolidate a critical and enlightened view of the complex relationships between genetics and epigenetics.

Lastly, the editors and authors would like to express their gratitude to the organizations that generously supported the writing of this handbook. Their contribution was essential in making this work possible: the CNRS, the GDR3E (Epigenetics, Ecology and Evolution), the Labex CeMEB (ANR-10-LABX-04-01) and the public library of Collioure (France).

February 2025

1

A Brief Conceptual History of Epigenetics, and More Besides

Arnaud POCHEVILLE
CRBE, Université de Toulouse, France

1.1. Introduction

Epigenetics concerns the expression of genetic information – or, more generally, non-genetic (biological) information altogether.

A polymorphous concept with multiple origins, sometimes used very liberally or absent where it is expected, epigenetics is deeply rooted in understanding "generation", in other words, the way in which living beings are formed.

There are two main groups of answers to the question of generation: either the form is already there, pre-existing its manifestation in the organized being; or it is created as it grows.

The history of epigenetics is the history of a dialectic between these two groups, always in tension, always on the verge of resolution, always in transformation.

Epigenetics in Ecology and Evolution,
coordinated by Christoph GRUNAU and Stéphane MAURY.

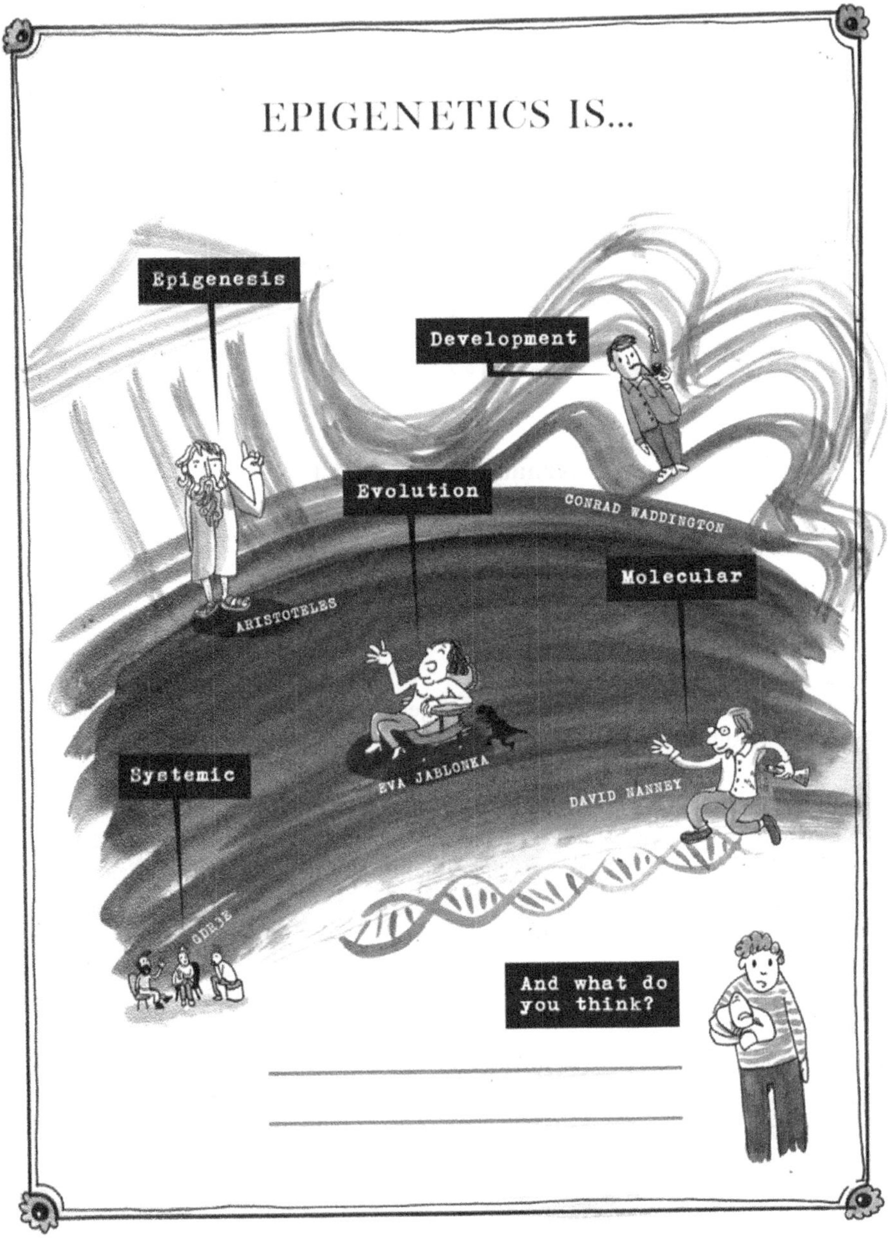

Figure 1.1. *Epigenetics has many faces. Its definition depends on the context in which we find ourselves, and it may yet evolve. Follow us in Chapter 1 to find out why. Illustration by Annette Köhn*

1.2. The birth of the term: from epigenesis to pangenesis

The term "epigenetics" appeared in the 17th century as a noun, "epigenesis", in a treatise by the English physician and anatomist William Harvey (1651, *Exercitatio* 45). In it, Harvey outlined his work on the causes of animal generation, responding in particular to Aristotle (330–322 BCE) (Aristotle 1887) and his own master, Fabricius (1621). Harvey distinguishes two possible modes of generation: either the material of the future animal is already present and receives its form like wax imprinted by a seal, all parts being formed simultaneously, which he calls metamorphosis; or the material is added as the form is born, like a potter's pot, the parts of the animal coming into being from one another, which he calls epigenesis. For Harvey, the generation of the chick from the egg is clearly the result of epigenesis, not metamorphosis. It is an idea that is tempting to see already discussed, albeit ambiguously, by Aristotle; however, it is important to bear in mind the differences in frameworks and debates between eras (Lennox 2006; Goy 2018; Bolduc 2021).

In the 18th century, another vision of generation gained ground: mechanistic preformationism (Bowler 2001, pp. 26–28). Preformationism assumes that form pre-exists generation. From an empirical point of view, the idea is justified by the observation (thanks to the use of the microscope) of microscopic organs and organisms, which suggest the possible existence in the living world of structures of infinitesimal size and therefore undetectable to our senses (Malpighi 1666; Malebranche 1674, pp. 41–42; see Gasking 1967, pp. 45–46, quoted by Roe 1981, pp. 83–87). From a philosophical point of view, preformation allows the formation of a living being to be reduced to simple growth, a mechanism set in motion at conception, at a time when the mechanistic vision is gaining traction (Bowler 2001, pp. 26–28; Wolfe 2016, 2017). The most extreme form of this vision (known as pre-existing germs) assumes that all living beings pre-exist their generation, nested like nesting dolls within each other since Creation, such as aphids whose parthenogenesis and nested generations are demonstrated by a proponent of preformation (Bonnet 1745, 1762, Chapter 1, 1770, pp. 190–191). This vision resonates with a Christian point of view, which in turn justifies the comprehensibility of the world (God lays down natural laws) (Roe 1981, p. 111; Bowler 2001, pp. 29–31). Conversely, preformationists see epigenesis as synonymous with chance, chaos and indeterminacy (and hence incomprehensibility). Advocates of atheism favor epigenetic theories, as well as those of spontaneous generation, which dispense with the idea of Creation (Buffon 1749; see Roe 1981, pp. 16–18; Spallanzani 1765; d'Holbach 1770; Bowler 2001, pp. 29–36).

Mechanistic preformationism fell into disfavor at the turn of the 18th and 19th centuries (Roe 1981, p. 150; Bowler 2001, p. 32; Wolfe 2016, 2017). We would prefer an epigenesis whose mechanistic vision is supplemented by the idea of forces specific to living organisms, which organize them (e.g. Maupertuis 1745, Chapter XI–XII, 1756a, Chapter LXV–LXVII; see Ibrahim 2019; Wolff 1759, 1768; Blumenbach 1780; see Bowler 2001, pp. 40–42; Dupont 2007; Wolfe 2019; Schmitt 2021). This vision gives rise to mystical trends (Naturphilosophie: Schelling 1797; Oken 1809; Bowler 2001, pp. 41–42), as well as to so-called teleomechanist trends, which aim to naturalize the notion of finality in living organisms, embryogenesis appearing to be such a process oriented towards a finality, the formed organism (Kant 1790; Lenoir 1982; Bowler 2001, p. 42; Huneman 2006; Moreno and Mossio 2015). The vision that emerges is that of a homology between the individual and the lineage, with parents shaping their children in the same way a craftsman produces an artifact, in a process akin to copying the parent, part by part – there is therefore, said in today's terms, inheritance of acquired traits (Bowler 2001, p. 23, 38).

This is a vision carried forward by Jean-Baptiste de Lamarck (1809a, Chapter II) in particular, who produced a veritable physicalist theory of the transformation of living beings (and of their spontaneous generation). For him, living beings become organized and complex throughout life and across generations by virtue of movements of fluids, caused by the use of certain parts of the organism whose needs, according to circumstances, are felt (Lamarck 1809b, Chapter VI). The use of a part strengthens it, and the acquired form is transmitted to descendants: this is a law that Lamarck considers empirically evident – the same one that posterity has associated with the neck of the giraffe (Lamarck 1809a, Chapter VII, p. 235). This posterity, sometimes simplifying and very unfaithful, remained influential until the middle of the 20th century, particularly in France, where it resonated with nationalist impulses that were sporadically reactivated by the international context (Boesiger 1980; Limoges 1980; Sapp 1987, pp. 124–128). However, we must be wary of reducing Lamarck to transformism through the inheritance of acquired traits (Bowler 2001, p. 37; Corsi 2001): his intuitions on the self-organization of what we would today call open systems are more reminiscent of the physics of out-of-equilibrium systems (Nicolis and Prigogine 1977; Bejan and Lorente 2004), and, in biology, organizational approaches to viability (Moreno and Mossio 2015).

In the Germanic world, teleomechanism underwent influential developments, culminating in the formulation of the law of parallelism – in the course of its development, the embryo passed through the hierarchy of animal forms, from fish to reptile to mammal, with man at the top (Meckel 1821, 1828 *contra* Von Baer 1828; Russell 1916; Temkin 1950; Oppenheimer 1967; Bowler 2001, pp. 42–44);

a law which, when interpreted as the product of evolutionary history, forms the basis of recapitulation theory – the developing embryo recapitulates its evolutionary history (Haeckel 1866). Recapitulation theory, such as Lamarck's transformism, invites us to understand the transmission of traits between generations as analogous to memory (Haeckel 1876; Gould 1977; Robinson 1979; Bowler 2001, p. 52). However, teleomechanist embryologists are not so much interested in trait transmission as in how the "potentialities" of a fertilized egg develop (Lenoir 1982, Chapter 3; Bowler 2001, p. 44).

Meanwhile, heredity was emerging as a subject of study in its own right, particularly in the medical field, with the study of hereditary diseases (Lopez-Beltran 2007)[1]. In the 18th century, the notion of heredity could be prefigured by the study of noble diseases (such as gout), supposedly acquired through lifestyle (delicious food) and transmissible diseases, which resonated with the pressure put on the aristocratic family (Pomata 2003, p. 150; Müller-Wille and Rheinberger 2007, p. 11, 19). Mendelian-looking studies, such as that by Pierre Louis Moreau de Maupertuis (1756b, letter xiv, pp. 275–278) on the prevalence of polydactyly in a Berlin family, were in fact interpreted within a framework of epigenesis with the transmission of acquired traits (Bowler 2001, pp. 33–35; Terrall 2007, pp. 265–268). In the 19th century, the social tropism for the study of hereditary diseases changed (Müller-Wille and Rheinberger 2007, p. 19). Among other things, the aim was to improve hygiene: the poorer classes and their share of diseases, supposedly caused by the environment and being more or less hereditary, were perceived as the new threat to social order (Pomata 2003; Cartron 2007; Müller-Wille and Rheinberger 2007, p. 19). Other fields of study also fed into these reflections, such as animal and plant hybridization experiments (aimed at testing fixism and the possibility of establishing new species, along Linnaeus' lines), natural history and taxonomy (concerned with the constancy of traits), the maintenance and improvement of plants and animals for breeding and collection (their transport over great distances making it possible to better distinguish the intrinsic characteristics of

1. The resemblance between parents and children has been documented since Antiquity. Although it is not the cornerstone of any concept of life, it figures in discussions of generation as a fact to be explained. For some primary sources, see Hippocrates (ca. 440–360 BCE) (Hippocrates 1842), Herodotus (ca. 425 BCE) (Herodotus 1850, par. 180), Aristotle (ca. 330–332 BCE) (Aristotle 1850, par. 180). C.) (Aristotle 1887, Book i, Chapter xi, § 1–6, p. 23), Plato (ca. 375 BCE) (Plato 1933, book v), Pliny the Elder (ca. 77 CE) (Pliny the Elder 1848, Book vii, Chapter x), Plutarch (ca. 72–126 CE) (Plutarch 1844, § 21), Galen (ca. 165–175 CE) (Galen 1822, p. 603–604, 1992, p. 235), and, closer to us (Fernel and Plancy 1656; Paré 1971, XIII, p. 45; Montaigne 2001, XXXVII, p. 1188). For some secondary sources, see Lesky (1950), cited by Preus (1977); (Boylan 1984, 1986; Bonnard 2015; Giorgianni and Provenza 2015; Loviconi 2019).

varieties from the effects of the environment), and anthropology, which studies human diversity, in particular skin color (Müller-Wille and Rheinberger 2007, pp. 19–23). In the 1830s, French physiologists came up with a noun corresponding to the adjective “hereditary”, which had been used to describe certain diseases since Antiquity – a noun that until then had only been used in the legal sphere, and which has now taken root in the study of living organisms: heredity (Lopez-Beltran 2007). With the study of heredity, the point of interest is no longer the material mechanism of reproduction (it would be a matter of understanding how a fellow human being is conceived), but the “transmission” of properties between parents and descendants (e.g. knowing the laws of transmission of hereditary diseases would suffice) (Sapp 1987, Chapter 2; Bowler 2001, p. 129; Müller-Wille and Rheinberger 2007).

At the same time as Darwin was proposing his theory of pangenesis, Gregor Mendel was conducting his now-famous hybridization experiments on peas (Mendel 1866, 1901). Mendel’s work was part of a long line of hybridization experiments aimed at obtaining new horticultural varieties or, with a more theoretical ambition, testing the possibility of creating new species, within a framework inherited from Linné where genera were fixed (Linné 1760; Koelreuter 1761; Gärtner 1849; Olby 1979; Bowler 2001, p. 97). It is within this framework that Mendel distinguishes between constant hybrids (in which traits do not segregate, and which can give rise to new species), and variable hybrids (in which traits segregate, according to precise ratios, with some traits being dominant and others recessive) (Callender 1988). Such phenomena had been observed before, but no one had undertaken a systematic study of trait genealogy across generations (Bowler 2001, p. 99). Mendel never mentions particulate inheritance; his categories, far from being materialistic, instead evoke the philosophy of Aristotle in which he was trained: the notion of trait is reminiscent of essence; the dominant/recessive notions are reminiscent of actual/potential (Kalmus 1983; quoted by Bowler 2001, p. 104). The fact that Mendel was not interested – contrary to what early geneticists claimed – in establishing universal laws of heredity may explain the obscure trajectory of his work (Olby 1979, 1985; Sandler and Sandler 1985; Callender 1988). Indeed, his name seems to have been used by researchers as much as a rhetorical argument (to provide a historical anchor in a priority dispute) as a scientific one (Bowler 2001, pp. 113–116).

Box 1.1. *Mendel, a historical digression*

The term *heredity* was borrowed by Charles Darwin and his followers to name the transmission of individual variations in his theory of evolution by natural selection (Darwin 1837, 1838; Spencer 1864, Chapter viii; Keller 2010, p. 21). This theory is based on the notion of hereditary variation, which can either be blind (not directed towards adaptation) or acquired (as in Lamarck). Darwin backs up his theory of evolution with a theory of reproduction, but the latter remains dependent on its century – not to say the previous century: it involves “gemmules“, germs of infinitesimal size that come from all parts of the parent organism and reconstitute these parts in the child organism (Darwin 1865, 1868, Chapter XXVII, p. 375, n. 29; Olby 1963; Bowler 2001, pp. 54–64). Darwin called this the "theory of pangenesis": from pan, "all", in Greek. However, it was not the near obsolescence of the pangenesis hypothesis that stood in the way of Darwin's contemporaries accepting the principle of natural selection, but rather their attachment – even by his most enthusiastic supporters – to a "neo-Lamarckian" vision of evolution, in which selection by blind variation is insufficient to explain living things (Bowler 1983, pp. 27–28; Bowler 2001, pp. 54–64).

1.3. From pangene to classical gene

A fervent reader of Darwin, and nonetheless a good critic, was his cousin Francis Galton (Galton 1871a, 1871b; Bowler 2001, p. 64). From the outset, Galton was wary of pangenesis; he developed a theory of heredity (known as ancestral heredity), which defines the transmission of a particulate material that remains stable through generations: the ancestors of all previous generations participate in the material inherited by an individual, following a geometric progression that ensures that the most recent ancestors contribute the most (Galton 1886, pp. 260–261, 1889, p. 148, 1897 *contra* Johannsen 1911, p. 138; see Sapp 1987, p. 39; Gayon 1992, p. 138; Bulmer 1998; Bowler 2001, p. 67). This stability of hereditary material represents "hard inheritance", as opposed to "soft inheritance", which allows the transmission of acquired traits (Mayr and Provine 1998, p. xi; Bowler 2001, p. 68). Hard heredity introduces a decoupling between the transmission of traits and the generation of individuals. From a methodological point of view, this decoupling enables Galton to forego physiology and approach the question of heredity at a population level through statistics (Bowler 2001, p. 69). Galton's notion of hard heredity resonated with his social concerns: Galton founded eugenics, a utopian movement which, concerned about the effects of civilization on human beings, aimed to improve them, particularly (but not only) through the selection of their reproducers (Galton 1883, p. 4, 24–25, n. 1; Gayon 1992, pp. 168–180; Bowler 2001, pp. 66, 70–72, Chapter 8; Gayon 2004). This being said, some advocates of soft heredity defended similar programs, while others, supporters of hard heredity, opposed them (Sapp 1987, Chapter 6; Bowler 2001, Chapter 8).

The development of a notion of hard heredity was also nurtured by developments in cytology. Cell theory, which posits the cell as the basic unit of life, developed in the 1830s, nurtured by a teleomechanist view (Lenoir 1982, Chapter 3, Chapter 5), with many biologists accepting it by the 1850s in a crystallized version according to which the cell cannot be formed from non-cellular material – notably in contrast to the pangenesis hypothesis (Bowler 2001, p. 74, 85). Moreover, it seems clear since the discovery of the mammalian ovum (Von Baer 1827) that the embryo grows from a single cell, raising the question of cellular differentiation (Bowler 2001, pp. 78–79). Improved microscopy and new dyeing techniques began to bring attention to the cell nucleus around the 1870s, with work on phenomena such as the fusion of gamete nuclei during fertilization (Hertwig 1875), chromosome migration during mitosis (Flemming 1879; Paweletz 2001) and the reduction in number of chromosomes during meiosis (Van Beneden 1883), which are suggestive of copying processes at work in reproduction (Bowler 2001, pp. 85–86). Embryologists' attention gradually turned away from speculative edifices such as recapitulation theory, and instead towards experimentation and the search for physiological mechanisms (Bowler 2001, p. 77).

It was in this context that August Weismann (1891, p. 190) became a key supporter of hard heredity, in a research program aimed at synthesizing Darwinian theory and the cytology of his day. After initially accepting the inheritance of acquired characteristics (Weismann 1882), Weismann rejected it on the basis of both theoretical and philosophical arguments (the theory of natural selection is sufficient to explain adaptation), as well as cytological arguments (no mechanism for the transmission of acquired traits is known; moreover, it seems impossible to translate a trait into a hereditary determinant) (Weismann 1893b, Chapter XIII, 1893a, pp. 107–108, 1904, p. 63). Weismann assumes that in multicellular organisms, only the germ line is immortal and involved in the transmission of traits, and that the somatic line cannot alter it in a directed way (Bowler 2001, p. 88). Traits are assumed to be predetermined by chromosomes, and differentiation occurs through an asymmetrical distribution of chromosomes between daughter cells (ibid., p. 75). This position was notoriously characterized as nuclear preformationism, reactivating a cleavage with the proponents of epigenesis (Hertwig 1894; Wilson 1896, pp. 327–331) – it should be remembered, however, that in preformation, the form pre-exists and only grows; in Weismann, it is the potential of a form that pre-exists (Bowler 2001, pp. 75–76, 81). Defending his position with a vigor that tended towards dogmatism, Weismann exacerbated tensions with advocates of soft heredity, who called him a "neo-Darwinian", so as to distinguish his position from that of Darwin himself (Romanes 1888; Bowler 2001, p. 89). The late Weismann was greatly influenced by his contemporary Hugo De Vries (Bowler 2001, p. 90). From Darwin's pangenesis, De Vries retains the idea of particles (which he calls "pangenes") responsible for traits; but for him, these remain in the cell (they are supposed to group around chromosomes and leave the nucleus for trait expression). All cells possess them in their entirety, even in latent form, and their number is perhaps limited, as traits develop through their interactions: heredity can thus take place without mixing, and new traits, even new species, appear through mutation (Vries 1889, sec. iv; Bowler 2001, pp. 90–91).

The year 1900 saw the rediscovery of Mendel's laws – or rather, the reinterpretation of Mendel's work in light of the gradually emerging framework of Mendelian genetics (Correns 1900; Tschermak 1900; Vries 1900; see Bowler 2001, pp. 113–116). Embryologists such as Thomas Morgan, frustrated by the failure of their experimental techniques to explain trait formation, turned to hybridization experiments that only studied transmission (Bowler 2001, p. 78, 82). These early Mendelians produced a vision of transmissible variation that was discrete, underpinning a saltatory, non-selective vision of evolution in direct opposition to that of the neo-Darwinians (notably the so-called biometricians), for whom evolution is gradual and based on the natural selection of continuous variation, fiercely opposing Mendelism (Gayon 1992, Chapter VII–VIII). This was also the time when, in tandem with developments in cytology, the chromosomal theory of heredity developed, with Mendelian overtones: chromosomes would be the medium for determining traits (Sutton 1903; Boveri 1904). Many people resisted this theory, including leading geneticists such as Morgan (who was converted to it, in a particularly fruitful Drosophila genetic mapping program), as well as William Bateson and Wilhelm Johannsen, for whom the hereditary determinants of traits needed to be conceived holistically, in terms of energy levels or waves in the cell (Bateson and Mendel 1902, pp. 2–3; Morgan 1910, 1913; Morgan et al. 1915; Johannsen 1923; see Coleman 1970; Allen 1974; Darden 1977; Roll-Hansen 1978; Allen 1978, cited by Morange 2020, Chapter 26; Bowler 2001, pp. 120–121, 129–131; Roll-Hansen 2014). It was, however, Bateson who founded the first Mendelian school of genetics in Great Britain, and Johannsen, who is credited with establishing the concepts of gene (a contraction of the pangene, the determinant of hereditary traits), genotype (the set of genes) and phenotype (a measurable trait) (Johannsen 1909, p. 113, 1911, pp. 132–134). These conceptual distinctions were part of a conquering movement to base the study of heredity and evolution on experimentation, and to exclude approaches that were deemed to be more descriptive or speculative, such as morphological or biometric studies (Sapp 1987, pp. 21–22, 36–41). In so doing, Johannsen was careful not to forge hypotheses concerning the physical support of hereditary material: for him, genetics did not study genotypes, but genotypic differences (Johannsen 1911, p. 133; see Wanscher 1975; Roll-Hansen 2014). Even though the chromosomal and Mendelian theory of heredity was fairly generally accepted in the community around 1915, particularly across the Atlantic (Morgan et al. 1915; Sapp 1987, pp. xiii, 46–48, 221), where it founded what would later be called classical genetics (Bowler 2001,

p. 128), the Mendelian gene remained a mainly abstract entity – not a speculation, but a computational tool used to describe the results of hybridization experiments (East 1912, p. 634; Morgan 1935, p. 315; see Stent 1970, p. 910; Bowler 2001, p. 125; Rheinberger et al. 2015). Thus, crystallized, classical genetics put aside the question of trait formation to focus solely on their transmission (Sapp 1987, Chapter 2; Bowler 2001, p. 129).

The American school of genetics founded by Morgan rapidly gained in power: it was an opportunity to secure posts and funding, with the promise of applications in both agriculture and public health, with the eugenics program (Sapp 1987, pp. 46–47; Bowler 2001, pp. 132–137). In evolutionary biology, the development of population genetics reconciled Mendelism with neo-Darwinism: continuous transmissible variation – which was dear to biometricians – could be supported by the interaction of multiple Mendelian genes, each causing small differences (Fisher 1918; Fisher 1930; Wright 1931; Haldane 1932; see also Yule 1902). However, classical genetics did not totally reign supreme. The chromosomal theory hardly shuts down the paradox, in multicellular organisms, of cellular differentiation taking place under the control of genes that are assumed to be identical, and active, in all cells (Sapp 1987, p. 17). Cases of non-Mendelian heredity have been reported, particularly in plants (Baur 1908; Correns 1909a, 1909b; Sapp 1987, p. 26, 73–74). Ultimately, as they themselves admit, geneticists have failed in their attempts to produce new species in the laboratory (Sapp 1987, p. 17, 21–22).

The Germans and the French, in particular, resisted the objects of study of classical genetics: for them, the topic of interest related to trait formation; Mendelians studied the transmission of trivial traits, with no biological relevance (Sapp 1987, Chapter 3, Chapter 5; Bowler 2001, pp. 143–152). Embryologists notably saw the cytoplasm of the egg cell as the essential element determining the properties of the organism, and encouraged research into cytoplasmic heredity in open opposition to the nuclear, then chromosomal, theory of heredity (*contra* Morgan et al. 1915; Morgan 1926; see Sapp 1987, p. xiii, 3, 27, 56). Research on cellular organelles experienced renewed interest (Sapp 1987, p. 25); the cell appears, beyond the resolution of the microscope, as an ordered system of particles (whose size can be supramolecular, and called "colloids") where chemical reactions of incalculable diversity take place (Wilson 1923; Sapp 1987, p. 18, 213–214), and which can form a "field" of metabolic gradients (Sapp 1987, p. 144); the cytoplasm is supposed to harbor "plasmagenes" (*plasmatischen Gene*), self-replicating microscopic elements that are differentially distributed between daughter cells (Winkler 1924), or to constitute a reactive, integrated, transmissible system that conditions trait expression, known as the "plasmon" (Wettstein 1926), which is the plasma complement of the "genome" (Winkler 1920, p. 165), in other words, the set of chromosomes (Sapp 1987, pp. 72–86, 2003, pp. 174–175).

In Europe, many geneticists are actively involved in cytoplasmic heredity research (Sapp 1987, p. 14, 54–55, 69, 73, Chapter 4, Chapter 5). Across the Atlantic, the differences between schools are more powerful: when a biologist of Jewish descent such as Victor Jollos, known for his discovery of *Dauermodifikationen* (environmentally induced variations transmissible over hundreds of vegetative generations in paramecia (Jollos 1920, 1921; Hämmerling 1929)) fled Germany for the US in 1934, he found it impossible to valorize his work and find a job again (Sapp 1987, pp. 62–65; Bowler 2001, pp. 148–149).

1.4. Classical epigenetics

It is in this context of resistance to classical genetics (as well as to the neo-Darwinism of this period, which reduces evolution, in its most extreme form, to dynamics of gene frequencies (e.g. Dobzhansky 1937, p. 11)) that Conrad Waddington coined a word to designate this study that was put aside by early geneticists: recalling the notion of epigenesis – perhaps, it must be said, rather vaguely (see Koltzoff 1935; Waddington 1956a; Wightman 1956) – he called the study of developmental processes that link genotype and phenotype "epigenetics" (Waddington 1942); see also Waddington 1939, pp. 154–156; Waddington 1957, p. ix, 151; cited by Haig 2004; Nicoglou and Merlin 2017). He calls this complex of developmental processes *epigenotype*. The reference to process is not random: Waddington is a reader of Alfred North Whitehead, the English mathematician and philosopher who built a philosophy of process (Whitehead 1925, 1929; Boisseau 2023). Waddington also suggests visualizing the development of an organism through a metaphor: that of a ball rolling down a landscape of hills and valleys. The genotype determines the landscape, which Waddington would later call the epigenetic landscape, in other words, the developmental potentials that are or are not actualized in the phenotype during development by the organism's interaction with its environment (Waddington 1939, 1940, 1957). Here again, these categories of potential and actual are the tacit and classic Western heritage of Aristotelian categories (ca. 350 BCE), see (Aristotle 1879, 1986; Cohen and Reeve 2021). Waddington created these concepts at a time when the notion of a gene remained abstract: it was a difference causing differences, the material support for this causality not being clearly identified at the time. As a result, Waddington's concepts bear the mark of this abstraction, making them highly flexible in terms of their material interpretation.

While the notion of the epigenetic landscape met with some success, particularly in the field of morphogenesis (Berry and Searle 1963; Herring 1993; cited by Haig 2004), Waddington's concepts of epigenetics and epigenotype did not catch on: apart from Waddington himself, hardly anyone referred to them, and these words were later independently reinvented to name different concepts. Waddington's real success was posthumous: from the 1990s and especially 2010, a new generation of evolutionary thinkers sought to extend evolutionary theory to reduce the explanatory share allocated to gene dynamics alone (Jablonka and Lamb 2002; Slack 2002; West-Eberhard 2003; Jablonka and Lamb 2005, pp. 63–64; Pigliucci and Müller 2010; see Morange 2009; Gilbert 2012; Haig 2012; Jablonka and Lamm 2012; Morange 2020, pp. 328–329). This new generation is re-appropriating Waddington's concepts (in what may look like a quest for historical legitimacy), and references to his work are multiplying.

1.5. Molecular epigenetics

Around the 1940s, biology became molecular, with the development of new techniques (crystallography, electrophoresis, etc.) and new models (fungi, bacteria, viruses, algae, protozoa) (Sapp 1987, pp. 88–89; Bowler 2001, p. 175; Rheinberger et al. 2015). Geneticists began to focus on the gene control of physiology and turned to microorganisms, which freed them from the complexities of cellular differentiation (Sapp 1987, p. 88). It was at this point that work on the *Neurospora* fungus suggested that each step of a metabolic pathway is under the control of a single gene (Beadle and Tatum 1941; Sapp 1987, p. 133), a result often summarized (somewhat anachronistically) as the "one gene – one enzyme" hypothesis (Morange 2020, Chapter 2). Work on bacterial transformation and bacteriophages suggests that the material support for genes is DNA (Avery et al. 1944; Hershey and Chase 1952; see Bowler 2001, p. 175; Morange 2020, Chapter 4). This vision is not self-evident: for many, DNA is not complex enough to carry genes; proteins are better candidates; on the other hand, the generality of results obtained on viruses and bacteria is questionable (Bowler 2001, pp. 174–175; Morange 2020, pp. 19–20, 35–39). However, when James Watson and Francis Crick proposed that the structure of DNA is a double helix, on the basis of X-ray diffraction patterns obtained by Rosalind Franklin and Raymond Wilkins, their model made it possible to explain both how genetic information can be encoded in DNA (via base sequence) and how this information can be replicated (via base complementarity) (Franklin and Gosling 1953; Watson and Crick 1953; Wilkins et al. 1953; see Bowler 2001, p. 176; Morange 2020, Chapter 11). The notion of a genetic code matching DNA and protein sequences quickly took root, and biologists tried to approach it from both the information-theoretic side (as a code to be broken) and the biochemical side – the latter ultimately being successful (Sarkar 1996)[2].

The domestication of microorganisms and the development of hybridization methods were accompanied by a broadening notion of heredity: the heredity of microorganisms became a model of the cellular heredity expressed during differentiation in multicellular organisms (viruses were also seen as vectors of heredity) (Sapp 1987, p. 89). However, this domestication of microorganisms ushered the discovery of many apparently non-Mendelian phenomena: for proponents of cytoplasmic heredity, genetics was on the eve of a major revolution that could mark the end of Mendelian supremacy and place the cytoplasm at the center of cellular functioning (Ephrussi 1953; see Waddington 1953; Sapp 1987, pp. 89, 166), thereby renewing the question of the evolutionary importance of the cytoplasm (Waddington 1940, p. 53; quoted by Sapp 1987, p. 101). A series of works are interpreted as marking the independence (if not pre-eminence) of non-nuclear cellular phenomena from the nucleus, involving, for example, the cell cortex or basal bodies (Just 1932, 1939; Lwoff 1990). In addition to this scientific turmoil, there was a political aspect: in the context of the Cold War, supporters of Lysenkoism (a scientific ideology that was then dominant in the USSR, defending a holistic vision of soft heredity) seized on work on cytoplasmic heredity to discredit Morgan's work on chromosomal Mendelism, generating a controversy discussed daily in Western newspapers and directly interfering with the development of research programs (Sapp 1987, Chapter 6, p. 172, 183). Various models of cytoplasmic heredity were proposed, involving plasmagenes (Lwoff 1949; Sonneborn and Beale 1949; Lwoff 1950; Morange 2020, p. 145) or metabolic feedback loops generating environmentally inducible and stable states (Wright 1945; Delbrück 1949; see Novick and Weiner 1957; Jablonka and Lamb 1995, pp. 82–83; Morange 2020, p. 327). Subsequent research was established around the transmission of supramolecular cellular structures, such as cortical structures in paramecia (Beisson and Sonneborn 1965; Nanney 1968), and cytoplasmic genes, now understood as nucleic acids (Sager 1972).

---

2. Subsequently, half a century of molecular biology would shatter the dream of a simple correspondence between Mendelian genes and determined genomic regions: phenomena such as non-coding DNA, overlapping reading frames, alternative splicing of transcripts, protein splicing, to name but a few, would lead us to instead consider the gene as a materially fragmented object, "a union of genomic sequences encoding a coherent set of potentially overlapping functional products" (Gerstein et al. 2007, p. 677; see Rheinberger et al. 2015).

At the height of research into cytoplasmic heredity, the cytoplasm was presented, in opposition to the nucleus, as a highly responsive interface that could explain cellular differentiation. However, this "geographical" opposition was soon to be dissolved by the conceptual changes introduced by molecular biology (Sapp 1987, pp. 192–200). By the end of the 1950s, it had become clear that genes could be regulated: nuclear phenomena could therefore explain differentiation (ibid., p. 193). This molecularization of genetics gave rise to the a concept of epigenetics, proposed by Nanney (1958a).

Initially, a supporter of cytoplasmic heredity, Nanney took on the vision of cytoplasmic heredity as the only possible explanation of cellular differentiation (Sapp 1987, p. 199). For Nanney, the whole question of heredity, posed in "geographical" terms (nucleus/cytoplasm, soma/germen), is badly put forward, both conceptually and empirically (Nanney 1957, p. 143; quoted by Sapp 1987, p. 200). A "totalitarian" vision of genetic mechanisms, in which genes are the only relevant determinants of traits, must be instead contrasted with a more interactive ("democratic") vision, in which biological organization is perpetuated as a stationary state by the interaction of its constituents (Nanney 1957; quoted by Haig 2004). At a conference on extrachromosomal heredity organized by Boris Ephrussi at the Centre de Génétique Moléculaire that he founded in Gif-sur-Yvette (Sapp 1987, Chapter 5), Nanney proposed that, in addition to a "library of specificities" produced by "genetic control systems", there are "auxiliary mechanisms", determining which "specificities" will be expressed in each cell, which he calls "epigenetic systems" (Nanney 1958a, p. 712). Although he refers to Waddington (1956b) after the fact, it seems that Nanney coined the term *epigenetics* independently, having first thought of calling these mechanisms "paragenetic" (Ephrussi 1958, p. 46; Haig 2004). Inflecting his concept, Nanney then proposed to see epigenetic systems as signal interpretation systems (that is, interpreting genetic signals), whose causal ability is constrained by the information contained in the genetic library, and which, he admitted, probably do not lead to irreversible changes (Nanney 1958b; quoted by Haig 2004). The terminology was defended by Ephrussi (1958, p. 47), who, from the perspective of a taxonomy of cytoplasmic heredity, took Nanney's criticism of a "geographical" taxonomy to heart. Ephrussi thus distinguishes genetic mechanisms (based on the "transmission of particles carrying their own structural information") from epigenetic mechanisms ("involving functional states of the nucleus"). In a commentary, Joshua Lederberg (1958, p. 384) embraces the concept, but not the term, which, he notes, is already used in a very different sense by Waddington (1952). Lederberg (1958, p. 385) prefers to distinguish between nucleic information (dependent on the sequence of nucleotides in a nucleic acid), epinucleic information (dependent on other aspects of the nucleic acid such as the presence of polyamines or polypeptides) and extranucleic information (dependent on molecules and reaction cycles that are not directly connected to the nucleic acid). Ironically, our current concepts are closer to Lederberg, and our terminology is from Nanney (Haig 2004).

The dichotomy between genetics and epigenetics has taken root: by the mid-1960s, it seemed sufficiently well known that it no longer required a definition in the literature (Cahn and Cahn 1966; Markert 1968; cited by Haig 2004). It has applications in a variety of fields, such as the biology of cancer (proposed as the result of an epigenetic mutation) or research into isozymes, enzymes with different sequences that catalyze the same reactions (Haig 2004). The term *epigenotype* was reinvented, this time to designate the set of self-replicating regulatory mechanisms that characterize the various cell lineages – under Waddington's approving view, who noted that he had already coined the term, with a different meaning (Abercrombie 1967; cited by Haig 2012). However, the early successes of epigenetic terminology were modest and, with the exception of isozymes, the association of what we would now call epigenetic research with the term itself remains anecdotal (see Morange 2002).

1.6. Epigenetics without knowing it, or Mr. Jourdain's epigenetics

Biology is first and foremost involved in epigenetics without knowing it, "à la Monsieur Jourdain"[3], in a disciplinary permeability context that is reminiscent of the intricacy of the mechanisms studied.

Barbara McClintock can be considered one of the first molecular epigeneticists before the word. Known today for her discovery of transposons, she conceived of them as elements controlling gene expression, in a highly organicist vision of biological functioning (McClintock 1948, 1951). However, two decades elapsed between her first publications and the genetic community's acceptance of her results – her results, but not so much her vision, organicism being cast aside (Keller 1983, Chapter 8, p. 193). Another heavyweight of genomic organicism, Richard Goldschmidt (1916, 1954), was similarly marginalized (Allen 1974; Keller 1983, p. 70, 98). While transposition can be seen today as a genetic phenomenon, this has not always been the case: Michel Morange (2002) notes that the dedicatee of the book *Epigenetic mechanisms of gene regulation* is McClintock herself (Russo et al. 1996). In a reverse move, Gary Felsenfeld (2014) counts the work of geneticist Hans

3. M. Jourdain, in Molière's *Le Bourgeois gentilhomme*, is delighted to discover, during a philosophy lesson, that he has been writing prose for 40 years without knowing it (1670, Act II, Scene IV).

Muller (1930) on chromosomal rearrangements caused by X-ray irradiation as precursors of epigenetics (see Carlson 1971).

In France – where, as we have said, a certain tradition exists that focuses on trait formation rather than transmission – Elie Wollman and François Jacob proposed a model similar to McClintock's, based on their work on bacterial conjugation, in which cellular differentiation (and, in a later version, regulation alone) would be under the control of genome modifications caused by mobile elements, "episomes" (Jacob and Wollman 1959, 1961; see Morange 2002). Work by Jacques Monod and other colleagues suggests other models of non-genetic differentiation, this time through the induction of enzyme formation by substrate molecules (such as the induction of lactose permease by intracellular lactose in *Escherichia coli*), with different stable states that can be achieved in the same environment (Monod 1956; Cohen and Monod 1957; see Thieffry 1996; cited by Morange 2002). Such models culminate in Jacob and Monod's operon model, in which a repressor (an allosteric protein encoded by a regulatory gene) binds to an inducer, with the conformational change of the allosteric repressor (and thus its binding capacity) depending on the presence of regulatory ligands (Jacob et al. 1960; Jacob and Monod 1961; see Morange 2002, 2020, p. 154). The allosteric model also provided the basis for the first model to explain the possible infectivity of a protein, the prion (Griffith 1967; Morange 2002).

The operon model is a geneticist's answer to the question of how traits are formed (and no longer simply transmitted); it provides the opportunity to forge a powerful metaphor, that of the genetic program: gene regulation would be comparable to the execution of a series of instructions (Jacob and Monod 1961, p. 354; Morange 2002, p. 55). This metaphor became fashionable in the late 1960s, and Jacob (1970) developed it further in *La Logique du vivant*: unlike a computer program, the genetic program requires its own outputs to be executed. For Jacob, the genetic program resolves the old unease between epigenesis and preformation: the egg does not contain the description of the future organism, but the instructions enabling it to develop in time and space (Jacob 1978, p. 249; quoted by Morange 2002, p. 55). The genetic program metaphor would go on to be severely criticized on philosophical and theoretical grounds, with its apparent determinism hardly relating to the improvisational capacity of biological systems (Morange 2002; Longo and Tendéro 2008; Pocheville 2018), but it only really fell out of favor from the 2000s onwards, in the post-genomic era – a time when the epigenetic reprogramming metaphor was gaining ground (Holliday 1990; Reik et al. 1993, 2001; Morgan et al. 2005; Suvà et al. 2013; see Brandt 2010).

The operon model is based on three assumptions: genetic material is stable during cell functioning (an assumption dating back to De Vries and Morgan); gene expression is controlled at the transcriptional level; and, lastly, this control is achieved by the binding of specific regulatory proteins to DNA (Morange 1997). While this vision of regulation has been successful to the point of turning to "mania" (ibid., p. 384), it has also led to reactions: for a whole section of the community, the operon cannot constitute a general explanation of regulation and differentiation, particularly in eukaryotes, where there is a lack of proof of its existence (Morange 1997, 2002, 2013). A series of alternatives have been proposed, such as the control of gene expression at the translation level (thanks to the stability of certain RNAs), at the DNA sequence level through gene amplification or reverse transcription, or at the enzymatic activity level (Morange 1997). Among these alternatives, those concerned with the regulation of transcription via structural and chemical modifications of DNA, histones and cytosine methylations, will be of particular interest to us.

The history of research into histones and DNA methylations is "long and tortuous", with the two fields developing in a remarkable lack of communication (Morange 2013, p. 453). The hypothesis that histones can inhibit gene activity was first speculatively proposed in the early 1950s (Stedman and Stedman 1950; Morange 2013, p. 451, 2020, p. 329). Histone modifications (acetylation and methylation) were described 10 years later, with the first in vitro results confirming histone inhibitory activity (Huang and Bonner 1962; Allfrey et al. 1963), as well as the role of acetylation in lifting this inhibition: the histone can function as a specific switch of gene activity (Allfrey and Mirsky 1964; Allfrey et al. 1964; Georgiev 1969; see Paik et al. 2007, p. 148; Morange 2013, p. 451, 2020, p. 329; Felsenfeld 2014, p. 5). However, these initial results were received with skepticism (Morange 2013, p. 451). Eric Davidson, for example, considered that in eukaryotes, inhibition of gene activity could not be achieved by specific repressors, the encoding of which would require too many genes; he preferred a model of gene regulation in which histones play a structural and global inhibitory role (the default state of the genome being inactivity), with activation by specific activators that he proposed to identify with RNAs (Davidson 1968, pp. 313–325; Morange 2013, p. 452). This model, subsequently developed with Roy Britten, has the advantage of predicting switches between global patterns of gene expression, which seem more plausible than cascades of local regulation to explain differentiation (Britten and Davidson 1969, p. 351; Morange 2002, p. 55, 2020, p. 268; García and Suárez 2010). Histones then appear to have a non-specific inhibitory activity; moreover, Morange (2013, p. 452) notes that the discovery of the nucleosome (Kornberg 1974; Olins and Olins 1974) and its subsequent characterization (Richmond et al. 1984) focused attention on the role of histones in chromatin structure (Kornberg 1977; McGhee and Felsenfeld 1980), with the suggested effect of their modifications on transcription remaining unknown (Holt 1985; Grunstein 1990) – at least until the 1990s, when proteins that recognize these modifications and act on transcription began to be discovered (Turner 1991; Wolffe 1994; Pazin and Kadonaga 1997; Kuo and Allis 1998; Kouzarides 1999; see also Felsenfeld 2014, p. 5; Morange 2020, p. 330).

As for cytosine methylations (5mC: 5-methylcytosine) of nucleic acids, results indicating their existence marked the first half of the 20th century (Wheeler 1904; Johnson and Coghill 1925; Hotchkiss 1948; see Mattei et al. 2022). The 1950s saw confirmation of their presence in variable quantities in plant and animal DNA (Wyatt 1950, 1951; Mattei et al. 2022), and as early as the 1960s, their potential role in eukaryotes was discussed: 5mCs could protect DNA (Srinivasan and Borek 1964) or, on the contrary, induce specific mutations that would participate in the differentiation process (Scarano et al. 1967) – this was a hypothesis, however, that was incompatible with recent results from cloning experiments (Gurdon 1962; see Morange 2013, p. 452). However, it is in prokaryotes that the first results flourished, thanks in particular to the in-depth description of the restriction-modification system (R-M system) in bacteria, a system in which so-called "restriction" enzymes cut DNA at specific sites via recognition of particular sequences (generally four to eight nucleotides long) (Luria and Human 1952; Luria 1953; Lederberg 1957; Gold et al. 1963; Arber 1965; Arber and Linn 1969; Morange 2013, p. 452, 2020, p. 331; Mattei et al. 2022). A bacterium possessing this system can digest the DNA of an infecting virus, the bacterial DNA being protected at the restriction sites by species-specific methylations (Arber 1965; Mattei et al. 2022). In addition to the role this system played in early genetic engineering experiments in the early 1970s (Jackson et al. 1972) and in the study of methylations themselves (Bird 1978; Bird and Southern 1978), it offers a possible analogy with the putative role of methylases in eukaryotic development (Morange 2013, p. 452, 2020, pp. 181, 331): the latter could recognize specific sites and control gene expression during differentiation, a mechanism that could explain the recently discovered inactivation of the X chromosome in mammals (a hypothesis that was later ruled out) (Lyon 1968; Lock et al. 1987; Morange 2013; Mattei et al. 2022, p. 12), as well as DNA loss during development in certain eukaryotes or senescence and apoptosis (Adams 1973; Holliday and Pugh 1975; Riggs 1975; Sager and Kitchin 1975; see Morange 2013, 2020, pp. 331–332; Mattei et al. 2022). That being said, this hypothetical ability of methylases to recognize specific sites has not been confirmed by subsequent studies (Morange 2013). While the precise role of methylations in the regulation of gene expression remains unknown (Felsenfeld and McGhee 1982; Doerfler 1983; cited by Morange 2013), converging studies based on improved locus-specific 5mC detection methods suggest that 5mCs are antagonistic to expression (Southern 1975; Waalwijk and Flavell 1978; Kuo et al. 1979; Mandel and Chambon 1979; McGhee and Ginder 1979; see Mattei et al. 2022). During the 1980s, technical advances (such as the possibility of inserting methylated or unmethylated DNA fragments into living cells) led to a more precise role for 5mC and, at least for mammals, a consensus emerged that 5' promoter methylation inhibits transcription (Mattei et al. 2022). However, the absence of 5mC in model organisms such as the yeast *Saccharomyces cerevisiae*, the nematode *Caenorhabditis elegans* and the vinegar fly *Drosophila melanogaster*, leaves the community dubious as to the generality of the mechanism (Morange 2013, 2020, p. 333; Mattei et al. 2022). It was at this point that Robin Holliday (1987), in a well-received speculative article, revived the question of the heritability of cellular states via methylations, and their possible role in evolution, formulating it in an explicitly epigenetic framework (for which he credited Waddington) that would briefly catalyze the association in molecular biology between "epigenetics" and heritable methylations (Holliday 1994; Haig 2004; Morange 2020, p. 333). It was in the 1990s that the outlines of a paradigm were established: 5mCs were shown to be essential for mammalian development (Li et al. 1992, 1993; Lei et al. 1996; Okano et al. 1999); their roles in genomic imprinting, X chromosome inactivation (Lock et al. 1987) and the repression of transposable elements during embryonic development (Walsh et al. 1998) were better understood; this was also when thale cress (*Arabidopsis thaliana*) became a leading model organism for epigenetic studies (Mattei et al. 2022). The mid-1990s also saw "omics" approaches flourish (proteomics, transcriptomics, etc.) which, on the model of genomics, focus on the study of all the components of a certain type in a living material (cell, organelle, tissue, etc.). The study of methylations is part of this movement: bisulfite sequencing, published in 1992 (Frommer et al. 1992), rapidly became the method of choice in a growing number of sequencing studies – some of which claimed to be "epigenomic".

1.7. Post-genomic epigenetics: epigenomics

In the early 2000s, several genome sequences began to be assembled, including those of Arabidopsis, mice and humans. The field was excited to enter the genomic era (Képès and Dardel 2002; Roses 2003), an era of abundant sequences that would pave the way for fine applications such as personalized medicine; but it was also, paradoxically, an era that would be referred to retrospectively as postgenomic – a "post" that initially only had temporal meaning ("after" the publication of the human genome) (Képès and Dardel 2002, p. 235), but which would soon be used in a roundabout way to signify a move beyond genomics, in a rhetorical pursuit for disruption aimed at singling out practices whose reality may show more continuity (Green 2016, p. 84). The era of genomics actually began two decades earlier, with the first sketches of the Human Genome Project in the mid-1980s (the project was officially launched in 1990) (Morange 2020, Chapter 27), and the assembly of several genomes (phage, murine and human mitochondria, *Marchantia* and tobacco chloroplasts, *Haemophilus influenzae* bacteria, *Saccharomyces cerevisiae* yeast), with published genomes accumulating at an exponential rate ever since (Fiers et al. 1976, 1978; Sanger et al. 1977; Anderson et al. 1981; Bibb et al. 1981; Ohyama et al. 1986; Shinozaki et al. 1986; Goffeau et al. 1996; Képès and Dardel 2002, p. 10; Smith 2016). The gradual completion of the Human Genome Project (with the publication of the very first draft sequence in 2000 (White House 2000)), however, is bittersweet: beyond its technical success, the project does not seem to be fulfilling all the – almost messianic – promises made by some of its leaders (Collins 1999; see Longo 2021). The field of molecular biology is always on the lookout for new territory: the situation is conducive to seeking explanations for cellular phenomena other than genetic sequences. The use of epigenetic terminology is exploding (Deichmann 2016), and epigenetics is becoming the focal point of the 21st-century biology (Figure 1.2).

From a mapping point of view, the publication of genomes (and, in the mid-2000s, the emergence of new sequencing and bioinformatics assembly techniques) paved the way for a broadening of methylation mapping, from the locus scale to that of the genome: these were the first methylomes, or genome-wide DNA methylation landscapes, the first maps being drawn up in the mid-2000s for *A. thaliana* and humans (Lippman et al. 2004; Tran et al. 2005; Weber et al. 2005; Rollins et al. 2006; Zhang et al. 2006; cited by Mattei et al. 2022). The mapping movement is moving towards both greater resolution (towards the base), and a more global description (towards the genome), an element which, alongside mechanistic results, will be essential to a better understanding of the dynamic character of methylations (Mattei et al. 2022). The field of epigenetics (like that of Morgan-style genetics before it) quickly thrives in topographical metaphors: the epigeneticist imagines herself as an explorer of a landscape made up of canyons, valleys and islands, which must be mapped, with each type of epigenetic modification representing a "layer" of this map (e.g. Mattei et al. 2022, p. 17). The inadequacy of one description (of the genome, for example) immediately calls for another (methylome, acetylome, etc.), until we obtain the most complete and precise description of the nucleic material. These metaphors can cause strange slippages, with some authors establishing a bridge between the methylation landscape, which concerns the methylation level profile along the sequence, and Waddington's epigenetic landscape, which concerns the potential curves that a developing organism can follow (Jenkinson et al. 2017).

From a mechanistic point of view, we are witnessing major advances in the biology of methylation readers, writers and erasers, which is an important step in the pursuit for explanations of methylation dynamics during differentiation and development (Mattei et al. 2022). It appears that most promoters are not regulated by 5mCs, which disrupts the previous promoter-centric paradigm and invites a more global study of genome function (Weber et al. 2007; Borgel et al. 2010; see Mattei et al. 2022, p. 21). Ultimately, these mechanistic advances will benefit from the development of epigenome-editing tools, with Cas9's ability to recognize sequences being used to produce targeted DNA methylations (Jinek et al. 2012; Hsu et al. 2014; Liu et al. 2016; cited by Mattei et al. 2022).

The early 2000s finally saw the convergence of research into cytosine methylation and histone modification, with results increasingly pointing to an interaction between these mechanisms (Tamaru and Selker 2001; Ooi et al. 2007; Weber et al. 2007; see Mattei et al. 2022). This point in time saw the hypothesis of a histone code (re)emerge, an epigenetic code of gene activity: histone modification, patterns (acetylation, methylation, phosphorylation, ubiquitination) would mark active or potentially active genes (Tsanev and Sendov 1971; Turner 2000; Jenuwein and Allis 2001; see Rando 2012; Morange 2020, p. 337). It has become clear that non-coding RNAs, particularly small RNAs, are involved in a whole range of epigenetic mechanisms, including cytosine methylations and histone modifications, and are of interest in phenomena such as transposon activity, X chromosome inactivation, parental imprinting or paramutation (Costa 2008), in other words, the heritable modification of one allele caused by the other allele at the same locus (Chandler 2007). Conversely, since the mid-2000s, there has been a renewed interest in epigenetic modifications of messenger RNAs and long non-coding RNAs, first discovered in the 1970s, with the latter emerging as an additional layer of epigenetic regulation of gene expression (Liu and Pan 2015). Epigenetic terminology, which until then only involved cytosine methylations, now encompasses all transcriptional effects, whether they are inherited or not, including histone modifications, followed by mechanisms involving RNAs (Haig 2012).

The meaning behind the term "epigenetics" has changed significantly over its history (Jablonka and Lamb 2002; Van de Vijver et al. 2002; Nicoglou and Merlin 2017). Two main families of definitions can be identified.

Developmental epigenetics: epigenetics is the science of epigenesis, in other words, a developmental process in which form is not predetermined at conception (see, for example, Wilson 1896, pp. 327–330; Koltzoff 1935, pp. 8–9; Waddington 1939, pp. 154–155; Berry and Searle 1963). Epigenetics is the science of the development, in particular, within the framework of classical and then molecular genetics, of mechanisms linking genotype and phenotype (Waddington 1942, 1952, 1968; Løvtrup 1974; Medawar and Medawar 1983, pp. 113–114; Holliday 1990, p. 329; Hall 2013, pp. 89–91; see Jablonka and Lamb 2002).

Molecular epigenetics: epigenetics is the science of factors modifying gene expression, whose states can be inherited through mitosis (which can explain differentiation) (Nanney 1958a, 1958b; Davis 1964, see, for example Cahn and Cahn 1966, p. 107; Markert 1968, p. 1908; Rosen 1984, p. 320; Ptashne 2007, R233) or even meiosis (which can produce soft heredity) (Holliday 1987). These factors can be identified with nucleic acid methylations (model of Holliday and Pugh 1975; Riggs 1975; terminology of Holliday 1987) – a model that can explain the putative reversibility of an "epigenetic program" at meiosis (Holliday 1990, pp. 335–336) – or even other nuclear (Holliday 1994, definition 2, p. 454) or at least chromatin modifications (Bird 2007, p. 398). In a strong sense, heredity is required through mitosis (Holliday 1994, definition 1) and/or meiosis (Russo et al. 1996). Epigenetics is then the study of heritable variations through mitosis or meiosis that do not depend on variations in nucleic acid sequence (Russo et al. 1996; Berger et al. 2009, p. 781; see Jablonka and Lamb 2002, p. 88). In a weak sense, a simple (causal or consequential) biologically functional relationship of these factors with gene expression is sufficient (factors limited to chromatin modifications (Bird 2007, p. 398)).

Epigenetics, like genetics before it, is moving towards an increasingly material identification of its object of study, at the cost of an increasing relaxation (*contra* Ptashne 2007, R234) of various functional criteria (heritability, gene expression) whose relationship to material factors can be ambiguous (Bird 2007, p. 397). There is a seamless shift from epigenesis to the study of chromatin, firstly by identifying development – and in particular differentiation – with a cascade of modifications in gene expression, and then by shifting towards the material identification of the factors causing these modifications – factors whose study, however structural, is then of its own interest (see Jablonka and Lamb 2002, p. 89). "Epigenetics", as a noun designating a discipline, only really came into use at the very beginning of the 21st century, once this material identification had been established (Figure 1.2).

Box 1.2. *Definitions of epigenetics*

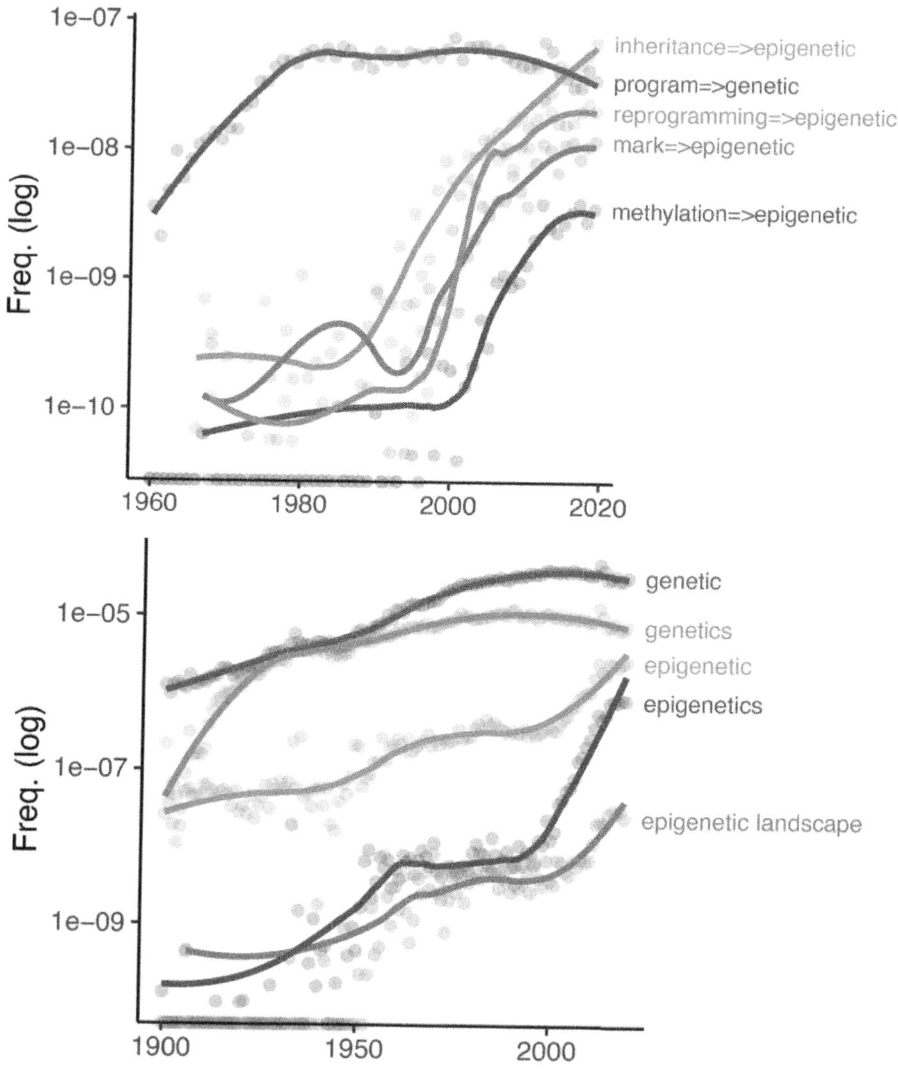

Figure 1.2(A). *Google Ngram representing the evolution of the frequency (in log) of some terms related to epigenetics, in Google's English-language corpus (texts prior to 2019). For a color version of this figure, see www.iste.co.uk/grunau/epigenetics.zip*

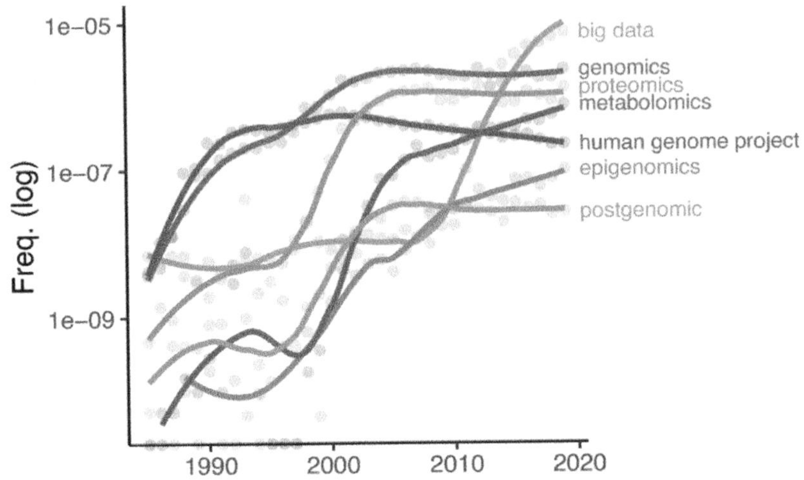

Figure 1.2(B). *Google Ngram representing the evolution of the frequency (in log) of some terms related to epigenetics, in Google's English-language corpus (texts prior to 2019) (continued). For a color version of this figure, see www.iste.co.uk/grunau/epigenetics.zip*

COMMENT ON FIGURE 1.2. – *The Ngram can be used to illustrate a major trend, but is no substitute for in-depth textual study; in particular, low frequencies are sensitive to false positives (e.g. document dating errors). The symbol ⇒ signifies a qualifying link (e.g. "epigenetic ⇒ mark" tracks the frequency of cases where "epigenetic" links together with "mark", as in the expression epigenetic methylation mark). "Epigenetic" as an adjective linking epigenesis has remained in use since Harvey, while the noun "epigenetics" designating a discipline only saw its use take off in the 2000s, when "omics" methods flourished and the use of genetic terminology decreased. From the 1990s onwards, the adjective "epigenetic" came into exponential use to describe mechanisms relating to chromatin.*

From a philosophical point of view, while approaches continue to move towards greater comprehensiveness (both in terms of the number of molecular species and the exhaustiveness of their sampling), the vision remains reductionist: we seek to explain the behavior of the whole by describing its parts.

1.8. Developmental, ecological and evolutionary epigenetics

While molecular biology and evolutionary biology have traditionally not interacted with one another very much – each assuming they can proceed with their research without a detailed knowledge of the other field – epigenetics is the site of encounters, resonances and even outright changes (Mayr 1961; see Morange 2012, 2020, Chapter 23). Morange (1997, p. 383, n. 63) notes that from the very beginnings of molecular epigenetics, similar criticisms were directed at models of gene regulation and the neo-Darwinian scheme of evolution, with the same "heretical" mechanisms (reverse transcription, transposition) and observations (abundance of repeated DNA, control of gene expression at a level other than transcription) being mobilized as putative explanations of evolutionary or developmental changes.

Two theoretical biologists were quick to take the measure of the changes that molecular biology encourages us to introduce into our vision of evolution: in a cutting-edge work, Eva Jablonka and Marion Lamb formulate a very precise proposal to go beyond the neo-Darwinian vision of a genome that would be a passive medium of information, evolving through the natural selection of blind variations (in other words, "random" mutations on an individual scale, which would be selected on a population scale) (Jablonka and Lamb 1995; see also Jablonka and Lamb 1989; Jablonka et al. 1992). For Jablonka and Lamb, the knowledge acquired over the last half-century in both genetics and epigenetics (of which they have a very inclusive, but still molecular, understanding that does not include cultural mechanisms (Jablonka and Lamb 1995, Chapter 4)) encourages us, on the contrary, to consider the genome as an active response system to environmental challenges, and to complement our vision of heredity by considering epigenetic heredity (Jablonka and Lamb 2005). Not only are epigenetic variations not necessarily blind to the challenges of the environment (the constant change of marks allows adaptation to fluctuating environments that genetic evolution alone would not allow, judged to be too slow) and mark patterns, if they are heritable, can be selected on a population scale (Lachmann and Jablonka 1996), but Jablonka and Lamb further argue that results obtained in bacteria suggest that epigenetic mechanisms can lead to directed (i.e. non-blind) genetic mutation on an individual scale: genes involved in responding to environmental challenges would be the target of epigenetic mechanisms that can be mutagenic, thus leading to increased variability at loci of physiological interest in response to the environment – a mechanism that Jablonka and Lamb call "mutational assimilation" (Jablonka and Lamb 1995, pp. 167–171; see Merlin 2010; Pocheville and Danchin 2017), in reference to genetic assimilation (Pigliucci et al. 2006). They thus advocate for a neo-Lamarckian vision of evolution, i.e. one that encompasses both neo-Darwinian mechanisms (selection of blind variation) and mechanisms of directed (non-blind) variation, mechanisms which they moreover assume may have been the result of earlier neo-Darwinian evolution (Jablonka and Lamb 1995, p. 25).

These theses were initially met with a rather hostile, if not totally hostile, response in the evolutionist community, who considered Lamarckism to be a long-dead and buried historical curiosity, with its occasional revivals to be the result of a misunderstanding of the explanatory power of neo-Darwinism (see Jablonka and Lamb 1995, p. vii; Griesemer 1998; Gouyon et al. 2007; Haig 2007; Dickins and Dickins 2008; Griesemer 2011; Bonduriansky 2012, box 1; Morange 2013, p. 453, 2020, p. 56). They would, by contrast, be regularly reinvented in molecular biology, as knowledge of evolutionary biology debates does not seem to have percolated deeply into this field (e.g. Costa 2008; Hilbricht et al. 2008; Martienssen 2008; Koonin and Wolf 2009; Schmidt and Kornfeld 2016; Veigl 2017; Wang et al. 2017; but see Koonin and Wolf 2016).

By the mid-2000s, the attitude of the evolutionary community began to change: a new generation of theorists saw epigenetics (and all developmental mechanisms in general) as an integral part of the evolutionary picture; no longer as a mere product of evolution, but as well as a producer of change, essential for understanding evolutionary trajectories (Odling-Smee et al. 2003; West-Eberhard 2003; Lehmann 2008; Helanterä and Uller 2010; Pigliucci and Müller 2010; Gissis and Jablonka 2011; Bonduriansky 2012; Sultan 2015; Bonduriansky and Day 2018; Pocheville 2019). In developmental biology, Waddington's epigenetics was exhumed and, from the 2010s onwards, became the spearhead of a whole series of theoretical (Bhattacharya et al. 2011; Wang et al. 2011; Huang 2012; Jaeger and Monk 2014; Matsushita and Kaneko 2020) and philosophical studies (Jamniczky et al. 2010; Baedke 2013; Allen 2015; Nicoglou and Merlin 2017; Baedke 2018; Fabris 2018; Nicoglou 2018); some experimental biologists are even getting into the game of explicitly measuring epigenetic landscapes (Banerji et al. 2013; Li and Wang 2013; Wang et al. 2014; Zwiessele and Lawrence 2017; Shi et al. 2020; Schiebinger 2021). The current perspective involves a re-coupling of developmental and evolutionary phenomena, which are no longer seen as separable, to be studied in isolation, but constituting one and the same phenomenon, in a vision that is ultimately very close to that of the 19th century before the establishment of neo-Darwinism (Pocheville 2019).

In ecology, the community seems to have taken more quickly to the epigenetic aspect, both conceptually and empirically (Bossdorf et al. 2008; Rey et al. 2020), although the deployment of epigenetic questions and methods still trails behind genetic approaches (but see Joly and Grunau 2018).

The central dogma of molecular biology is one of a pair of hypotheses formulated by Crick (1958), to specify the causal relationships existing between nucleic acids and proteins in present-day living beings (Crick considers that he is only making the dominant scheme of his day explicit). The first hypothesis, known as the Sequence Hypothesis, is that the sequence of a nucleic acid is a (simple) code for the sequence of an associated protein. The second is the "central dogma": information is only transferred from nucleic acids (to other nucleic acids or proteins), never from proteins (Figure 1.3). *Information* here means sequential information, in other words, precise determination of the sequence (of nucleic bases for DNA and RNA, of amino acids for proteins) (Crick 1958, 1970). The central dogma thus simply postulates the de facto non-existence (but not the impossibility in principle) of reverse translation, i.e. the reading of a protein sequence and its translation into a nucleic sequence[4], as well as protein replication (Figure 1.3). The reason for this hypothesis is that no trace of such machinery was known at the time, and it did not appear to be really necessary for life – the degeneracy of the genetic code, sometimes cited as an additional argument, was only hypothetical at the time (Sarkar 1996, pp. 192–199; Morange 2020, p. 122).

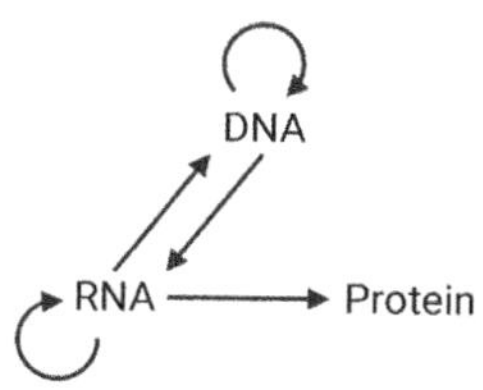

Figure 1.3. *The central dogma of molecular biology, in its contemporary version. From Crick (1970, Figure 2), with deletion of a hypothetical arrow between DNA and protein, and a confirmed arrow between RNA and DNA. Created with BioRender.com*

Many biologists consider that the discoveries of biology over the last half-century should bring about the refutation of the central dogma (e.g. Anon 1970; Temin 1976; Monk 1995; Koonin 2012). In particular, some epigenetic mechanisms seem to be interpretable as transfers, on an individual scale, of some information from the environment to DNA (Monk 1995; Koonin 2012). These arguments are based on a misunderstanding of the central dogma, and in particular of the notion of sequential information (Thieffry and Sarkar 1998; Rosenberg 2006; Ogryzko 2008; Saade and Ogryzko 2014; Steele 2016; Wang and Allard 2022). Clearly, transfers of sequence information from a protein are, by definition, not at work in mechanisms where the source of sequence information is RNA (*contra* Anon 1970; Temin 1976; Dhanasekaran et al. 2013), prion folding (a conformational transfer between proteins) (*contra* Bussard 2005; Koonin 2012), or the modulation, albeit functional, of epigenetic marks without modification of the nucleic sequence (*contra* Monk 1995) – unless 5-methylcytosine is interpreted as a fifth base (Griffith and Mahler 1969, p. 580; Millar et al. 2003) or histone

4. The possibility of reverse translation in vivo has occasionally been discussed (Cook 1977), particularly in the field of the origins of life (Nashimoto 2001; Vitas and Dobovišek 2017).

modifications as part of a code (Tsanev and Sendov 1971; Jenuwein and Allis 2001; but see Morange 2020, p. 337), in which case these modifications constitute edits to an extended alphabet sequence, both nucleic and epinucleic. Nor, strictly speaking, is there any transfer of information between sequences in mutagenic–epigenetic mechanisms (see, for example, Chahwan et al. 2010; Sassa et al. 2016), even in the case of mutational assimilation, mutation being assumed to amount to sequence noise (see Jablonka and Lamb 1995; Griesemer 2002; Merlin 2010; Pocheville and Danchin 2017). On the other hand – although not part of reverse translation or protein replication – the mechanisms of correction, editing, alternative splicing and protein splicing do represent cases of sequence information modification made possible by proteins (see Thieffry and Sarkar 1998; Stotz 2006; Morange 2008; Griffiths et al. 2015; Pocheville 2018).

This misunderstanding is encouraged by another misunderstanding: the association of the central dogma with neo-Darwinism (Figure 1.4). Under a reading of biology in which DNA represents the only relevant support of heredity (Mayr 1961), does the central dogma not signify the impossibility of directional modification of hereditary information on an individual scale (Maynard Smith 1993, pp. 79–81; see Lamb 2011, pp. 118–119)? In other words, does it not express the impossibility of heredity of acquired traits (Sarkar 1996, p. 196; Meloni 2015; Ashe et al. 2021)? Hence, do discoveries about the heredity of environmentally-induced epigenetic modifications not upset this sealed-off heredity postulated by the central dogma (Monk 1995; Watson and Goodman 2002; Koonin 2012; González-Recio et al. 2015; Chakravarty and Jarosz 2018)? There is additional confusion here regarding the level at which information transfers take place: the central dogma concerns cellular functioning, not heredity, in other words, information transfers between generations.

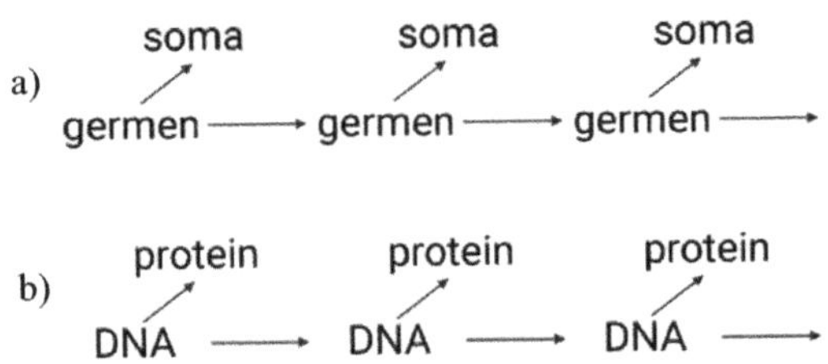

Figure 1.4. *The reinterpretation of Weismannism (a) in terms of the central dogma (b) according to Maynard Smith (1993, Figure 8, 1st edition, 1958). Maynard Smith notes that the validity of the central dogma alone would not establish the impossibility of heredity of acquired traits (ibid., p. 81), a point not all his commentators have taken on board. Created with BioRender.com*

These misunderstandings contribute to the perception of the central dogma as a great theoretical principle, whose refutation should lead to a revolution. In fact, it is a mechanistic hypothesis with a remarkably limited scope, aimed solely at organizing molecular biology research in its day (see also Morange 2008).

Box 1.3. *Epigenetics and the central dogma of molecular biology*

1.9. Epigenetics and ethics

By offering a new layer of biological causality, epigenetics encourages us to revisit ethical debates that were previously centered on genetics (Chiapperino 2018; Soulier 2019). A certain deterministic, fatalistic vision of the human biological condition (genes are fixed for the life of an individual) is succeeded by a richer vision, which takes up elements of previous determinism (which may be genetic or epigenetic, or even social), while mixing it with a certain voluntarism, due to the supposed responsiveness of epigenetic material to the biological and social environment, a responsiveness that puts the agency of each individual back at the center of ethical questioning: individuals are no longer condemned by their genes (according to a certain vision of genetics), but empowered towards themselves and future generations (see Meloni and Müller 2018). A whole series of philosophical and ethical questions can then be put forward in renewed terms: for example, will epigenetics offer an unexpected way out of the nature/nurture debates (Guthman and Mansfield 2013; Lock 2013; Moore 2017; see Dupras et al. 2019)? Under what conditions will the importation of epigenetic causality be liberating or stigmatizing, for example, if the empowerment of putative future mothers' behavior from childhood leads to increased control of women's bodies (Richardson et al. 2014; Richardson 2015; Kenney and Müller 2018; Dupras et al. 2019; Richardson 2021)? Could epigenetics lead to a new form of biological reification of certain ethnic or social groups, or even to a new epi-eugenics (Pickersgill et al. 2013; Juengst et al. 2014; Meloni 2016; Saulnier and Dupras 2017; see Dupras et al. 2019)? The weight of these questions and the penetration of epigenetics into the public sphere resonate with a long-term ethical impetus (ethics is a field that has been growing at an accelerated pace for three centuries) to produce a plethora of work, the pace of which has been increasing exponentially since the 2000s (Dupras et al. 2019; Figure 1).

1.10. Conclusion

The question of generation, in the sense of the formation of living beings, has led to the opposition, and sometimes articulation, of two families of schemes: in an epigenetic scheme, formation is contingent, taking place during growth, interacting with the environment; in a preformationist (or, today, genetic) scheme, formation is determined by constraints that pre-exist conception. The family resemblances are both profound and tenuous: Bonnet's preformation is as distantly related to our contemporary vision of developmental genes as Harvey's epigenesis is to a Waddington's conception of development; preformation and epigenesis, in their various historical meanings, would be "unthinkable today" (Bowler 2001, p. 32).

Each scheme, whether epigenetic or preformationist, responds to the alternative scheme of its time, in a conjugal relationship that is always tumultuous and renewed, and which, notwithstanding the proponents of extreme positions, is characterized by a remarkable fluidity that reflects the intricacy of the objects studied. In the 17th and 18th centuries, epigenesis was first conceived as a model for the formation of a living being by accretion, in opposition to a mechanistic preformationism that put aside the question of formation and reduced it to simple growth. In the 19th century, as cytology developed, the question was revisited in cellular and then nuclear terms. When the hypothesis of material structures in the cell nucleus determining traits crystallized, this vision was referred to as "nuclear preformationism", and was opposed or articulated by explicitly epigenetic considerations that encouraged us to re-examine the role of the cytoplasm in development and heredity. In between the 19th and 20th centuries, by focusing on the transmission of traits, Mendelian and then classical genetics developed a theory of hard heredity that allowed it to be separated from embryology (dealing with the formation of traits), a separation against which the proponents of soft heredity, whose scheme is based on intergenerational epigenesis, continue to argue. In response to classical genetics, the adjective "epigenetic" was used to simply mean "pertaining to epigenesis" or "non-genetic" – a usage that persists today, notably in a certain understanding of epigenetic heredity (Morange 2020, p. 327). A particular case in point is Waddington, for whom "epigenetic" took on the meaning of "developmental", and "epigenetics" the meaning of "developmental science" (or developmental genetics), in a quest to reconcile genetics and embryology – as well as overcome a certain neo-Darwinism of his day. In the mid-20th century, molecular biology reshuffled the deck. The production of a molecular gene concept and the specification of gene regulation models led Nanney to translate the relics of the dualism between cytoplasm and nucleus, into a dualism between structural information and modulation of the expression of this information. "Epigenetics" then took on the meaning of gene regulation, a regulation that could eventually be transmitted through mitosis or even meiosis. Epigenetics produces a profusion of terms modeled on genetics, in a lexical derivation whose slow pace – and perhaps capacity for attraction – suggests the long semantic drift of epigenetics from very different concepts and status to the quasi-"genetics bis" it is today (Figure 1.5). In fact, the use of epigenetic terminology boomed at the beginning of the 21st century, in the post-genomic era of "omic" abundance, when biology became eager for explanations of living phenomena other than genetic sequences. Epigenetics became the focal point of 21st-century biology. In molecular biology, epigenetics began to designate – to the dismay of biologists less liberal with this term (Ptashne 2007; see Pearson 2008) – the molecular field of close interactions with DNA, with molecular epigenetics crystallizing as a field revolving around 5mCs and histones as their relationships began to be better understood (before expanding by phagocytosis of other fields, such as small RNA research).

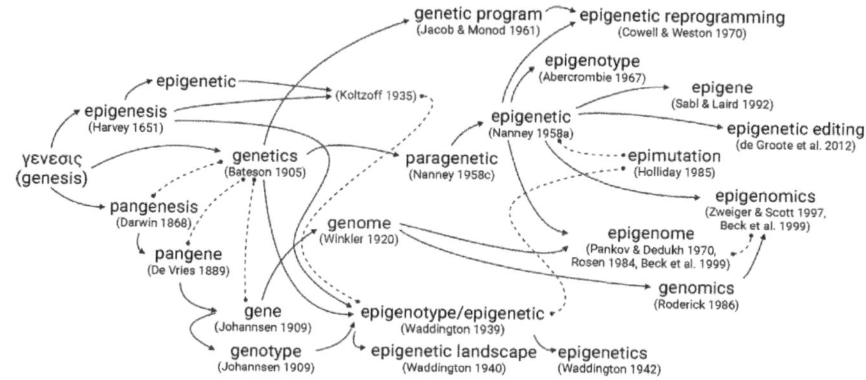

Figure 1.5. *Terminological linkages of epigenetics*

COMMENT ON FIGURE 1.5.– *The arrows represent inkages that are certain (ubiquitous term or direct quotation), the dotted links uncertain linkages. Many terms appear to have been invented several times, and some inventions or linkages may have escaped the author's notice. This tree does not represent conceptual linkage, which is partially decoupled from terminology. Nanney (1958c): see Ephrussi (1958) and Haig (2004). Koltzoff (1935): see Morange (2011). Holliday (1985): see Holliday (1987) citing Waddington (1956b), although a linkage via Nanney (1958a) also seems plausible. Cowell and Weston (1970): see also Lederberg (1966, 1967, cell reprogramming), Nirenberg (1967, cell programming), Gurdon and Woodland (1968, cytoplasmic chromosome reprogramming), Tsanev and Sendov (1971, epigenetic code and cell reprogramming) and Brandt (2010, review). Created with BioRender.com.*

Several components of meaning thus emerge that trace the contours of epigenetics more than they define it, none of which are entirely necessary or independent of the others: the non-genetic (or the non-nucleic, the non-sequential, etc., in an abundance of negative definitions), soft heredity, development, gene regulation (whether heritable or not through cell division), or even what we might call, after Lederberg, the epinucleic, or perhaps more aptly the "perinucleic" (the molecular domain of close interactions with nucleic acids). Some of these components of meaning are functional (epigenetic would be anything that produces traits, or, in a more restricted sense, heritable traits, and which does not rely on the DNA sequence); others are material (epigenetic would be anything that clings to DNA). The terminological drift that epigenetics has undergone is thus accompanied by a constant change in meaning, with some approaches being explicitly contradictory, concerning the requisite of heritability in particular (e.g. Bird 2007; *contra* Deans and Maggert 2015). Faced with this polysemy of usage, the scientist's reflex is to forge new definitions – with however low chances of changing usage – which the philosopher-lepidopterist is quick to pin down. Polysemy, ubiquitous in science, can just as easily lead to misunderstandings as to fruitful friction in our attempts to apprehend a vague reality. Putting concepts back into their historical context can give them meaning back – but not *their* meaning back, as history cannot provide access to its raw object in all its irreducible, rich ephemerality. In this respect, epigenetics, like genetics, has had many false ancestors – those venerable predecessors politely quoted without much reading – and the historical quest always seems to have to be the rewriting of a myth. However, the myth here is performative: these false ancestors, living large in their paper existence, end up attracting new readings and substantially influencing the field. The mischievous epigeneticist might see this as the equivalent of "libraries of specificities" transposed into a new reading frame and expressed anew[5].

1.11. References

Abercrombie, M. (1967). General review of the nature of differentiation. In *Ciba Foundation Symposium – Cell Differentiation*. Wiley Online Library, 3–17.

Adams, R.L.P. (1973). Delayed methylation of DNA in developing sea urchin embryos. *Nature New Biology*, 244(131), 27–29.

Allen, G.E. (1974). Opposition to the Mendelian-chromosome theory: The physiological and developmental genetics of Richard Goldschmidt. *Journal of the History of Biology*, 7(1), 49–92.

---

5. Figures were produced with BioRender and R (version 4.2.2) using the ngramr package (Carmody 2023) and Google Books Ngram Viewer, corpus in 2019 (see: http://books.google.com/ngrams). The author thanks the participants of the Collioure workshop (October 15–18, 2021), in particular the editors of the present work for their proofreading and encouragement, as well as Francesca Merlin for nucleating the project. Michel Morange very kindly sent the author a reprint of his article "The relations between genetics and epigenetics". This chapter bears the mark of discussions with Emmanuelle Maciel on the naturalist epigeneticist and with Alexandra Soulier on the biosocial turn. Alexandra Soulier very kindly contributed to the bibliographical research on the creation of epigenetic terms (Figure 1.5) and passed on her best references on ethics. Alexis Boisseau and Jonathan Racine very kindly shared their unpublished work on Waddington. Charles Wolfe very kindly shared his best references on epigenesis and epigenetics. Clémentine Guillot very kindly improved Figure 1.4. Alexandra Soulier, Charles Wolfe and Clémentine Guillot very kindly reviewed an earlier version of this chapter. The author thanks Lou de Barrelles and Osean van Lilivil for their support.

Allen, G.E. (1978). *Thomas Hunt Morgan: The Man and His Science*. Princeton University Press, Princeton.

Allen, M. (2015). Compelled by the diagram: Thinking through C.H. Waddington's epigenetic landscape. *Contemporaneity: Historical Presence in Visual Culture*, 4, 119.

Allfrey, V.G. and Mirsky, A.E. (1964). Structural modifications of histones and their possible role in the regulation of RNA synthesis. *Science (New York, NY)*, 144(3618), 559.

Allfrey, V.G., Littau, V.C., Mirsky, A.E. (1963). On the role of histones in regulating ribonucleic acid synthesis in the cell nucleus. *Proceedings of the National Academy of Sciences*, 49(3), 414–421.

Allfrey, V.G., Faulkner, R., Mirsky, A.E. (1964). Acetylation and methylation of histones and their possible role in the regulation of RNA synthesis. *Proceedings of the National Academy of Sciences*, 51(5), 786–794.

Anderson, S., Bankier, A.T., Barrell, B.G., de Bruijn, M.H.L., Coulson, A.R., Drouin, J., Eperon, I.C., Nierlich, D.P., Roe, B.A., Sanger, F. et al. (1981). Sequence and organization of the human mitochondrial genome. *Nature*, 290(5806), 457–465. doi: 10.1038/290457a0.

Anon (1970). Central dogma reversed. *Nature*, 226(5252), 1198–1199.

Arber, W. (1965). Host-controlled modification of bacteriophage. *Annual Reviews in Microbiology*, 19(1), 365–378.

Arber, W. and Linn, S. (1969). DNA modification and restriction. *Annual Review of Biochemistry*, 38(1), 467–500.

Aristotle (1879). *Métaphysique*, J. Barthélemy-Saint-Hilaire (trans.). Librairie Germer-Baillière et Cie, Paris.

Aristotle (1887). *Traité de la génération des animaux d'Aristote*, J. Barthélemy-Saint-Hilaire (trans.). Hachette et Cie, Paris.

Aristotle (1986). *La métaphysique*. J. Tricot (trans.). Bibliothèque des Textes Philosophiques, Paris.

Ashe, A., Colot, V., Oldroyd, B.P. (2021). How does epigenetics influence the course of evolution? *Philosophical Transactions of the Royal Society B*, 376(1826), 20200111.

Avery, O.T., MacLeod, C.M., McCarty, M. (1944). Studies on the chemical nature of the substance inducing transformation of pneumococcal types: Induction of transformation by a desoxyribonucleic acid fraction isolated from pneumococcus type III. *Journal of Experimental Medicine*, 79(2), 137–158.

Baedke, J. (2013). The epigenetic landscape in the course of time: Conrad Hal Waddington's methodological impact on the life sciences. *Studies in History and Philosophy of Science Part C: Studies in History and Philosophy of Biological and Biomedical Sciences*, 44(4), 756–773.

Baedke, J. (2018). *Above the Gene, Beyond Biology: Toward a Philosophy of Epigenetics.* University of Pittsburgh Press, Pittsburgh.

Banerji, C.R.S., Miranda-Saavedra, D., Severini, S., Widschwendter, M., Enver, T., Zhou, J.X., Teschendorff, A.E. (2013). Cellular network entropy as the energy potential in Waddington's differentiation landscape. *Scientific Reports*, 3(1), 1–7.

Bateson, W. and Mendel, G. (1902). *Mendel's Principles of Heredity: A Defence, with a Translation of Mendel's Original Papers on Hybridisation.* Cambridge University Press, Cambridge.

Baur, E. (1908). Das Wesen und die Erblichkeitsverhältnisse der "Varietates albomarginatae hort." von Pelargonium zonale. *Zeitschrift für induktive Abstammungs-und Vererbungslehre*, 1(1), 330–351.

Beadle, G.W. and Tatum, E.L. (1941). Genetic control of biochemical reactions in Neurospora. *Proceedings of the National Academy of Sciences*, 27(11), 499–506.

Beck, S., Olek, A., Walter, J. (1999). From genomics to epigenomics: A loftier view of life. *Nature Biotechnology*, 17(12), 1144–1144.

Beisson, J. and Sonneborn, T.M. (1965). Cytoplasmic inheritance of the organization of the cell cortex in *Paramecium aurelia. Proceedings of the National Academy of Sciences*, 53(2), 275–282.

Bejan, A. and Lorente, S. (2004). *La loi constructale.* L'Harmattan, Paris.

Berger, S.L., Kouzarides, T., Shiekhattar, R., Shilatifard, A. (2009). An operational definition of epigenetics. *Genes & Development*, 23(7), 781–783.

Berry, R.J. and Searle, A.G. (1963). Epigenetic polymorphism of the rodent skeleton. In *Proceedings of the Zoological Society of London.* Wiley Online Library, 577–615.

Bhattacharya, S., Zhang, Q., Andersen, M.E. (2011). A deterministic map of Waddington's epigenetic landscape for cell fate specification. *BMC Systems Biology*, 5, 85.

Bibb, M.J., Van Etten, R.A., Wright, C.T., Walberg, M.W., Clayton, D.A. (1981). Sequence and gene organization of mouse mitochondrial DNA. *Cell*, 26(2), 167–180.

Bird, A.P. (1978). Use of restriction enzymes to study eukaryotic DNA methylation: II. The symmetry of methylated sites supports semi-conservative copying of the methylation pattern. *Journal of Molecular Biology*, 118(1), 49–60.

Bird, A.P. (2007). Perceptions of epigenetics. *Nature*, 447(7143), 396–398.

Bird, A.P. and Southern, E.M. (1978). Use of restriction enzymes to study eukaryotic DNA methylation: I. The methylation pattern in ribosomal DNA from *Xenopus laevis*. *Journal of Molecular Biology*, 118(1), 27–47.

Blumenbach, J.F. (1780). Über den Bildungstrieb (*Nisus formativus*) und seinen Einfluß auf die Generation und Reproduction. *Göttingisches Magazin der Wissenschaften und Litteratur*, 1(5), 247–266.

Boesiger, E. (1980). Evolutionary biology in france at the time of the evolutionary synthesis. In *The Evolutionary Synthesis: Perspectives on the Unification of Biology*, Mayr, E. and Provine, W.B. (eds). Harvard University Press, Cambridge, 309–321.

Boisseau, A. (2023). La causalité du développement selon C.H. Waddington. PhD Thesis, Université Jean Jaurès, Toulouse.

Bolduc, G. (2021). *Préformation et épigenèse en développement. Naissance de l'embryologie expérimentale.* Les Presses de l'Université de Montréal, Montreal, 652.

Bonduriansky, R. (2012). Rethinking heredity, again. *Trends in Ecology & Evolution*, 27(6), 330–336.

Bonduriansky, R. and Day, T. (2018). *Extended Heredity: A New Understanding of Inheritance and Evolution.* Princeton University Press, Princeton.

Bonnard, J.-B. (2015). Il paraît en effet que les fils ressemblent aux pères. In *Penser et représenter le corps dans l'Antiquité*, Prost, F. and Wilgaux, J. (eds). Presses Universitaires de Rennes, Rennes, 307–318.

Bonnet, C. (1745). *Traité d'insectologie ou Observations sur les pucerons. Partie 1.* Durand, Paris.

Bonnet, C. (1762). *Considérations sur les corps organisés.* Marc-Michel Rey, Amsterdam.

Bonnet, C. (1770). *La palingénésie philosophique, ou Idées sur l'état passé et sur l'état futur des êtres vivans. Tome premier.* Claude Philibert and Barthelemi Chirol, Geneva.

Borgel, J., Guibert, S., Li, Y., Chiba, H., Schübeler, D., Sasaki, H., Forné, T., Weber, M. (2010). Targets and dynamics of promoter DNA methylation during early mouse development. *Nature Genetics*, 42(12), 1093–1100.

Bossdorf, O., Richards, C.L., Pigliucci, M. (2008). Epigenetics for ecologists. *Ecology Letters*, 11(2), 106–115.

Boveri, T. (1904). *Ergebnisse über die Konstitution der chromatischen Substanz des Zellkerns.* Verlag von Gustav Fischer in Jena, Jena.

Bowler, P.J. (1983). *The Eclipse of Darwinism: Anti-Darwinian Evolution Theories in the Decades Around 1900.* JHU Press, Baltimore.

Bowler, P.J. (2001). *The Mendelian Revolution: The Emergence of Hereditarian Concepts in Modern Science and Society.* Continuum International Publishing Group, London.

Boylan, M. (1984). The galenic and hippocratic challenges to Aristotle's conception theory. *Journal of the History of Biology*, 17(1), 83–112.

Boylan, M. (1986). Galen's conception theory. *Journal of the History of Biology*, 19(1), 47–77.

Brandt, C. (2010). The metaphor of "nuclear reprogramming": 1970's cloning research and beyond. In *The Hereditary Hourglass. Genetics and Epigenetics, 1868–2000*, Barahona, A., Suarez-Díaz, E., Rheinberger, H.-J. (eds). Max Planck Institute for the History of Science, Berlin, 85–96.

Britten, R.J. and Davidson, E.H. (1969). Gene regulation for higher cells: A theory. *Science*, 165(3891), 349–357.

Buffon, G.-L.L. (1749). *Histoire naturelle, générale et particulière, avec la description du Cabinet du Roy. Tome 2*. Imprimerie Royale, Paris.

Bulmer, M. (1998). Galton's law of ancestral heredity. *Heredity*, 81(5), 579–585.

Bussard, A.E. (2005). A scientific revolution? The prion anomaly may challenge the central dogma of molecular biology. *EMBO Reports*, 6(8), 691–694.

Cahn, R.D. and Cahn, M.B. (1966). Heritability of cellular differentiation: Clonal growth and expression of differentiation in retinal pigment cells in vitro. In *Proceedings of the National Academy of Sciences*, 55(1), 106–114.

Callender, L.A. (1988). Gregor mendel: An opponent of descent with modification. *History of Science*, 26(1), 41–75.

Carlson, E.A. (1971). An unacknowledged founding of molecular biology: HJ Muller's contributions to gene theory, 1910–1936. *Journal of the History of Biology*, 4(1), 149–170.

Carmody, S. (2023). Ngramr: Retrieve and Plot Google n-Gram Data. Version 1.9.3.

Cartron, L. (2007). Degeneration and "alienism" in early nineteenth-century France. In *Heredity Produced: At the Crossroads of Biology, Politics, and Culture, 1500-1870*, Müller-Wille, S. and Rheinberger, H.-J. (eds). MIT Press, Cambridge.

Chahwan, R., Wontakal, S.N., Roa, S. (2010). Crosstalk between genetic and epigenetic information through cytosine deamination. *Trends in Genetics*, 26(10), 443–448.

Chakravarty, A.K. and Jarosz, D.F. (2018). More than just a phase: Prions at the crossroads of epigenetic inheritance and evolutionary change. *Journal of Molecular Biology*, 430(23), 4607–4618.

Chandler, V.L. (2007). Paramutation: From maize to mice. *Cell*, 128(4), 641–645.

Chiapperino, L. (2018). Epigenetics: Ethics, politics, biosociality. *British Medical Bulletin*, 128(1), 49–60.

Cohen, G.N. and Monod, J. (1957). Bacterial permeases. *Bacteriological Reviews*, 21(3), 169–194.

Cohen, S.M. and Reeve, C.D.C. (2021). Aristotle's metaphysics. In *The Stanford Encyclopedia of Philosophy*, Zalta, E.N. (ed.). Stanford University Press, Stanford.

Coleman, W. (1970). Bateson and chromosomes: Conservative thought in science. *Centaurus*, 15(3), 228–314.

Collins, F.S. (1999). Medical and societal consequences of the human genome project. *New England Journal of Medicine*, 341(1), 28–37.

Cook, N.D. (1977). The case for reverse translation. *Journal of Theoretical Biology*, 64(1), 113–135.

Correns, C. (1900). G. Mendel's Regel über das Verhalten der Nachkommenschaft der Rassenbastarde. *Berichte der deutschen botanischen Gesellschaft*, 18, 158–167.

Correns, C. (1909a). Vererbungsversuche mit blass(gelb)grünen und buntblättrigen Sippen bei Mirabilis Jalapa, Urtica pilulifera und Lunaria annua. *Zeitschriftfur induktive Abstammungs- und Vererbungslehre*, 1, 291–329.

Correns, C. (1909b). Zur Kenntnis der Rolle von Kern und Plasma bei der Vererbung. *Zeitschrift für Induktive Abstammungs und Vererbungslehre*, 2(1), 331–340.

Corsi, P. (2001). *Lamarck. Genèse et enjeux du transformisme, 1770–1830*. CNRS editions, Paris, 444.

Costa, F.F. (2008). Non-coding RNAs, epigenetics and complexity. *Gene*, 410(1), 9–17.

Cowell, L.A. and Weston, J.A. (1970). An analysis of melanogenesis in cultured chick embryo spinal ganglia. *Developmental Biology*, 22(4), 670–697.

Crick, F. (1958). *On Protein Synthesis. Symposium of the Society for Experimental Biology XII*. Academic Press, New York.

Crick, F. (1970). Central dogma of molecular biology. *Nature*, 227(5258), 561–563.

Darden, L. (1977). William Bateson and the promise of Mendelism. *Journal of the History of Biology*, 10(1), 87–106.

Darwin, C.R. (1837). Notebook B: Transmutation of species (1837–1838). CUL-DAR121. Transcribed by K. Rookmaaker.

Darwin, C.R. (1838). Notebook D: Transmutation of species (7-10.1838). CUL-DAR123. Transcribed by K. Rookmaaker.

Darwin, C.R. (1865). Manuscript 51.

Darwin, C.R. (1868). *The Variation of Animals and Plants Under Domestication*, 1st edition, volume 2. John Murray, London.

Davidson, E.H. (1968). *Gene Activity in Early Development*, 1st edition. Academic Press, New York.

Davis, B.D. (1964). Theoretical mechanisms of differentiation. *Medicine*, 43(6), 639.

Deans, C. and Maggert, K.A. (2015). What do you mean, "epigenetic"? *Genetics*, 199(4), 887–896.

Deichmann, U. (2016). Epigenetics: The origins and evolution of a fashionable topic. *Developmental Biology*, 416(1), 249–254.

Delbrück, M. (1949). Discussion sur T.M. Sonneborn et G.H. Beale: Influence des gènes, des plasmagènes et du milieu dans le déterminisme des caractères antigéniques chez Paramecium aurelia (variété 4). In *Unités biologiques douées de continuité génétique*, Sonneborn, T.M. and Beale, G.H. (eds). Éditions du Centre National de la Recherche Scientifique, Paris, 33–34.

Dhanasekaran, K., Kumari, S., Kanduri, C. (2013). Noncoding RNAs in chromatin organization and transcription regulation: An epigenetic view. In *Epigenetics: Development and Disease*, Kundu, T.K. (ed.). Springer Netherlands, Dordrecht, 343–372.

Dickins, T. and Dickins, B. (2008). Mother nature's tolerant ways: Why non-genetic inheritance has nothing to do with evolution. *New Ideas in Psychology*, 26(1), 41–54.

Dobzhansky, T. (1937). *Genetics and the Origin of Species*. Columbia University Press, New York.

Doerfler, W. (1983). DNA methylation and gene activity. *Annual Review of Biochemistry*, 52(1), 93–124.

Dupont, J.-C. (2007). Pre-kantian revival of epigenesis: Caspar Friedrich Wolff's de formatione intestinorum (1768–1769). In *Understanding Purpose: Kant and the Philosophy of Biology*, Huneman, P. (ed.). University of Rochester Press, New York.

Dupras, C., Saulnier, K.M., Joly, Y. (2019). Epigenetics, ethics, law and society: A multidisciplinary review of descriptive, instrumental, dialectical and reflexive analyses. *Social Studies of Science*, 49(5), 785–810.

East, E.M. (1912). The Mendelian notation as a description of physiological facts. *The American Naturalist*, 46(551), 633–655.

Ephrussi, B. (1953). *Nucleo-Cytoplasmic Relations in Micro-Organisms. Their Bearing on Cell Heredity and Differentiation*. Clarendon Press, Oxford.

Ephrussi, B. (1958). The cytoplasm and somatic cell variation. *Journal of Cellular and Comparative Physiology*, 52(S1), 35–53.

Fabricius, H. (1621). *De Formatione Ovi et Pulli*. J. Prevotius, Padua.

Fabris, F. (2018). Waddington's processual epigenetics and the debate over cryptic variability. In *Everything Flows: Towards a Processual Philosophy of Biology*, Nicholson, D.J. and Dupré, J. (eds). Oxford University Press, Oxford, 246–263.

Felsenfeld, G. (2014). A brief history of epigenetics. *Cold Spring Harbor Perspectives in Biology*, 6(1), a018200–a018200.

Felsenfeld, G. and McGhee, J. (1982). Methylation and gene control. *Nature*, 296(5858), 602–603.

Fernel, J. and Plancy, G. (1656). *Universa Medicina*, O. van Heurne (trans.). Trajecti ad Rhenum, Utrecht.

Fiers, W., Contreras, R., Duerinck, F., Haegeman, G., Iserentant, D., Merregaert, J., Min Jou, W., Molemans, F., Raeymeakers, A., Van den Berghe, A. et al. (1976). Complete nucleotide sequence of bacteriophage MS2 RNA: Primary and secondary structure of the replicase gene. *Nature*, 260(5551), 500–507.

Fiers, W., Contreras, R., Haegemann, G., Rogiers, R., Van de Voorde, A., Van Heuverswyn, H., Van Herreweghe, J., Volckaert, G., Ysebaert, M. (1978). Complete nucleotide sequence of SV40 DNA. *Nature*, 273(5658), 113–120.

Fisher, S.R.A. (1918). The correlation between relatives on the supposition of Mendelian inheritance. *Philosophical Transactions of the Royal Society of Edinburgh*, 52, 399–433.

Fisher, R.A. (1930). *The Genetical Theory of Natural Selection*. Oxford Clarendon Press, Oxford.

Flemming, W. (1879). Ueber das Verhalten des Kerns bei der Zelltheilung, und über die Bedeutung mehrkerniger Zellen. *Archiv für pathologische Anatomie und Physiologie und für klinische Medicin*, 77(1), 1–29.

Franklin, R.E. and Gosling, R.G. (1953). Molecular configuration in sodium thymonucleate. *Nature*, 171(4356), 740–741.

Frommer, M., McDonald, L.E., Millar, D.S., Collis, C.M., Watt, F., Grigg, G.W., Molloy, P.L., Paul, C.L. (1992). A genomic sequencing protocol that yields a positive display of 5-methylcytosine residues in individual DNA strands. *Proceedings of the National Academy of Sciences*, 89(5), 1827–1831.

Galen, C. (1822). *Claudii Galenii. Opera omnia. Tomus IV*, D.C.G. Kühn (trans.). Car. Cnobloch, Lipsiae.

Galen, C. (1992). *Galen: On Semen*. P. De Lacy (trans.). Akademie Verlag, Berlin.

Galton, F. (1871a). Experiments in pangenesis, by breeding from rabbits of a pure variety, into whose circulation blood taken from other varieties had previously been largely transfused. *Proceedings of the Royal Society of London*, 19(123), 393–410.

Galton, F. (1871b). Pangenesis. *Nature*, 4(79), 5–6.

Galton, F. (1883). *Inquiries into Human Faculty and Its Development*. Macmillan, New York.

Galton, F. (1886). Regression towards mediocrity in hereditary stature. *The Journal of the Anthropological Institute of Great Britain and Ireland*, 15, 246–263.

Galton, F. (1889). *Natural Inheritance*. Macmillan and Co., London.

Galton, F. (1897). The average contribution of each several ancestor to the total heritage of the offspring. *Proceedings of the Royal Society of London*, 91(369), 401–413.

García, V. and Suárez, E. (2010). Switches and batteries: Two models of gene regulation and a note on the historiography of 20th century biology. In *The Hereditary Hourglass. Genetics and Epigenetics, 1868–2000*, Barahona, A., Suarez- Díaz, E., Rheinberger., H.-J. (eds). Max Planck Institute for the History of Science, Berlin, 59–84.

Gärtner, C.F.V. (1849). *Versuche und Beobachtungen über die Bastarderzeugung im Pflanzenreich: mit Hinweisung auf die ähnlichen Erscheinungen im Thierreich*. Gedruckt bei K.F. Hering.

Gasking, E.B. (1967). *Investigations Into Generation, 1651–1828*. Johns Hopkins Press, Baltimore.

Gayon, J. (1992). *Darwin et l'après-Darwin*. Éditions Kimé, Paris.

Gayon, J. (2004). Eugénisme. In *Dictionnaire de la pensée médicale*, Lecourt, D., Delaporte, F., Bourgeois, T. (eds). Presses Universitaires de France, Paris, 450–457.

Georgiev, G.P. (1969). Histones and the control of gene action. *Annual Review of Genetics*, 3(1), 155–180.

Gerstein, M.B., Bruce, C., Rozowsky, J.S., Zheng, D., Du, J., Korbel, J.O., Emaunuelsson, O., Zhang, Z.D., Weissman, S., Snyder, M. (2007). What is a gene, post-ENCODE? History and updated definition. *Genome Research*, 17(6), 669.

Gilbert, S.F. (2012). Commentary: 'The Epigenotype' by C.H. Waddington. *International Journal of Epidemiology*, 41(1), 20–23.

Giorgianni, F. and Provenza, A. (2015). An entry for a 'dictionary of genetics' generation and aspects of heredity from the presocratics to galen: The main notions and the technical terminology. *Medicina Nei Secoli: Journal of History of Medicine and Medical Humanities*, 27(3), 1111–1158.

Gissis, S.B. and Jablonka, E. (2011). *Transformations of Lamarckism: From Subtle Fluids to Molecular Biology*. MIT Press, Cambridge.

Goffeau, A., Barrell, B.G., Bussey, H., Davis, R.W., Dujon, B., Feldmann, H., Galibert, F., Hoheisel, J.D., Jacq, C. et al. (1996). Life with 6000 genes. *Science*, 274(5287), 546–567.

Gold, M., Hurwitz, J., Anders, M. (1963). The enzymatic methylation of RNA and DNA, II. On the species specificity of the methylation enzymes. *Proceedings of the National Academy of Sciences*, 50(1), 164–169.

Goldschmidt, R.B. (1916). Genetic factors and enzyme reaction. *Science*, 43(1099), 98–100.

Goldschmidt, R.B. (1954). Different philosophies of genetics. *Science*, 119(3099), 703–710.

González-Recio, O., Toro, M.A., Bach, A. (2015). Past, present, and future of epigenetics applied to livestock breeding. *Frontiers in Genetics*, 6, 305.

Gould, S.J. (1977). *Ontogeny and Phylogeny*. Harvard University Press, Cambridge.

Gouyon, P.-H., Henry, J.-P., Arnould, J. (2007). *Gene Avatars: The Neo-Darwinian Theory of Evolution*. Springer Science & Business Media, Berlin.

Goy, I. (2018). Was Aristotle the 'father' of the epigenesis doctrine? *History and Philosophy of the Life Sciences*, 40(2), 28.

Green, S. (2016). What is the 'post' in postgenomics? *Metascience*, 25(1), 83–86.

Griesemer, J. (1998). The case for epigenetic inheritance in evolution. *Journal of Evolutionary Biology*, 11(2), 193–200.

Griesemer, J. (2002). What is "epi" about epigenetics? *Annals of the New York Academy of Sciences*, 981(1), 97–110.

Griesemer, J. (2011). The relative significance of epigenetic inheritance in evolution: Some philosophical considerations. In *Transformations of Lamarckism: From Subtle Fluids to Molecular Biology*, Gissis, S.B. and Jablonka, E. (eds). MIT Press, Cambridge, 331–344.

Griffith, J.S. (1967). Nature of the scrapie agent: Self-replication and scrapie. *Nature*, 215(5105), 1043–1044.

Griffith, J.S. and Mahler, H.R. (1969). DNA ticketing theory of memory. *Nature*, 223(5206), 580–582.

Griffiths, P.E., Pocheville, A., Calcott, B., Stotz, K., Kim, H., Knight, R. (2015). Measuring causal specificity. *Philosophy of Science*, 82(4), 529–555.

de Groote, M.L., Verschure, P.J., Rots, M.G. (2012). Epigenetic editing: Targeted rewriting of epigenetic marks to modulate expression of selected target genes. *Nucleic Acids Research*, 40(21), 10596–10613.

Grunstein, M. (1990). Histone function in transcription. *Histone Function in Transcription*, 6, 643–678.

Gurdon, J.B. (1962). The developmental capacity of nuclei taken from intestinal epithelium cells of feeding tadpoles. *Journal of Embryology and Experimental Morphology*, 10(4), 622–640.

Gurdon, J.B. and Woodland, H.R. (1968). The cytoplasmic control of nuclear activity in animal development. *Biological Reviews*, 43(2), 233–267.

Guthman, J. and Mansfield, B. (2013). The implications of environmental epigenetics: A new direction for geographic inquiry on health, space, and nature-society relations. *Progress in Human Geography*, 37(4), 486–504.

Haeckel, E. (1866). *Generelle Morphologie der Organismen: allgemeine Grundzüge der organischen Formen-Wissenschaft, mecha- nisch begründet durch die von Charles Darwin reformirte Descendenz-Theorie. Allgemeine Entwickelungsgeschichte der Organismen*. Georg Reimer, Berlin.

Haeckel, E. (1876). *Die Perigenesis der Plastidule oder die Wellenerzeugung der Lebenstheilchen*. Georg Reimer, Berlin.

Haig, D. (2004). The (dual) origin of epigenetics. In *Cold Spring Harbor Symposia on Quantitative Biology*, Stillman, B. and Stewart., D. (eds). CSHL Press, New York, 67–70.

Haig, D. (2007). Weismann rules! OK? epigenetics and the lamarckian temptation. *Biology & Philosophy*, 22(3), 415–428.

Haig, D. (2012). Commentary: The epidemiology of epigenetics. *International Journal of Epidemiology*, 41(1), 13–16.

Haldane, J.B.S. (1932). *The Causes of Evolution*. Longmans, London.

Hall, B.K. (2013). *Evolutionary Developmental Biology*. Springer Science & Business Media, Berlin.

Hämmerling, J. (1929). *Dauermodifikationen*. Handbuch der Vererbungswissenschaft, Borntraeger.

Harvey, W. (1651). *Exercitationes de generatione animalium*. Du Gard, London.

Helanterä, H. and Uller, T. (2010). The Price equation and extended inheritance. *Philosophy & Theory in Biology*, 2. 1–17.

Henry Ford Hospital (1955). *International Symposium on Cardiovascular Surgery*. Saunders, Detroit.

Herodotus (1850). *Livre IV: Melpomène*, Larcher (trans.). Charpentier, Paris.

Herring, S.W. (1993). Formation of the vertebrate face epigenetic and functional influences. *American Zoologist*, 33(4), 472–483.

Hershey, A.D. and Chase, M. (1952). Independent functions of viral protein and nucleic acid in growth of bacteriophage. *The Journal of General Physiology*, 36(1), 39–56.

Hertwig, O. (1875). *Beiträge zur kenntniss der bildung, befruchtung und theilung des thierischen eies...* Wilhelm Engelmann, Leipzig.

Hertwig, O. (1894). *Präformation oder Epigenese? Grundzüge einer Entwicklungstheorie der Organismen*. Gustav Fischer, Jena.

Hilbricht, T., Varotto, S., Sgaramella, V., Bartels, D., Salamini, F., Furini, A. (2008). Retrotransposons and siRNA have a role in the evolution of desiccation tolerance leading to resurrection of the plant *Craterostigma plantagineum*. *New Phytologist*, 179(3), 877–887.

Hippocrates (1842). *Oeuvres*. Société encyclographique des sciences médicales, Brussels.

d'Holbach, P.-H.T. (1770). *Système de la nature, ou des lois du monde physique et du monde moral*. London [Online]. Available at: https://gallica.bnf.fr/ark:/12148/btv1b8626110p.

Holliday, R. (1985). The significance of DNA methylation in cellular aging. In *Molecular Biology of Aging*, Woodhead, A.D., Blackett, A.D., Hollaender, A. (eds). Springer, Boston, 269–283.

Holliday, R. (1987). The inheritance of epigenetic defects. *Science*, 238(4824), 163–170.

Holliday, R. (1990). DNA methylation and epigenetic inheritance. *Philosophical Transactions of the Royal Society of London. B, Biological Sciences*, 326(1235), 329–338.

Holliday, R. (1994). Epigenetics: An overview. *Developmental Genetics*, 15(6), 453–457.

Holliday, R. and Pugh, J.E. (1975). DNA modification mechanisms and gene activity during development: Developmental clocks may depend on the enzymic modification of specific bases in repeated DNA sequences. *Science*, 187(4173), 226–232.

von Holt, C. (1985). Histones in perspective. *BioEssays*, 3(3), 120–124.

Hotchkiss, R.D. (1948). The quantitative separation of purines, pyrimidines, and nucleosides by paper chromatography. *J. Biol. Chem*, 175(1), 315–332.

Hsu, P.D., Lander, E.S., Zhang, F. (2014). Development and applications of CRISPR-Cas9 for genome engineering. *Cell*, 157(6), 1262–1278.

Huang, S. (2012). The molecular and mathematical basis of Waddington's epigenetic landscape: A framework for post-Darwinian biology? *BioEssays*, 34(2), 149–157.

Huang, R.-C.C. and Bonner, J. (1962). Histone, a suppressor of chromosomal RNA synthesis. *Proceedings of the National Academy of Sciences*, 48(7), 1216–1222.

Huneman, P. (2006). Naturalising purpose: From comparative anatomy to the 'adventure of reason'. *Studies in History and Philosophy of Science Part C: Studies in History and Philosophy of Biological and Biomedical Sciences*, 37(4), 649–674.

Ibrahim, A. (2019). Maupertuis : un système de la nature ? In *Philosophies de la nature*, Bloch, O. (ed.). Éditions de la Sorbonne, Paris, 67–77.

Jablonka, E. and Lamb, M.J. (1989). The inheritance of acquired epigenetic variations. *Journal of Theoretical Biology*, 139(1), 69–83.

Jablonka, E. and Lamb, M.J. (1995). *Epigenetic Inheritance and Evolution: The Lamarckian Dimension*. Oxford University Press, Oxford.

Jablonka, E. and Lamb, M.J. (2002). The changing concept of epigenetics. *Annals of the New York Academy of Sciences*, 981(1), 82–96.

Jablonka, E. and Lamb, M.J. (2005). *Evolution in Four Dimensions: Genetic, Epigenetic, Behavioral, and Symbolic Variation in the History of Life*. MIT Press, Cambridge.

Jablonka, E. and Lamm, E. (2012). Commentary: The epigenotype–a dynamic network view of development. *International Journal of Epidemiology*, 41(1), 16–20.

Jablonka, E., Lachmann, M., Lamb, M.J. (1992). Evidence, mechanisms and models for the inheritance of acquired characters. *Journal of Theoretical Biology*, 158(2), 245–268.

Jackson, D.A., Symons, R.H., Berg, P. (1972). Biochemical method for inserting new genetic information into DNA of Simian Virus 40: Circular SV40 DNA molecules containing lambda phage genes and the galactose operon of Escherichia coli. *Proceedings of the National Academy of Sciences*, 69(10), 2904–2909.

Jacob, F., Perrin, D., Sánchez, C., Monod, J. (1960). L'opéron : groupe de gènes à expression coordonnée par un opérateur. *Comptes Rendus Hebdomadaires des Séances de l'Académie des Sciences*, 250(9), 1727–1729.

Jacob, F. (1970). *La logique du vivant. Une histoire de l'hérédité*. Gallimard, Paris.

Jacob, F. (1978). The Leeuwenhoek Lecture, 1977. Mouse teratocarcinoma and mouse embryo. *Proceedings of the Royal Society of London. Series B. Biological Sciences*, 201(1144), 249–270.

Jacob, F. and Monod, J. (1961). Genetic regulatory mechanisms in the synthesis of proteins. *Journal of Molecular Biology*, 3(3), 318–356.

Jacob, F. and Wollman, E.L. (1959). *La sexualité des bactéries*. Masson, Paris.

Jacob, F. and Wollman, E.L. (1961). *Sexuality and the Genetics of Bacteria*. Academic Press, New York.

Jaeger, J. and Monk, N. (2014). Bioattractors: Dynamical systems theory and the evolution of regulatory processes. *The Journal of Physiology*, 592(11), 2267–2281.

Jamniczky, H.A., Boughner, J.C., Rolian, C., Gonzalez, P.N., Powell, C.D., Schmidt, E.J., Parsons, T.E., Bookstein, F.L., Hallgrímsson, B. (2010). Rediscovering Waddington in the post-genomic age: Operationalising Waddington's epigenetics reveals new ways to investigate the generation and modulation of phenotypic variation. *Bioessays*, 32(7), 553–558.

Jenkinson, G., Pujadas, E., Goutsias, J., Feinberg, A.P. (2017). Potential energy landscapes identify the information-theoretic nature of the epigenome. *Nature Genetics*, 49(5), 719–729.

Jenuwein, T. and Allis, C.D. (2001). Translating the histone code. *Science*, 293(5532), 1074–1080.

Jinek, M., Chylinski, K., Fonfara, I., Hauer, M., Doudna, J.A., Charpentier, E. (2012). A programmable dual-RNA–guided DNA endonuclease in adaptive bacterial immunity. *Science*, 337(6096), 816–821.

Johannsen, W. (1909). *Elemente der exakten Erblichkeitslehre.* Gustav Fischer, Jena.

Johannsen, W. (1911). The genotype conception of heredity. *The American Naturalist*, 45(531), 129–159.

Johannsen, W. (1923). Some remarks about units in heredity. *Hereditas*, 4, 133–141.

Johnson, T.B. and Coghill, R.D. (1925). Researches on pyrimidines. C111. The discovery of 5-methyl-cytosine in tuberculinic acid, the nucleic acid of the tubercle bacillus1. *Journal of the American Chemical Society*, 47(11), 2838–2844.

Jollos, V. (1920). Experimentelle Vererbungsstudien an Infusorien. *Zeitschrift für Induktive Abstammungs-und Vererbungslehre*, 24(1), 77–97.

Jollos, V. (1921). *Experimentelle Protistenstudien. I. Untersuchungen über Variabilität und Vererbung bei Infusorien.* Verlag von Gustav Fischer, Jena.

Joly, D. and Grunau, C. (2018). *Prospective épigénétique, écologie & évolution. Une prospective de l'Institut Ecologie et Environnement.* CNRS, Paris.

Juengst, E.T., Fishman, J.R., McGowan, M.L., Settersten Jr, R.A. (2014). Serving epigenetics before its time. *Trends in Genetics*, 30(10), 427–429.

Just, E.E. (1932). On the origin of mutations. *The American Naturalist*, 66(702), 61–74.

Just, E.E. (1939). *The Biology of the Cell Surface.* P. Blakiston's Son & Company, Incorporated, Philadelphia.

Kalmus, H. (1983). The scholastic origins of Mendel's concepts. *History of Science*, 21(1), 61–83.

Kant, I. (1790). *Critique de la faculté de juger*. Vrin, Paris.

Keller, E.F. (1983). *A Feeling for the Organism, 10th Anniversary Edition: The Life and Work of Barbara McClintock.* Macmillan, New York.

Keller, E.F. (2002). *The Century of the Gene.* Harvard University Press, Cambridge.

Keller, E.F. (2010). *The Mirage of a Space between Nature and Nurture*. Duke University Press, Durham/London.

Kenney, M. and Müller, R. (2018). Of rats and women: Narratives of motherhood in environmental epigenetics. In *The Palgrave Handbook of Biology and Society*, Meloni, M., Cromby, J., Fitzgerald, D., Lloyd, S. (eds). Palgrave Macmillan, London, 799–830.

Képès, F. and Dardel, F. (2002). *Bioinformatique. Génomique et post-génomique*, 1st edition. Éditions de l'école polytechnique, Palaiseau.

Koelreuter, J.G. (1761). *Vorläufige Nachricht von einigen das Geschlecht der Pflanzen betreffenden Versuchen und Beobachtungen*. Gleditsch, Leipzig.

Koltzoff, N.K. (1935). *Physiologie du développement et génétique*. Hermann & Cie, Paris.

Koonin, E.V. (2012). Does the central dogma still stand? *Biology Direct*, 7(1), 1–7.

Koonin, E.V. and Wolf, Y.I. (2009). Is evolution darwinian or/and Lamarckian? *Biology Direct*, 4(1), 42.

Koonin, E.V. and Wolf, Y.I. (2016). Just how Lamarckian is CRISPR-Cas immunity: The continuum of evolvability mechanisms. *Biology Direct*, 11(1), 9.

Kornberg, R.D. (1974). Chromatin structure: A repeating unit of histones and DNA. *Science*, 184(4139), 868–871.

Kornberg, R.D. (1977). Structure of chromatin. *Annual Review of Biochemistry*, 46(1), 931–954.

Kouzarides, T. (1999). Histone acetylases and deacetylases in cell proliferation. *Current Opinion in Genetics & Development*, 9(1), 40–48.

Kuo, M.-H. and Allis, C.D. (1998). Roles of histone acetyltransferases and deacetylases in gene regulation. *Bioessays*, 20(8), 615–626.

Kuo, M.T., Mandel, J.L., Chambon, P. (1979). DNA methylation: Correlation with DNase I sensitivity of chicken ovalbumin and conalbumin chromatin. *Nucleic Acids Research*, 7(8), 2105–2113.

Lachmann, M. and Jablonka, E. (1996). The inheritance of phenotypes: An adaptation to fluctuating environments. *Journal of Theoretical Biology*, 181(1), 1–9.

Lamarck, J.-B.D. (1809a). *Philosophie zoologique ou exposition des considérations relatives à l'histoire naturelle des animaux. Tome 1*. Dentu, Paris.

Lamarck, J.-B.D. (1809b). *Philosophie zoologique ou exposition des considérations relatives à l'histoire naturelle des animaux. Tome 2*. Dentu, Paris.

Lamb, M.J. (2011). Attitudes to soft inheritance in Great Britain, 1930s–1970s. In *The Transformation of Lamarckism: From Subtle Fluids to Molecular Biology*, Gissis, S.B. and Jablonka, E. (eds). MIT Press, Cambridge, 109–120.

Lederberg, S. (1957). Suppression of the multiplication of heterologous bacteriophages in lysogenic bacteria. *Virology*, 3(3), 496–513.

Lederberg, J. (1958). Genetic approaches to somatic cell variation: Summary comment. *Journal of Cellular and Comparative Physiology*, 52(S1), 383–401.

Lederberg, J. (1966). Experimental genetics and human evolution. *Bulletin of the Atomic Scientists*, 22(8), 4–11.

Lederberg, J. (1967). Dangers of reprogramming cells. *Science*, 158(3799), 313–313.

Lehmann, L. (2008). The adaptive dynamics of niche constructing traits in spatially subdivided populations: Evolving posthumous extended phenotypes. *Evolution*, 62(3), 549–566.

Lei, H., Oh, S.P., Okano, M., Jüttermann, R., Goss, K.A., Jaenisch, R., Li, E. (1996). De novo DNA cytosine methyltransferase activities in mouse embryonic stem cells. *Development*, 122(10), 3195–3205.

Lennox, J.G. (2006). The comparative study of animal development: From Aristotle to William Harvey's Aristotelianism. In *The Problem of Animal Generation in Early Modern Philosophy*, Smith, J.E.H. (ed.). Cambridge University Press, Cambridge.

Lenoir, T. (1982). *The Strategy of Life: Teleology and Mechanics in Nineteenth-Century German Biology*. University of Chicago Press, Chicago.

Lesky, E. (1950). *Die Zeugungs- und Vererbungslehren der Antike und ihr Nachwirken*. Abhandlungen der Geistes- und Sozialwis- senschaftlichen Klasse 19. Akademie der Wissenschaften und der Literatur, Mainz.

Li, C. and Wang, J. (2013). Quantifying Waddington landscapes and paths of non-adiabatic cell fate decisions for differentiation, reprogramming and transdifferentiation. *Journal of The Royal Society Interface*, 10(89), 20130787.

Li, E., Bestor, T.H., Jaenisch, R. (1992). Targeted mutation of the DNA methyltransferase gene results in embryonic lethality. *Cell*, 69(6), 915–926.

Li, E., Beard, C., Jaenisch, R. (1993). Role for DNA methylation in genomic imprinting. *Nature*, 366(6453), 362–365.

Limoges, C. (1980). A second glance at evolutionary biology in France. In *The Evolutionary Synthesis: Perspectives on the Unification of Biology*, Mayr, E. and Provine, W.B. (eds). Harvard University Press, Cambridge, 322–328.

Linné, C.V. (1760). *Disquisitio de sexu plantarum*.

Lippman, Z., Gendrel, A.-V., Black, M., Vaughn, M.W., Dedhia, N., McCombie, W.R., Lavine, K., Mittal, V., May, B., Kasschau, K.D. et al. (2004). Role of transposable elements in heterochromatin and epigenetic control. *Nature*, 430(6998), 471–476.

Liu, N. and Pan, T. (2015). RNA epigenetics. *Translational Research*, 165(1), 28–35.

Liu, X.S., Wu, H., Ji, X., Stelzer, Y., Wu, X., Czauderna, S., Shu, J., Dadon, D., Young, R.A., Jaenisch, R. (2016). Editing DNA methylation in the mammalian genome. *Cell*, 167(1), 233–247.

Lock, M. (2013). The epigenome and nature/nurture reunification: A challenge for anthropology. *Medical Anthropology*, 32(4), 291–308.

Lock, L.F., Takagi, N., Martin, G.R. (1987). Methylation of the Hprt gene on the inactive X occurs after chromosome inactivation. *Cell*, 48(1), 39–46.

Longo, G. (2021). Programming evolution: A crack in science. *Organisms. Journal of Biological Sciences*, 5(1), 5–16.

Longo, G. and Tendéro, P.-E. (2008). L'alphabet, la Machine et l'ADN : l'incomplétude causale de la théorie de la programmation en biologie moléculaire. In *Biologie du XXIe siècle : évolution des concepts fondateurs*, Miquel, P.-A. (ed.). DeBoeck, Brussels.

Lopez-Beltran, C. (2007). The medical origins of heredity. In *Heredity Produced: At the Crossroads of Biology, Politics, and Culture, 1500–1870*, Müller-Wille, S. and Rheinberger, H.-J. (eds). MIT Press, Cambridge.

Loviconi, L. (2019). Réflexions autour des maladies héréditaires dans les traités médicaux des XIVe et XVe siècles. *Annales de démographie historique*, 137(1), 49–73.

Løvtrup, S. (1974). *Epigenetics; a Treatise on Theoretical Biology*. Wiley, New York.

Luria, S.E. (1953). Host-induced modifications of viruses. *Cold Spring Harbor Symposia on Quantitative Biology*, 18, 237–244.

Luria, S.E. and Human, M.L. (1952). A nonhereditary, host-induced variation of bacterial viruses. *Journal of Bacteriology*, 64(4), 557–569.

Lwoff, A. (1949). Les organites doués de continuité génétique chez les Protistes. Unités biologiques douées de continuité génétique. In *Colloques Internationaux du Centre National de la Recherche Scientifique VIII*. Éditions du Centre National de la Recherche Scientifique, Paris, 7–24.

Lwoff, A. (1950). *Problems of Morphogenesis in Ciliates: The Kinetosomes in Development, Reproduction and Evolution*. Wiley, New York.

Lwoff, A. (1990). L'organisation du cortex chez les ciliés: un exemple d'hérédité de caractère acquis. *Comptes Rendus de l'Académie des Sciences de Paris, Sciences de la Vie*, 310(4), 109–111.

Lyon, M.F. (1968). Chromosomal and subchromosomal inactivation. *Annual Review of Genetics*, 2(1), 31–52.

Malebranche, N. (1674). *De la recherche de la vérité. Où l'on traite de la nature de l'esprit de l'homme, & de l'usage qu'il en doit faire pour éviter l'erreur dans les sciences. Tome I*. André Pralard, Paris.

Malpighi, M. (1666). *Dissertatio epistolica de formatione pulli in ovo*. J. Martyn, London.

Mandel, J.L. and Chambon, P. (1979). DNA methylation: Organ specific variations in the methylation pattern within and around ovalbumin and other chicken genes. *Nucleic Acids Research*, 7(8), 2081–2103.

Markert, C.L. (1968). Neoplasia: A disease of cell differentiation. *Cancer Research*, 28(9), 1908–1914.

Martienssen, R. (2008). Great leap forward? Transposable elements, small interfering RNA and adaptive lamarckian evolution. *The New Phytologist*, 179(3), 570–572.

Matsushita, Y. and Kaneko, K. (2020). Homeorhesis in Waddington's landscape by epigenetic feedback regulation. *Physical Review Research*, 2(2), 023083.

Mattei, A.L., Bailly, N., Meissner, A. (2022). DNA methylation: A historical perspective. *Trends in Genetics*, 38(7), 676–707.

Maupertuis, P.L.M.D. (1745). *Vénus physique, contenant deux dissertations, l'une sur l'origine des hommes et des animaux, et l'autre sur l'origine des noirs*. Jean-Martin Husson, Paris.

Maupertuis, P.L.M.D. (1756a). *Œuvres. Tome second.* New corrected and amended edition. Jean Marie Bruyset, Lyon.

Maupertuis, P.L.M.D. (1756b). *Œuvres de M. de Maupertuis. Tome second.* New corrected and amended edition. Jean Marie Bruyset, Lyon.

Maynard Smith, J. (1993). *The Theory of Evolution*, 3rd edition. Cambridge University Press, Cambridge/New York [1958].

Mayr, E. (1961). Cause and effect in biology. *Science*, 134, 1501–1506.

Mayr, E. and Provine, W.B. (1998). *The Evolutionary Synthesis: Perspectives on the Unification of Biology, with a New Preface*. Harvard University Press, Cambridge.

McClintock, B. (1948). Mutable loci in maize. *Carnegie Institution of Washington Yearbook*, 47, 155–169.

McClintock, B. (1951). Chromosome organization and genic expression. *Cold Spring Harbor Symposia on Quantitative Biology*, 16, 40.

McGhee, J.D. and Felsenfeld, G. (1980). Nucleosome structure. *Annual Review of Biochemistry*, 49(1), 1115–1156.

McGhee, J.D. and Ginder, G.D. (1979). Specific DNA methylation sites in the vicinity of the chicken β-globin genes. *Nature*, 280(5721), 419–420.

Meckel, J.F. (1821). *System der vergleichenden Anatomie*. Renger, Halle.

Meckel, J.F. (1828). *Traité général d'anatomie comparée*. Villeret, Paris.

Medawar, P.B. and Medawar, J.S. (1983). *Aristotle to Zoos: A Philosophical Dictionary of Biology*. Harvard University Press, Cambridge.

Meloni, M. (2015). *Heredity 2.0: The Epigenetics Effect*. Taylor & Francis, Abingdon-on-Thames.

Meloni, M. (2016). *Political Biology: Science and Social Values in Human Heredity from Eugenics to Epigenetics*. Springer, Berlin.

Meloni, M. and Müller, R. (2018). Transgenerational epigenetic inheritance and social responsibility: Perspectives from the social sciences. *Environmental Epigenetics*, 4(2), dvy019.

Mendel, G.J. (1866). Versuche über Pflanzenhybriden. *Verhandlungen des naturforschenden Vereines in Brünn*, 3–47.

Mendel, G.J. (1901). Experiments in plant hybridization (1865). *Read at the February 8th, and March 8th, 1865, Meetings of the Brünn Natural History Society*, W. Bateson (trans.), Brno.

Merlin, F. (2010). Evolutionary chance mutation: A defense of the modern synthesis' consensus view. *Philosophy & Theory in Biology*, 2.

Millar, D.S., Holliday, R., Grigg, G.W. (2003). Five not four: History and significance of the fifth base. In *The Epigenome: Molecular Hide and Seek*, Beck, S. and Olek, A. (eds). John Wiley & Sons, New York, 3–22.

Molière, J.-B. P. (1670). *Le Bourgeois gentilhomme, comédie-ballet*. Chambord.

Monk, M. (1995). Epigenetic programming of differential gene expression in development and evolution. *Developmental Genetics*, 17(3), 188–197.

Monod, J. (1956). Remarks on the mechanism of enzyme induction. In *Enzymes: Units of Biological Structure and Function*, Gaebler, O.H. (ed.). Academic Press, New York, 7–28.

Montaigne, M.D. (2001). *Les Essais*. Livre de Poche, Paris [1595].

Moore, D.S. (2017). The potential of epigenetics research to transform conceptions of phenotype development. *Human Development*, 60(2), 69–80.

Morange, M. (1997). The transformation of molecular biology on contact with higher organisms, 1960–1980: From a molecular description to a molecular explanation. *History and Philosophy of the Life Sciences*, 369–393.

Morange, M. (2002). The relations between genetics and epigenetics: A historical point of view. *Annals of the New York Academy of Sciences*, 981(1), 50–60.

Morange, M. (2008). What history tells us XIII. Fifty years of the central dogma. *Journal of Biosciences*, 33(2), 171.

Morange, M. (2009). What history tells us XVII. Conrad Waddington and the nature of life. *Journal of Biosciences*, 34(2), 195–98.

Morange, M. (2011). The attempt of Nikolai Koltzoff (Koltsov) to link genetics, embryology and physical chemistry. *Journal of Biosciences*, 36(2), 211–214.

Morange, M. (2012). *Les secrets du vivant. Contre la pensée unique en biologie*. La Découverte, Paris.

Morange, M. (2013). What history tells us XXXII. The long and tortuous history of epigenetic marks. *Journal of Biosciences*, 38(3), 451–454.

Morange, M. (2020). *The Black Box of Biology: A History of the Molecular Revolution*, M. Cobb (trans.). Harvard University Press, Cambridge.

Moreno, A. and Mossio, M. (2015). *Biological Autonomy: A Philosophical and Theoretical Enquiry*. Springer, Berlin.

Morgan, T.H. (1910). Chromosomes and heredity. *The American Naturalist*, 44(524), 449–496.

Morgan, T.H. (1913). Factors and unit characters in Mendelian heredity. *The American Naturalist*, 47(553), 5–16.

Morgan, T.H. (1926). Genetics and the physiology of development. *The American Naturalist*, 60(671), 489–515.

Morgan, T.H. (1935). The relation of genetics to physiology and medicine. *The Scientific Monthly*, 41(1), 5–18.

Morgan, T.H., Sturtevant, A.H., Bridges, C.B. (1915). *The Mechanism of Mendelian Heredity*. Henry Holt, New York.

Morgan, H.D., Santos, F., Green, K., Dean, W., Reik, W. (2005). Epigenetic reprogramming in mammals. *Human Molecular Genetics*, 14(1), R47–R58.

Muller, H.J. (1930). Types of visible variations induced by X-rays in Drosophila. *Journal of Genetics*, 22(3), 299–334.

Müller-Wille, S. and Rheinberger, H.-J. (2007). Heredity–the formation of an epistemic space. In *Heredity Produced: At the Crossroads of Biology, Politics, and Culture, 1500–1870*, Müller-Wille, S. and Rheinberger, H.-J. (eds). MIT Press, Cambridge.

Nanney, D.L. (1957). The role of the cytoplasm in heredity. In *The Chemical Basis of Heredity*, McElroy, W.D. and Glass, H.B. (eds). Johns Hopkins Press, Baltimore, 134–164.

Nanney, D.L. (1958a). Epigenetic control systems. *Proceedings of the National Academy of Sciences*, 44(7), 712–717.

Nanney, D.L. (1958b). Epigenetic factors affecting mating type expression in certain ciliates. In *Cold Spring Harbor Symposia on Quantitative Biology*. Cold Spring Harbor Laboratory Press, Cold Spring Harbor, 23, 327–335.

Nanney, D.L. (1958c). *Genetic and Paragenetic Systems* [apocryphal title]. Gif-sur-Yvette.

Nanney, D.L. (1968). Ciliate genetics: Patterns and programs of gene action. *Annual Review of Genetics*, 2(1), 121–140.

Nashimoto, M. (2001). The RNA/Protein symmetry hypothesis: Experimental support for reverse translation of primitive proteins. *Journal of Theoretical Biology*, 209(2), 181–187.

Nicoglou, A. (2018). Waddington's epigenetics or the pictorial meetings of development and genetics. *History and Philosophy of the Life Sciences*, 40(4), 1–25.

Nicoglou, A. and Merlin, F. (2017). Epigenetics: A way to bridge the gap between biological fields. *Studies in History and Philosophy of Science Part C: Studies in History and Philosophy of Biological and Biomedical Sciences*, 66, 73–82.

Nicolis, G. and Prigogine, I. (1977). *Self-Organization in Nonequilibrium Systems*. Wiley, New York.

Nirenberg, M.W. (1967). Will society be prepared? *Science*, 157(3789), 633–633.

Novick, A. and Weiner, M. (1957). Enzyme induction as an all-or-none phenomenon. *Proceedings of the National Academy of Sciences*, 43(7), 553–566.

Odling-Smee, J.F., Laland, K.N., Feldman, M.W. (2003). *Niche Construction: The Neglected Process in Evolution*. Princeton University Press, Princeton.

Ogryzko, V.V. (2008). Erwin Schroedinger, Francis Crick and epigenetic stability. *Biology Direct*, 3(1), 1–10.

Ohyama, K., Fukuzawa, H., Kohchi, T., Shirai, H., Sano, T., Sano, S., Umesono, K., Shiki, Y., Takeuchi, M., Chang, Z. et al. (1986). Chloroplast gene organization deduced from complete sequence of liverwort Marchantia polymorpha chloroplast DNA. *Nature* 322(6079), 572–574.

Okano, M., Bell, D.W., Haber, D.A., Li, E. (1999). DNA methyltransferases Dnmt3a and Dnmt3b are essential for de novo methylation and mammalian development. *Cell*, 99(3), 247–257.

Oken, L. (1809). *Lehrbuch der Naturphilosophie*. Friedrich Fromman, Jena.

Olby, R.C. (1963). Charles Darwin's manuscript of pangenesis. *The British Journal for the History of Science*, 1(3), 251–263.

Olby, R.C. (1979). Mendel no mendelian? *History of Science*, 17(1), 53–72.

Olby, R.C. (1985). *Origins of Mendelism*. University of Chicago Press, Chicago.

Olins, A.L. and Olins, D.E. (1974). Spheroid chromatin units (ν bodies). *Science*, 183(4122), 330–332.

Ooi, S.K., Qiu, C., Bernstein, E., Li, K., Jia, D., Yang, Z., Erdjument-Bromage, H., Tempst, P., Lin, S.-P., Allis, C.D. et al. (2007). DNMT3L connects unmethylated lysine 4 of histone H3 to de novo methylation of DNA. *Nature*, 448(7154), 714–717.

Oppenheimer, J.M. (1967). *Essays in the History of Embryology and Biology*. MIT Press, Cambridge.

Paik, W.K., Paik, D.C., Kim, S. (2007). Historical review: The field of protein methylation. *Trends in Biochemical Sciences*, 32(3), 146–152.

Pankov, E.Y. and Dedukh, N.V. (1970). Effect of exogenous RNA isolated from bones and embryonic tissues on growth of sarcoma M1. *Bulletin of Experimental Biology and Medicine*, 70(3), 1046–1048.

Paré, A. (1971). *Des monstres et prodiges*. Librairie Droz, Geneva.

Paweletz, N. (2001). Walther Flemming: Pioneer of mitosis research. *Nature Reviews Molecular Cell Biology*, 2(1), 72–75.

Pazin, M.J. and Kadonaga, J.T. (1997). What's up and down with histone deacetylation and transcription? *Cell*, 89(3), 325–328.

Pearson, H. (2008). Disputed definitions. Epigenetic. *Nature*, 455(7216), 1023–1028.

Pickersgill, M., Niewöhner, J., Müller, R., Martin, P., Cunningham-Burley, S. (2013). Mapping the new molecular landscape: Social dimensions of epigenetics. *New Genetics and Society*, 32(4), 429–447.

Pigliucci, M. and Müller, G.B. (2010). *Evolution: The Extended Synthesis*. MIT Press, Cambridge/London.

Pigliucci, M., Murren, C.J., Schlichting, C.D. (2006). Phenotypic plasticity and evolution by genetic assimilation. *Journal of Experimental Biology*, 209(12), 2362–2367.

Plato (1933). *La République*, É. Chambry (trans.). Les Belles Lettres, Paris.

Pliny the Elder (1848). *Histoire naturelle: oeuvre compète*. Dubochet, Paris.

Plutarch (1844). *Oeuvres morales. Tome troisième. Sur les délais de la justice divine*, Ricard (trans.). Lefèvre, Paris.

Pocheville, A. (2018). Biological information as choice and construction. philosophy of science. *Proceedings of the 2016 Biennial Meeting of the Philosophy of Science Association*, 85(5).

Pocheville, A. (2019). A Darwinian dream: On time, levels, and processes in evolution. In *Evolutionary Causation. Biological and Philosophical Reflections*, Uller, T. and Laland, K.N. (eds). MIT Press, Boston.

Pocheville, A. and Danchin, É. (2017). Genetic assimilation and the paradox of blind variation. In *Challenging the Modern Synthesis*, Walsh, D.M. and Huneman, P. (eds). Oxford University Press, Oxford.

Pomata, G. (2003). Comments on session III: Heredity and medicine. In *Conference: A Cultural History of Heredity II: 18th and 19th Centuries, Prepoint*. Max Planck Institute for the History of Science Berlin, Berlin, 145–152.

Preus, A. (1977). Galen's criticism of Aristotle's conception theory. *Journal of the History of Biology*, 10(1), 65–85.

Ptashne, M. (2007). On the use of the word 'epigenetic'. *Current Biology*, 17(7), R233–R236.

Rando, O.J. (2012). Combinatorial complexity in chromatin structure and function: Revisiting the histone code. *Current Opinion in Genetics & Development. Genome Architecture and Expression*, 22(2), 148–155.

Reik, W., Römer, I., Barton, S.C., Surani, M.A., Howlett, S.K., Klose, J. (1993). Adult phenotype in the mouse can be affected by epigenetic events in the early embryo. *Development*, 119(3), 933–942.

Reik, W., Dean, W., Walter, J. (2001). Epigenetic reprogramming in mammalian development. *Science*, 293(5532), 1089–1093.

Rey, O., Eizaguirre, C., Angers, B., Balthazar-Soares, M., Sagonas, K., Prunierm, J.G., Blanchet, S. (2020). Linking epigenetics and biological conservation: Towards a conservation epigenetics perspective. *Functional Ecology*, 34(2), 414–427.

Rheinberger, H.-J., Müller-Wille, S., Meunier, R. (2015). Gene. In *The Stanford Encyclopedia of Philosophy*, Zalta, E.N. (ed.). Metaphysics Research Lab, Stanford University Press, Stanford.

Richardson, S.S. (2015). Maternal bodies in the postgenomic order: Gender and the explanatory landscape of epigenetics. In *Postgenomics: Perspective on Biology after the Genome*, Richardson, S. and Stevens, H. (eds). Duke University Press, Durham, 210–231.

Richardson, S.S. (ed.) (2021). Introduction: The maternal imprint. In *The Maternal Imprint*. University of Chicago Press, Chicago, 1–25.

Richardson, S.S., Daniels, C.R., Gillman, M.W., Golden, J., Kukla, R., Kuzawa, C., Rich-Edwards, J. (2014). Society: Don't blame the mothers. *Nature*, 512(7513), 131–132.

Richmond, T.J., Finch, J.T., Rushton, B., Rhodes, D., Klug, A. (1984). Structure of the nucleosome core particle at 7 Å resolution. *Nature*, 311(5986), 532–537.

Riggs, A.D. (1975). X inactivation, differentiation, and DNA methylation. *Cytogenetic and Genome Research*, 14(1), 9–25.

Robinson, G. (1979). *A Prelude to Genetics: Theories of a Material Substance of Heredity, Darwin to Weismann*. Coronado Press, Lawrence.

Roderick, T.H. (1986). Private conversation at McDonald's Raw Bar, Bethesda; with Franck Ruddle, Victor McKusick and others, as reported in Kuska, Bob. 1998. "Beer, Bethesda, and Biology: How 'Genomics' Came Into Being. *JNCI: Journal of the National Cancer Institute*, 90(2), 93. doi: 10.1093/jnci/90.2.93.

Roe, S.A. (1981). *Matter, Life, and Generation: Eighteenth-Century Embryology and the Haller-Wolff Debate*. Cambridge University Press, Cambridge.

Roll-Hansen, N. (1978). Drosophila genetics: A reductionist research program. *Journal of the History of Biology*, 11(1), 159–210.

Roll-Hansen, N. (2014). Commentary: Wilhelm Johannsen and the problem of heredity at the turn of the 19th century. *International Journal of Epidemiology*, 43(4), 1007–1013.

Rollins, R.A., Haghighi, F., Edwards, J.R., Das, R., Zhang, M.Q., Ju, J., Bestor, T.H. (2006). Large-scale structure of genomic methylation patterns. *Genome Research*, 16(2), 157–163.

Romanes, G.J. (1888). Lamarckism versus Darwinism. *Nature*, 38, 413–413.

Rosen, R. (1984). Genomic control of global features in morphogenesis. In *Modelling of Patterns in Space and Time*, Jäger, W. and Murray, J.D. (eds). Springer, Berlin/Heidelberg, 318–330.

Rosenberg, A. (2006). Is epigenetic inheritance a counterexample to the central dogma? *History and Philosophy of the Life Sciences*, 28(4), 549–565.

Roses, A.D. (2003). The genome era begins... *Nature Genetics*, 33(3), 217–217.

Russell, E.S. (1916). *Form and Function (1982 Reprint)*. University of Chicago Press, Chicago.

Russo, V.E., Martienssen, R.A., Riggs, A.D. (1996). *Epigenetic Mechanisms of Gene Regulation*. Cold Spring Harbor Laboratory Press, Cold Spring Harbor.

Saade, E. and Ogryzko, V.V. (2014). Epigenetics: What it is about? *Biopolymers and Cell*, 30(1), 3–9.

Sabl, J.F. and Laird, C.D. (1992). Epigene conversion: A proposal with implications for gene mapping in humans. *American Journal of Human Genetics*, 50(6), 1171.

Sager, R. (1972). *Cytoplasmic Genes and Organelles*. Academic Press, Cambridge.

Sager, R. and Kitchin, R. (1975). Selective silencing of eukaryotic DNA: A molecular basis is proposed for programmed inactivation or loss of eukaryotic DNA. *Science*, 189(4201), 426–433.

Sandler, I. and Sandler, L. (1985). A conceptual ambiguity that contributed to the neglect of Mendel's paper. *History and Philosophy of the Life Sciences*, 3–70.

Sanger, F., Air, G.M., Barell, B.G., Brown, N.L., Coulson, A.R., Fiddes, J.C., Hutchison III, C.A., Slocombe, P.M., Smith, M. (1977). Nucleotide sequence of bacteriophage φX174 DNA. *Nature*, 265(5596), 687–695.

Sapp, J. (1987). *Beyond the Gene: Cytoplasmic Inheritance and the Struggle for Authority in Genetics*. Oxford University Press, New York.

Sapp, J. (2003). *Genesis: The Evolution of Biology*. Oxford University Press, New York.

Sarkar, S. (1996). Biological information: A skeptical look at some central dogmas of molecular biology. In *The Philosophy and History of Molecular Biology: New Perspectives*, Sarkar, S. (ed.). Kluwer Academic Publishers, Dordrecht, 187–232.

Sassa, A., Kanemaru, Y., Kamoshita, N., Honma, M., Yasuam, M. (2016). Mutagenic consequences of cytosine alterations site-specifically embedded in the human genome. *Genes and Environment*, 38(1), 1–6.

Saulnier, K.M. and Dupras, C. (2017). Race in the postgenomic era: Social epigenetics calling for interdisciplinary ethical safe-guards. *The American Journal of Bioethics*, 17(9), 58–60.

Scarano, E., Iaccarino, M., Grippo, P., Parisi, E. (1967). The heterogeneity of thymine methyl group origin in DNA pyrimidine isostichs of developing sea urchin embryos. *Proceedings of the National Academy of Sciences*, 57(5), 1394–1400.

Schelling, F.W.J.V. (1797). *Ideen zu einer Philosophie der Natur als Einleitung in das Studium dieser Wissenschaft*. Philipp Krüll, Landshut.

Schiebinger, G. (2021). Reconstructing developmental landscapes and trajectories from single-cell data. *Current Opinion in Systems Biology*, 27, 100351.

Schmidt, E. and Kornfeld, J.-W. (2016). Decoding Lamarck–transgenerational control of metabolism by noncoding RNAs. *Pflügers Archiv – European Journal of Physiology*, 468(6), 959–969.

Schmitt, S. (2021). Les théories buffoniennes de la génération, entre permanence et plasticité. In *La biologie au défi de l'histoire*, Loison, L. and Pradeu, T. (eds). Éditions Matériologiques, Paris.

Shi, J., Teschendorff, A.E., Chen, W., Chen, L., Li, T. (2020). Quantifying Waddington's epigenetic landscape: A comparison of single-cell potency measures. *Briefings in Bioinformatics*, 21(1), 248–261.

Shinozaki, K., Ohme, M., Tanaka, M., Wakasugi, T., Hayashida, N., Matsubayashi, T., Zaita, N., Chunwongse, J., Obokata, J., Yamaguchi-Shinozaki, K. et al. (1986). The complete nucleotide sequence of the tobacco chloroplast genome: Its gene organization and expression. *The EMBO Journal*, 5(9), 2043–2049.

Slack, J.M. (2002). Conrad Hal Waddington: The last renaissance biologist? *Nature Reviews Genetics*, 3(11), 889–895.

Smith, D.R. (2016). The past, present and future of mitochondrial genomics: Have we sequenced enough mtDNAs? *Briefings in Functional Genomics*, 15(1), 47–54.

Sonneborn, T.M. and Beale, S.H. (eds) (1949). Influence des gènes, des plasmagènes et du milieu dans le déterminisme des caractères antigéniques chez *Paramecium aurelia* (variété 4). In *Unités biologiques douées de continuité génétique*. Éditions du Centre National de la Recherche Scientifique, Paris, 25–36.

Soulier, A. (2019). Les recherches génomiques et post-génomiques dans le tournant biosocial des sciences médicales : aspects épistémologiques et éthiques. Research Project, CNRS, section 35, CID 53.

Southern, E.M. (1975). Detection of specific sequences among DNA fragments separated by gel electrophoresis. *Journal of Molecular Biology*, 98(3), 503–517.

Spallanzani, L. (1765). *Saggio di osservazioni microscopiche concernenti il sistema della generazione de' signori di Needham et Buffon*. Bartolomeo Soliani, Modena.

Spencer, H. (1864). *The Principles of Biology*. Williams and Norgate, London/Edinburgh.

Srinivasan, P.R. and Borek, E. (1964). Enzymatic alteration of nucleic acid structure: Enzymes put finishing touches characteristic of each species on RNA and DNA by insertion of methyl groups. *Science*, 145(3632), 548–553.

Stedman, E. and Stedman, E. (1950). Cell specificity of histones. *Nature*, 166(4227), 780–781.

Steele, E.J. (2016). Commentary: Past, present, and future of epigenetics applied to livestock breeding. Hard versus soft lamarckian inheritance mechanisms. *Frontiers in Genetics*, 7, 1–7.

Stent, G.S. (1970). DNA. *Daedalus*, 99(4), 909–937.

Stotz, K. (2006). Molecular epigenesis: Distributed specificity as a break in the central dogma. *History and Philosophy of the Life Sciences*, 28(4), 533–548.

Sultan, S.E. (2015). *Organism and Environment: Ecological Development, Niche Construction, and Adaptation*. Oxford University Press, Oxford.

Sutton, W.S. (1903). The chromosomes in heredity. *The Biological Bulletin*, 4(5), 231–250.

Suvà, M.L., Riggi, N., Bernstein, B.E. (2013). Epigenetic reprogramming in cancer. *Science*, 339(6127), 1567–1570.

Tamaru, H. and Selker, E.U. (2001). A histone H3 methyltransferase controls DNA methylation in *Neurospora crassa*. *Nature*, 414(6861), 277–283.

Temin, H.M. (1976). The DNA provirus hypothesis: The establishment and implications of RNA-directed DNA synthesis. *Science*, 192(4244), 1075–1080.

Temkin, O. (1950). German concepts of ontogeny and history around 1800. *Bulletin of the History of Medicine*, 24(3), 227–246.

Terrall, M. (2007). Speculation and experiment in enlightenment life sciences. In *Heredity Produced: At the Crossroads of Biology, Politics, and Culture, 1500–1870*, Müller-Wille, S. and Rheinberger, H.-J. (eds). MIT Press, Cambridge.

Thieffry, D. (1996). *Escherichia coli* as a model system with which to study cell differentiation. *History and Philosophy of the Life Sciences*, 163–193.

Thieffry, D. and Sarkar, S. (1998). Forty years under the central dogma. *Trends in Biochemical Sciences*, 23(8), 312–316.

Tran, R.K., Henikoff, J.G., Zilberman, D., Ditt, R.F., Jacobsen, S.E., Henikoff, S. (2005). DNA methylation profiling identifies CG methylation clusters in Arabidopsis genes. *Current Biology*, 15(2), 154–159.

Tsanev, R. and Sendov, B. (1971). Possible molecular mechanism for cell differentiation in multicellular organisms. *Journal of Theoretical Biology*, 30(2), 337–393.

Tschermak, E.V. (1900). Über künstliche Kreuzung bei Pisum sativum. *Berichte der deutschen botanischen Gesellschaft*, 18, 232–239.

Turner, B.M. (1991). Histone acetylation and control of gene expression. *J. Cell. Sci.*, 99(1), 13–20.

Turner, B.M. (2000). Histone acetylation and an epigenetic code. *Bioessays*, 22(9), 836–845.

Van Beneden, É. (1883). *Recherches sur la maturation de l'oeuf, la fécondation, et la division cellulaire*. Librairie Clemm, Paris.

Van de Vijver, G., Van Speybroeck, L., De Waele, D. (2002). Epigenetics: A challenge for genetics, evolution, and development? *Annals of the New York Academy of Sciences*, 981(1), 1–6.

Veigl, S.J. (2017). Use/disuse paradigms are ubiquitous concepts in characterizing the process of inheritance. *RNA Biology*, 14(12), 1700–1704.

Vitas, M. and Dobovišek, A. (2017). On a quest of reverse translation. *Foundations of Chemistry*, 19(2), 139–155.

Von Baer, K.E. (1827). *De ovi mammalium et hominis genesi*. Leopold Voss, Leipzig.

Von Baer, K.E. (1828). *Über entwickelungsgeschichte der thiere: beobachtung und reflexion. Tome 1*. Bei den gebrüdern Bornträger.

de Vries, H. (1889). *Intracellulare pangenesis*. G. Fischer, Jena.

de Vries, H. (1900). Das Spaltungsgesetz der Bastarde (Vorläufige Mitteilung). *Berichte der deutschen botanischen Gesellschaft*, 18, 83–90.

Waalwijk, C. and Flavell, R.A. (1978). DNA methylation at a CCGG sequence in the large intron of the rabbit β-globin gene: Tissue-specific variations. *Nucleic Acids Research*, 5(12), 4631–4642.

Waddington, C.H. (1939). *An Introduction to Modern Genetics*. Routledge, London.

Waddington, C.H. (1940). *Organisers and Genes*. Cambridge University Press, London.

Waddington, C.H. (1952). *The Epigenetics of Birds*. Cambridge University Press, London.

Waddington, C.H. (1953). Role of plasmagenes. *Nature*, 172, 784–785.

Waddington, C.H. (1956a). Embryology, epigenetics and biogenetics. *Nature*, 177(4522), 1241–1241.

Waddington, C.H. (1956b). *The Principles of Embryology*. Routledge, London.

Waddington, C.H. (1957). *The Strategy of the Genes*. Routledge, London.

Waddington, C.H. (1968). The basic ideas of biology. In *Towards a Theoretical Biology: An IUBS Symposium*, Waddington, C.H. (ed.). Edinburgh U.P, Edinburgh.

Walsh, C.P., Chaillet, J.R., Bestor, T.H. (1998). Transcription of IAP endogenous retroviruses is constrained by cytosine methylation. *Nature Genetics*, 20(2), 116–117.

Wang, H.D. and Allard, P. (2022). Challenging dogmas: How transgenerational epigenetics reshapes our views on life. *Journal of Experimental Zoology Part A: Ecological and Integrative Physiology*, 337(1), 70–74.

Wang, J., Zhang, K., Xu, L., Wang, E. (2011). Quantifying the Waddington landscape and biological paths for development and differentiation. *Proceedings of the National Academy of Sciences*, 108(20), 8257–8262.

Wang, P., Song, C., Zhang, H., Wu, Z., Tian, X.-J., Xing, J. (2014). Epigenetic state network approach for describing cell phenotypic transitions. *Interface Focus*, 4(3).

Wang, Y., Liu, H., Sun, Z. (2017). Lamarck rises from his grave: Parental environment-induced epigenetic inheritance in model organisms and humans. *Biological Reviews*.

Wanscher, J.H. (1975). An analysis of Wilhelm Johannsen's genetical genotype "term" 1909–1926. *Hereditas*, 79(1), 1–4.

Watson, J.D. and Crick, F.H.C. (1953). Molecular structure of nucleic acids: A structure for deoxyribose nucleic acid. *Nature*, 171(4356), 737–738.

Watson, R.E. and Goodman, J.I. (2002). Epigenetics and DNA methylation come of age in toxicology. *Toxicological Sciences*, 67(1), 11–16.

Weber, M., Davies, J.J., Wittig, D., Oakeley, E.J., Haase, M., Lam, W.L., Schübeler, D. (2005). Chromosome-wide and promoter-specific analyses identify sites of differential DNA methylation in normal and transformed human cells. *Nature Genetics*, 37(8), 853–862.

Weber, M., Hellman, I., Stadler, M.B., Ramos, L., Pääbo, S., Rebhan, M., Schübeler, D. (2007). Distribution, silencing potential and evolutionary impact of promoter DNA methylation in the human genome. *Nature Genetics*, 39(4), 457–466.

Weismann, A. (1882). *Studies in the Theory of Descent. Translated and Edited, with Notes, by Raphael Meldola. With a Prefatory Notice by Charles Darwin*, volume 2. Sampson Low/Marston/Searle/Rivington, London.

Weismann, A. (1891). *Essays Upon Heredity and Kindred Biological Problems*, volume 1. Clarendon Press, Oxford.

Weismann, A. (1893a). The all-sufficiency of natural selection: A reply to herbert spencer. *Contemporary Review*, 309–38, 556–610.

Weismann, A. (1893b). *The Germ-Plasm a Theory of Heredity*. Charles Scribner's Sons, New York.

Weismann, A. (1904). *The Evolution Theory Vol. 2*, Thomson, A.J. (trans.). Edward Arnold, London.

West-Eberhard, M.J. (2003). *Developmental Plasticity and Evolution*. Oxford University Press, New York.

von Wettstein, F. (1926). Über plasmatische Vererbung, sowie Plasma-und Genwirkung. *Nachrichten von der Gesellschaft der Wissenschaften zu Göttingen, Mathematisch-Physikalische Klasse*, 250–281.

Wheeler, H.L. (1904). Researches on pyrimidine derivatives: 5-Methylcytosine. *Journal of the American Chemical Society*, 31, 591–606.

White House (2000). White House press release. *Office of the Press Secretary*, Washington, D.C.

Whitehead, A.N. (1925). *Science and the Modern World*. Macmillan, London.

Whitehead, A.N. (1929). *Process and Reality*. Simon and Schuster, New York.

Wightman, W.P. (1956). Embryology, epigenetics and biogenetics. *Nature*, 177(4522), 1240–1241.

Wilkins, M.H.F., Stokes, A.R., Wilson, H.R. (1953). Molecular structure of nucleic acids: Molecular structure of deoxypentose nucleic acids. *Nature*, 171(4356), 738–740.

Wilson, E.B. (1896). *The Cell in Development and Inheritance.* Macmillan, New York.

Wilson, E.B. (1923). The physical basis of life. *Science*, 57(1471), 277–286.

Winkler, H. (1920). *Verbreitung und Ursache der Parthenogenesis im Pflanzen- und Tierreiche.* Gustav Fischer, Jena.

Winkler, H. (1924). Über die Rolle von Kern und Protoplasma bei der Vererbung. *Zeitschrift fur induktive Abstammungs- und Vererbungslehre*, 33, 238–253.

Wolfe, C.T. (2016). *Materialism: A Historico-Philosophical Introduction.* Springer, Berlin.

Wolfe, C.T. (2017). Varieties of vital materialism. In *The New Politics of Materialism: History, Philosophy, Science*, Ellenzweig, S. and Zammito, J.H. (eds). Routledge/Taylor & Francis Group, London/New York.

Wolfe, C.T. (2019). *La philosophie de la biologie avant la biologie: une histoire du vitalisme.* Classiques Garnier, Paris.

Wolff, C.F. (1759). *Theoria Generationis.* Hendel, Halle.

Wolff, C.F. (1768). *De formatione intestinorum.* Hendel, Halle.

Wolffe, A.P. (1994). Transcription: In tune with the histones. *Cell*, 77(1), 13–16.

Wright, S. (1931). Evolution in Mendelian populations. *Genetics*, 16(2), 97.

Wright, S. (1945). Genes as physiological agents: General considerations. *American Naturalist*, 79(783), 289–303.

Wyatt, G.R. (1950). Occurrence of 5-methyl-cytosine in nucleic acids. *Nature*, 166(4214), 237–238.

Wyatt, G.R. (1951). Recognition and estimation of 5-methylcytosine in nucleic acids. *Biochemical Journal*, 48(5), 581.

Yule, G.U. (1902). Mendel's laws and their probable relations to intra-racial heredity (continued). *New Phytologist*, 1(10), 222–238.

Zhang, X., Yazaki, J., Sundaresan, A., Cokus, S., Chan, S.W.-L., Chen, H., Henderson, I.R., Shinn, P., Pellegrini, M., Jacobsen, S.E., Ecker, J.R. (2006). Genome-wide high-resolution mapping and functional analysis of DNA methylation in Arabidopsis. *Cell*, 126(6), 1189–1201.

Zweiger, G. and Scott, R.W. (1997). From expressed sequence tags to 'epigenomics': An understanding of disease processes. *Current Opinion in Biotechnology*, 8(6), 684–687.

Zwiessele, M. and Lawrence, N.D. (2017). Topslam: Waddington landscape recovery for single cell experiments. *BioRxiv*, 057778.

2

Molecular Players of Epigenetic Information

Natacha BIES ETHEVE[1], Séverine CHAMBEYRON[2] and Frédéric BANTIGNIES[2]

[1] *LGDP, Université de Perpignan Via Domitia, France*

[2] *Institut de Génétique Humaine, CNRS, Université de Montpellier, France*

2.1. Introduction

The carriers of epigenetic information are chromosomal (DNA, histones and other associated proteins) and extrachromosomal (regulatory RNAs). These carriers undergo specific molecular modifications and are found at precise locations in the genome. These modifications or epigenetic marks are put in place by nuclear proteins with specific enzymatic activity (writers), which are recruited to their active sites by transcription factors, regulatory RNAs or even marks that have already been deposited.

Epigenetic marks are genuine "signposts" found at the DNA level, in the form of methylation, and at the histone level, mainly in the form of acetylation, methylation and phosphorylation. These modifications are recognized by chromatin factors (readers), which have a direct impact on gene expression, thus influencing cell functionality. These processes are reversible, so deposited epigenetic marks can be removed by other enzymes (erasers) and replaced, or not, by other writers.

Epigenetic modifications also affect chromatin topology, in other words, its fine organization within chromosomes and cell nuclei. Together, these epigenetic modifications form information vectors that can be stable or reversible, as well as conducive to the development of organisms and their environmental adaptation.

Epigenetics in Ecology and Evolution,
coordinated by Christoph GRUNAU and Stéphane MAURY. © ISTE Ltd 2025.

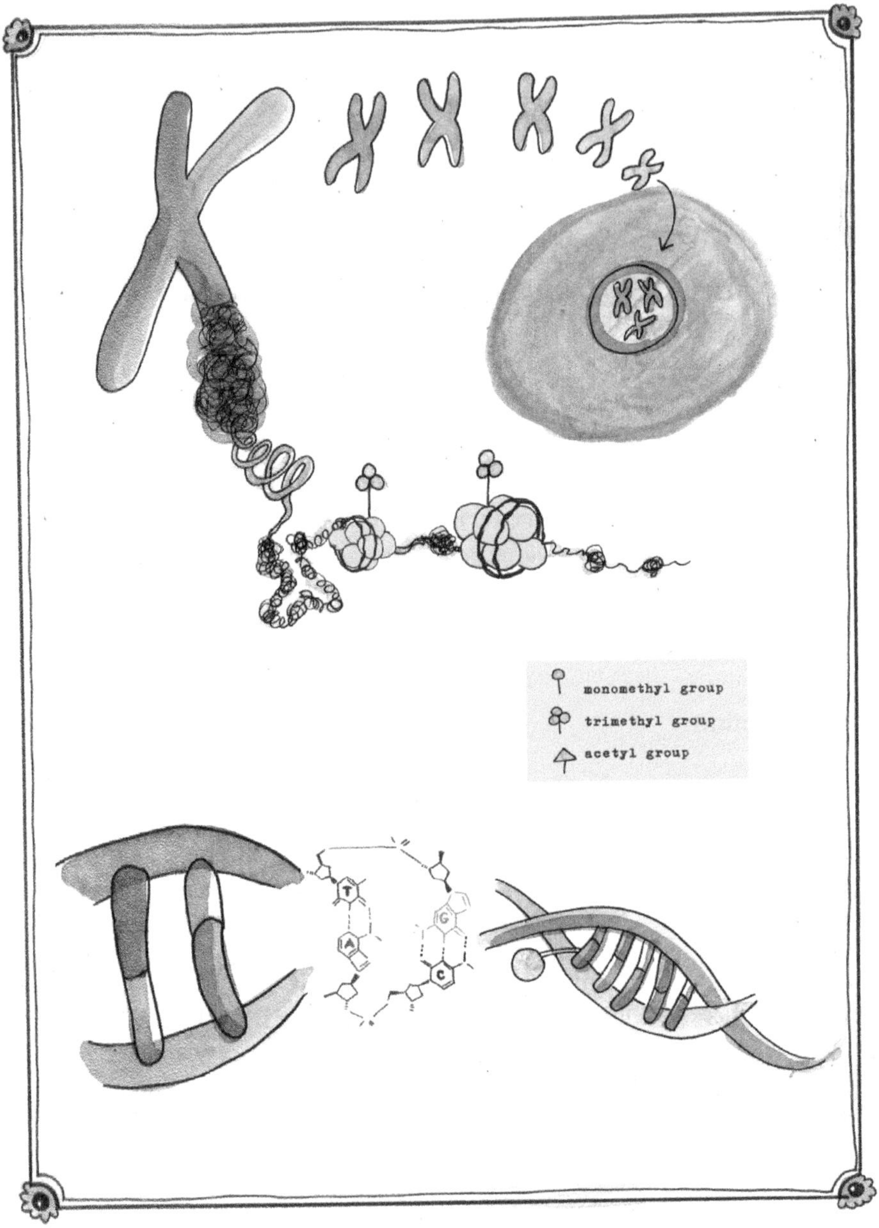

Figure 2.1. *Epigenetic information is physically stored in several types of molecules, as well as in the topological configuration adopted by DNA and the proteins that surround it. Chapter 2 is devoted to this subject. Illustration by Annette Köhn*

2.2. DNA methylation

DNA methylation was first demonstrated in tuberculosis bacillus DNA by Johnson and colleagues in 1925 (Johnson and Coghill 1925). It was in 1975 that Riggs, Holliday and Pugh proposed that this methylation, sometimes described as characterizing a "fifth nucleotide", influences gene expression and is transmitted through cell divisions (Meng et al. 2015). DNA methylation involves an epigenetic mark that acts on gene activity and thus controls cell specificity. This mark partly explains how, starting from the totipotent cells of the zygote, different tissues containing differentiated cells can divide and maintain a certain gene expression profile, thereby enabling organ development. DNA methylation is a mechanism that appears to predate the appearance of fungi, plants and animals; however, some model organisms, such as the yeast *Saccharomyces cerevisiae,* the worm *Caenorhabditis elegans* or the fly *Drosophila melanogaster,* have very little methylated DNA (He et al. 2011; Bartels et al. 2018). In the following sections, we look at the establishment of DNA methylation, its reversibility and its impact on genome regulation.

2.2.1. *The different types of DNA methylation*

DNA methylation is an active process involving the addition of a methyl group to certain nucleotide bases. The most commonly detected methylation is 5-methylcytosine, or 5mC (methylation at position 5 of a cytosine, one of the four nucleotide bases of the DNA molecule). 5mC is implemented by DNA methyl-transferase enzymes (abbreviated as DNMT), which are writers of DNA methylation. DNMTs are capable of transferring a methyl group from the substrate S-adenosyl-L-methionine (SAM) to a cytosine (Figure 2.2(a)). In humans, there are four DNMT proteins: DNMT1, which is involved in maintaining methylation on both DNA strands during DNA replication prior to cell division, DNMT2, whose role seems to be limited to transfer RNAs, and DNMT3a and 3b, which are capable of adding new DNA methylations. This is known as de novo methylation (Figure 2.2(b)). Numerous studies show that DNMT3a is responsible for maternal imprinting, while DNMT3b is involved in the methylation of pericentromeric regions that are rich in repeated sequences, as well as in the inactivation of the X chromosome during embryogenesis (Meng et al. 2015). DNMT dysfunction leads to abnormal and potentially lethal development, demonstrating the impact of DNA methylation on developmental control and the loss of cell totipotency[1].

1. Ability of an undifferentiated embryonic cell to differentiate into any specialized cell.

In vertebrates, it is often the cytosines directly followed by a guanosine ("CG" sequence), which are located in genomic regions known as "CpG islands" (p for the phosphate linking the two nucleotide bases), that are methylated. In general, 70–80% of CpG islands are methylated, and 60% of them are located in gene promoter regions and control their expression (Meng et al. 2015). In plants, in addition to CG methylations, cytosine methylations can also occur at CHG sites or asymmetric CHH sites (where H = A, T or C). These methylations are maintained by a MEthylTransferase MET1, orthologous to DNMT1, ChromoMeThylases such as CMT2 and CMT3, which methylate CHG and CHH through a pathway that is largely controlled by histone modifications, as well as by DRM1 and DRM2 (Domain Rearranged Methylase) (Stroud et al. 2014; Markulin et al. 2021). The latter enzyme is primarily involved in de novo methylation, under the control of a plant-specific pathway, the RdDM (RNA-directed DNA methylation) pathway, which involves small regulatory RNAs and plant-specific polymerase complexes (see section 2.5).

In prokaryotes (organisms without a nucleus, such as bacteria), DNA methylation can also take place on cytosines, in the form of 5mC or N4-methylcytosine (4mC), as well as, sometimes in large proportions, on the N6 position of adenines (6mA), in 5'-GATC-3' sequences (dam site). The latter methylation is implemented by dam methyltransferase, which recognizes hemimethylated DNA, as well as by TAMT-1 and MTA1c, which belong to the MT-A70 family (Liang et al. 2018). This mark, initially observed in bacteria, is also present in eukaryotes, in very low proportions, and mainly in insects and plants (Lyer et al. 2016; Liang et al. 2018). While all GATC sites are methylated in the bacterium *Escherichia coli*, in plants, less than 0.5% of adenines are generally methylated.

2.2.2. *DNA demethylation*

It was by using restriction enzymes2 that are sensitive to methylation that, as early as 1978, Bird and Southern demonstrated that CpGs could or could not be methylated, depending on the cell type analyzed (Bird and Southern 1978). More recently, DNA bisulfite treatments followed by high-throughput sequencing, luminometric analysis3 (LUMA for LUminometric Methylation Assay) or chromatin immunoprecipitation using methylation-recognizing antibodies, have made it possible to acquire genome-wide mapping of DNA methylations in a given cell type. Together, these analyses demonstrate that DNA methylation is a reversible process and can be eliminated by erasers.

2. Enzymes capable of cutting DNA at specific points determined by their sequence.

3. A molecular technique that uses the emission of small amounts of light by a specific enzymatic reaction as a measuring tool.

DNA demethylation is achieved by successive hydroxylations of methylcytosines via enzymes of the ten–eleven translocation (TET) cytosine dioxygenase family and by action of a thymine DNA glycosylase (TDG). In vertebrates, TET enzymes are capable of converting 5mC to 5-hydroxymethylcytosine (5hmC). Oxidation products of 5hmC such as 5-formylcytosine (5fC) and 5-carboxylcytosine (5caC) can also be detected (Meng et al. 2015). These last two forms of methylation are less numerous because they are unstable and often mark a transition before methylation loss, through the action of TDGs, which act as DNA methylation erasers (Figure 2.2(a)). In mammals, the control of DNA demethylation is particularly significant during germline development, fertilization (in the zygote) and early embryogenesis. Thus, during fertilization, a strong and rapid active demethylation is observed in the male pronucleus4, while the maternal genome is slowly and passively demethylated. Certain genes subject to parental imprinting, as well as repeated elements, are resistant to demethylation. DNA demethylation occurs up to the morula stage, and is followed by de novo methylation (Kota and Feil 2010; Miyoshi et al. 2016). We thus note the importance of methylation and demethylation mechanisms and their balance in the genome.

A more recently discovered second pathway for 5mC removal involves enzymes of the BER (Base Excision Repair) family, which are capable of excising methylated cytosine and replacing it with unmethylated cytosine (Meng et al. 2015). In plants, Demeter DNA Glycosylase (DME) and ROS1 (Repressor of Silencing 1) are the preferred excision enzymes (Markulin et al. 2021).

Adenine demethylation is carried out by proteins of the ALKBH1 family, which are enzymes that are also capable of demethylating 6mA RNA (Liang et al. 2018).

2.2.3. *The effects of DNA methylation*

2.2.3.1. *The effects of DNA methylation on gene activity*

DNA methylations recruit MeCPs (Methyl CpG binding proteins), which are readers that are often capable of binding histone-modifying proteins (see section 2.3.5). MeCPs include methyl-CpG binding domain (MBD) proteins, such as MBD1, MBD2 and MBD4, KAISO proteins such as ZBTB33, ZBTB4 and ZBTB38, as well as UHRF proteins (Meng et al. 2015). Interestingly, MeCP2, a transcriptional repressor, is more weakly bound to 5hmC than to 5mC, suggesting

4. Spermatocyte nucleus after fusion with the oocyte in the fertilization process, but before fusion with the oocyte nucleus (maternal germ cell).

that the functional effects depend on the type of DNA methylation, and will therefore not necessarily be the same (Meng et al. 2015).

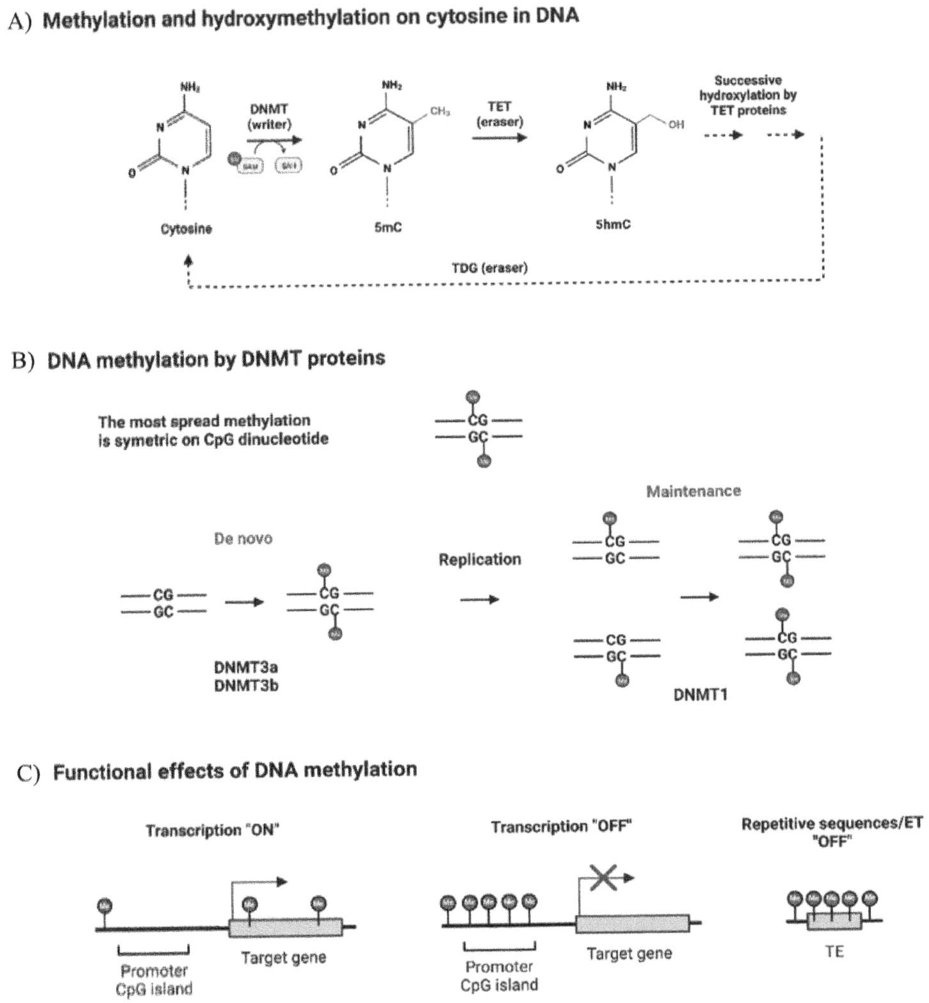

Figure 2.2. *DNA methylation. a) Principle of DNA methylation by DNMTs and demethylation initiated by TET proteins. b) Action of DNMT proteins for de novo methylation or its maintenance during DNA replication. c) Functional effects of methylation on gene transcription (ON/active and OFF/repressed) and transposable element (TE) repression. For a color version of this figure, see www.iste.co.uk/grunau/epigenetics.zip*

DNA methylation can be thought of as the placement of "biochemical tags" that act primarily on DNA compaction, and thus on the accessibility of different genes to the reading process that enables their transcription. DNA methylation will therefore affect gene expression, and consequently the production of the protein it encodes. The methylation/expression relationship can sometimes be complex, but in general, low methylation at the start of a gene promotes transcription (gene activation), while high methylation at this same point will repress it (Figure 2.2(c)). However, recent analyses show that the effects of DNA methylation can depend on the region that is methylated: the region that regulates gene expression (promoter), the coding region or exons (the region of the gene that will be translated) or the intergenic region. There are examples where high methylation, often in the coding sequence of genes, has no effect on their level of expression. In plants, the increasing degree of adenine methylation in the coding part of genes is generally associated with higher transcription (Liang et al. 2018; Zhang et al. 2018). These observations show the diversity of effects depending on the type of DNA methylation, as well as its location relative to gene sequences.

It is largely through DNA methylation that many genes involved in various cellular processes, development and reproduction are controlled. It is important to note that, in addition to development, DNA methylation can also be impacted by environmental factors (see Chapter 6). The heritability of these marks across generations, which is clearly demonstrated in plants, may allow some form of epigenetic inheritance.

2.2.3.2. *The effects of DNA methylation on RNA maturation*

In addition to playing a role in gene expression, DNA methylation is also involved in the control of alternative splicing of transcripts produced from gene activation (Shukla et al. 2011). An active gene produces a pre-mRNA that undergoes several maturation stages. One of these is alternative splicing5 (whether or not certain exons are retained), which varies according to different factors. Mature transcripts produced from the same gene will therefore not produce the same protein, and in some cases, protein function may be altered as a result. It was, in fact, by comparing transcript splicing in wild-type cells with DNA methylation-affected cells that an impact on alternative splicing versus constitutive splicing was observed (Yearim et al. 2015). More recently, using a DNA sequence modification (editing) technique to prevent methylation of intron sequence ends, it has been shown that transcript splicing can be altered (Shayevitch et al. 2018).

---

5. Splicing is one of the stages of maturation of messenger RNAs, in which parts of the transcript are retained (the "exons") and others are removed (the "introns").

2.2.3.3. *The effects of DNA methylation on genome protection*

Lastly, DNA methylation also protects genomes from mobile genetic elements such as bacteriophages in prokaryotes or transposons in eukaryotes. When infecting prokaryotic cells, bacteriophages are able to transfer their genetic material into the infected cells, enabling them to multiply and spread. A protective mechanism enables prokaryotic cells to recognize the unmethylated DNA of the bacteriophage, cleaving it and causing its degradation. This mechanism protects the integrity of the host's genetic heritage by preventing the integration of the pathogen's DNA into its genome. Transposons, on the other hand, represent truly mobile pieces of DNA, which are in a "silent or repressed" state thanks to their methylation levels (Figure 2.2(c)). A methylation deficit generally results in transposon activation, and thus in the possibility of random neo-insertions in the genome, which can sometimes impact the expression of certain genes and lead to genomic instability in the cell (see Chapter 3).

2.3. Histone modifications

We previously mentioned an epigenetic modification that takes place at the level of DNA, without affecting its sequence. The other modifications we are now going to describe relate to histone proteins, the major components of chromatin that enable DNA to be wound. We will describe three main types of histone marks: acetylation, methylation and phosphorylation, whose molecular functions are now fairly well established.

2.3.1. *The nucleosome*

Let us start by defining chromatin, the main constituent of chromosomes. It is found in all eukaryotes, including unicellular organisms such as the baker's yeast *Saccharomyces cerevisiae*. In 1882, biologist Walther Flemming named it after his affinity for dyes. Within nuclei, there are two main types of chromatin: euchromatin and heterochromatin, with different genomic and epigenetic traits.

In eukaryotes, DNA is never bare, but wrapped around highly basic proteins called histones, which have a high affinity for the acidic DNA molecule. Histones are extremely well-conserved proteins throughout evolution. A 146-base-pair (bp) fragment of DNA is wrapped around a histone octamer about twice. This is made up of four main histones H2A, H2B, H3 and H4, which combine to form two

H2A–H2B dimers and two H3–H4 dimers to form the nucleosome, which is considered the basic unit of chromatin (Figure 2.3(a)). The nucleosome thus provides the first level of DNA molecule winding, making a major contribution to its folding inside the nucleus. A succession of nucleosomes forms the chromatin fiber.

We will see in the next chapter how specific higher-order organizations contribute to additional folding of this fiber. In humans, chromatin enables 2 meters of DNA molecules to be wound into a nucleus that measures approximately 10 micrometers in diameter.

Chromosomes are thus found in the form of a chromatin fiber, and nucleosome density, nucleosome positioning, as well as its modifications, play a significant role in gene expression and other genome functions.

2.3.2. *Histone marks*

In 1997, the crystallographic structure of the nucleosome revealed N-terminal ends protruding from the nucleosome, particularly for the two histones H3 and H4 (Luger et al. 1997). These histone ends are also known as "histone tails" (Figure 2.3(a)). They are very poorly organized, but rich in basic amino acids, notably lysines (K). Exposed to the outside of nucleosomes, these lysine residues are the target of numerous post-translational modifications, such as acetylation (ac), methylation (me1, me2, me3, respectively, for monomethylation, dimethylation or trimethylation), phosphorylation (ph) or ubiquitination (ub).

As seen above for DNA methylation, we discover the notion of writers, readers and erasers here. Acetylation writers and erasers are called histone acetyltransferase (abbreviated as HAT) or lysine acetyltransferase (KAT), and histone deacetylase (HDAC), respectively. In the case of acetylation, the substrate used by writers to deposit the mark on a lysine is acetyl-CoA (AcCoA) (Figure 2.3(b)). Histone methylation writers and erasers are called lysine methyltransferase (KMT) and lysine demethylase (KDM), respectively. In this case, the nuclear substrate used is S-adenosyl-L-methionine (SAM) (Figure 2.3(b)). Arginines (R) located on histone H3 and H4 tails can also be the target of methylation. Phosphorylation writers and erasers are, respectively, called histone kinase (phosphate of an ATP molecule deposited on a serine) and histone phosphatase, and those of ubiquitination are, respectively, called ubiquitin ligase (E3) (ubiquitin deposited on a lysine) and deubiquitinase (DUB).

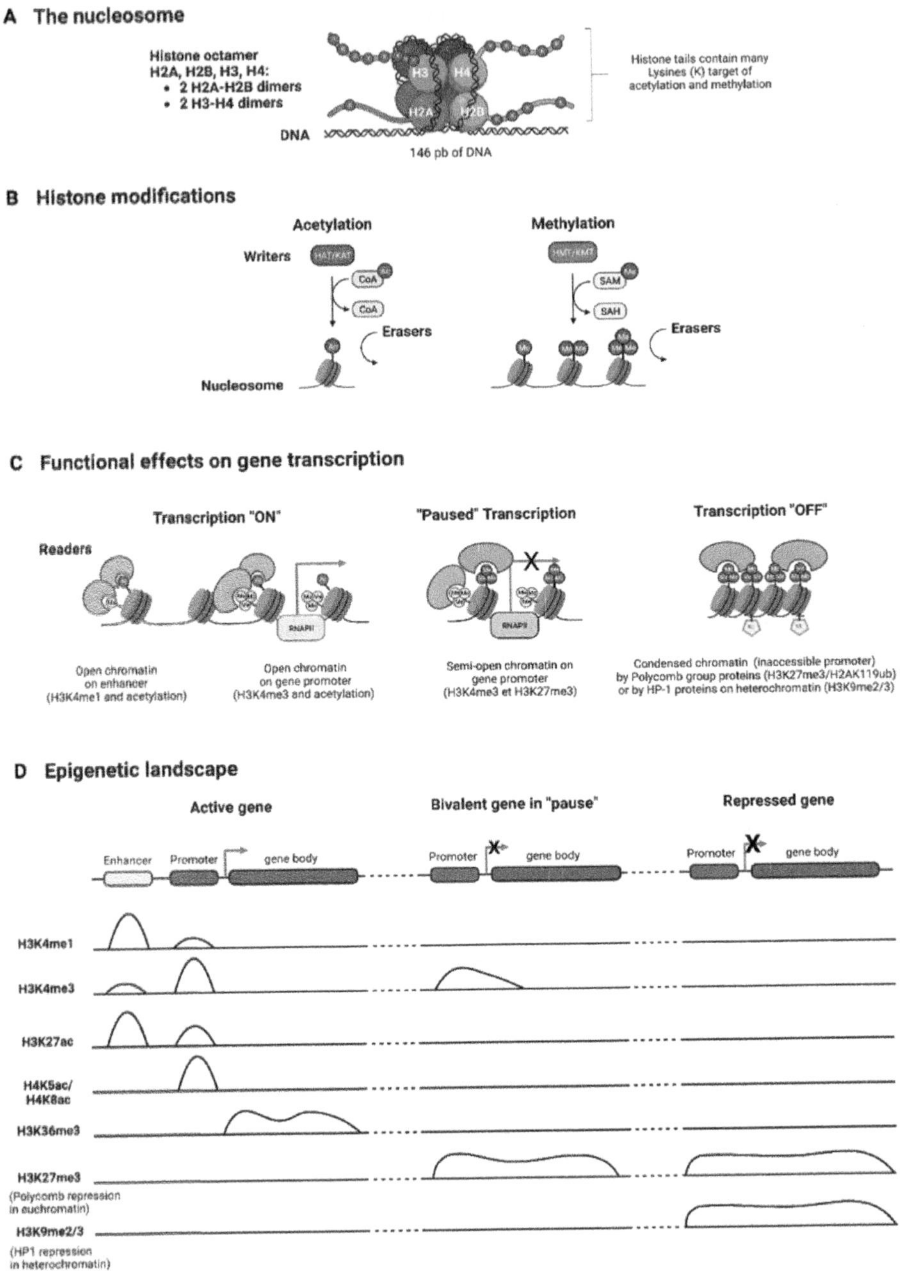

Figure 2.3. *Histone modifications and their functions. For a color version of this figure, see www.iste.co.uk/grunau/epigenetics.zip*

COMMENT ON FIGURE 2.3. – *A) The nucleosome and lysine-rich histone tails. B) Acetylation and methylation of histone tails. C) Functional effects on gene transcription. Three transcriptional states are shown: ON/active, "paused" and OFF/repressed. D) Epigenetic* landscapes *or profiles formed by specific histone marks along the chromatin fiber. Created with BioRender.com.*

As for readers, they read a specific mark thanks to particular protein domains (e.g. a chromodomain or PHD finger for the specific recognition of trimethylation; a bromodomain for the recognition of acetylated lysines, etc.) (Figure 2.3(c)). These chromatin players rarely act alone, but as multi-protein complexes, contributing to the very high specificity of epigenetic mechanisms and their function (Gardner et al. 2011).

Note that while most known modifications are found at the N-terminal ends of histones H3 and H4, there are also modifications in the core of histone proteins, such as the H3K79me3 or H2AK119ub marks (respectively, trimethylation on a lysine at position 79 or ubiquitination on a lysine at position 119 of the histone). Nevertheless, in all cases, the modifications are targeted at amino acid residues that are exposed to the external environment and therefore accessible to the writer enzymes that will deposit them, as well as the reader complexes that will read them.

There are also histone variants, some of which are known to have an epigenetic function (Henikoff et al. 2004). For example, the H3.3 and H2A.Z variants are found on active genomic regions, facilitating the concentration of active marks or chromatin remodeling. The macroH2A variant is associated with X chromosome inactivation (see section 2.3.5), as well as the H2A.X variant, whose phosphorylated form at serine 139 enables the recruitment of factors to repair DNA during double-strand breaks that may occur accidentally during cell life.

2.3.3. *Histone marks and their function*

Histone marks can dictate a functional code at chromatin level. Here, we will focus on marks that influence gene regulation. Along the chromatin fiber, epigenetic marks represent signals.

H3K4me3, H3K36me3, H3K79me3 (trimethylations on lysines in positions 4, 36, 79 of histone H3) and acetylations (such as H3K27ac, H4K5ac, H4K8ac) act as "green lights", enabling gene transcription and hence expressivity in a given cell type. The H3K4me1 mark and the H3K27ac mark can also be found on enhancers, acting as catalysts for gene transcription (Figure 2.3(c)). Acetylation has the interesting property of neutralizing the positive charge of lysines, thus reducing interactions between histones and DNA (negatively charged), making it more accessible to transcription factors.

Conversely, the H3K9me2/3 and H3K27me3 marks (dimethylations or trimethylations on histone H3 lysines 9 and 27) are gene repression marks, acting as "red lights" (Figure 2.3(c)). Genes bearing these marks will be inactive. Lastly, in some cases, there are "orange lights", with the simultaneous presence of active and repressive marks at particular chromatin sites, known as "bivalent domains or genes" (Figure 2.3(c)). Bivalent domains correspond to the coexistence of the active mark H3K4me3 and the repressive mark H3K27me3 at the promoter level of a gene, allowing the expression of this gene to be at a standstill or "paused", but ready to restart as soon as an external signal transits to the cell nucleus to activate the gene (e.g. if it is indispensable to the cell type to which the cell is destined).

In contrast to genetic modifications, epigenetic modifications are reversible, yet their stabilities are highly variable. Let us take a few examples to illustrate the different stabilities of marks. There are marks that only appear briefly during each cell division, such as the H3S10ph and H3S28ph phosphorylation marks, which are important for the ultra-condensation of chromosomes during each mitosis (cell division). In the case of DNA breaks, the histone variant H2A-X will accumulate at the point of break to enable efficient repair of the DNA molecule, and then disappear. The case of bivalent domains is typical of stable but relatively labile epigenetic marks. For example, a cell-specification gene possessing the double mark will be inactive in pluripotent stem cells. During differentiation, one of the two marks will disappear in favor of the other, for example, the active H3K4me3 mark will remain when the gene is essential for the cell type to which the cell is destined (such as a neural differentiation gene in neurons).

On the contrary, it will become inactive with the H3K27me3 mark, which will persist if the gene is dispensable in a given cell type (e.g. a muscle differentiation gene in neurons or epidermal cells). In this case, the marks are relatively stable, persisting over several generations of cells within a tissue or organ. There are also cases where marks are rapidly established during development and become very stable throughout the life of the organism. This is notably the case with Drosophila homeotic genes, which are stably inactive, covered with the repressive mark H3K27me3 in parts of the body where they must be kept repressed. However, they are covered by the H3K4me3, H3K36me3 and acetylation marks in parts where their function is indispensable, for example, the *Antennapedia* gene in the parts that will give legs, or the *Abdominal-B* gene in the posterior abdominal parts of Drosophila. In mammals, we can also cite the case of homeotic genes, as well as that of the inactivation of one of the X chromosomes in females, which is inactivated very early (blastocyst stage6) and remains very stably repressed throughout the life of the organism (see section 2.3.5). Lastly, certain marks could even contribute to the maintenance of a gene state across generations, such as the H3K27me3 mark in Drosophila (Ciabrelli et al. 2017; Zenk et al. 2017), representing a certain form of epigenetic heredity.

2.3.4. *Histone mark complexes*

Each histone mark will be associated with singular protein complexes, contributing to the high specificity of these mechanisms and their functionality at the chromatin level. At the euchromatin level, active methylation marks (notably H3K4me1, H3K4me3, H3K36me3) are deposited by SET-domain KMT proteins, notably the Trithorax group proteins in Drosophila and their mammalian counterpart, MLL (Mixed-Lineage Leukemia) proteins, which are part of COMPASS complexes. These complexes are often associated with acetylation HAT complexes that make chromatin more open and accessible. These active marks will enable the recruitment of chromatin remodeling factors, making chromatin compatible for transcription by RNA polymerase II (Schuettengruber et al. 2017). The H3K27me3 mark is the result of the KMT Enhancer of Zeste (E(Z) in Drosophila and EZH2 in humans), which belongs to the Polycomb Repressive Complex 2 (PRC2). This SET-domain writer is the sole owner of the repressive H3K27me3 mark in the genome. This mark is recognized by the chromodomain of the Polycomb protein (CBX in mammals) belonging to the PRC1 complex, which also contains the RING protein, an E3 ubiquitin ligase. PRC1 and PRC2 form an effective repression system in many organisms (including plants), notably by inducing chromatin compaction and less accessibility to transcription factors and RNA polymerase II (Schuettengruber et al. 2017). In order to maintain chromatin compaction, Polycomb complexes often act together with HDACs, which are responsible for histone deacetylation and chromatin closure.

6. Early stage of development corresponding to an embryo composed of approximately 200 cells, a stage that occurs before the embryo is implanted in the mother's uterus.

At the heterochromatin level, in particular pericentromeric heterochromatin, it is mainly the KMT SUV39H1/2 that registers the H3K9me2/3 mark, although other KMTs such as G9a and SETDB1 may also register this mark on other heterochromatin domains. The H3K9me3 mark is read by the chromodomain factor HP1 (Heterochromatin Protein 1), a highly conserved protein that is capable of carrying out oligomerization to induce the known compact structure of heterochromatin (Machida et al. 2018). At the heterochromatin level, we also find the H4K20me3 mark deposited by KMT SUV4-20H1/2 (Saksouk et al. 2015).

2.3.5. *Links between DNA methylation and histone modification*

We have just seen two major types of modification: DNA methylation and histone modifications. In mammals in particular, these two mechanisms of repression can be synergistic. Take the example of X chromosome inactivation: the presence of sex chromosomes X and Y induces a gene imbalance between the male and the female. The X chromosome is present in two copies in the female, but in a single copy in the male. To compensate for this difference, one of the X chromosomes is inactivated in the female. Most of what we know about how this process takes place in the early stages of development comes from the mouse model, which suggests that a series of epigenetic mechanisms lead to the stable inactivation of one of the X chromosomes. A non-coding RNA7 named *Xist* covers the inactive X chromosome, enabling the recruitment of HDAC enzymes (chromosome deacetylation), then the recruitment of the Polycomb repression system covering the chromosome with the H3K27me3 mark, followed by the recruitment of DNMT to methylate DNA, enabling the recruitment of other repression readers. The coexistence of these two repressive forms, DNA methylation and trimethylation of histone H3 lysine 27, results in an extremely stable repressive configuration of one of the X chromosomes ("Barr body" formation). In humans, the repression mechanism appears to be even more complex (Okamoto et al. 2011). It should be noted that one of the triggering players at the chromatin level is a long non-coding RNA. These players appear to play a significant role in the recruitment of chromatin effectors (see section 2.5.2.1 for further examples).

Another characteristic example is the coupling of the H3K9me2/3 mark and the SUV39H1,2/HP1 repression system with DNA methylation, in order to lock the

7. RNAs that are not translated into proteins, see section 2.5.

inactive form at the heterochromatin level. In plants, a typical example is the regulation of the FLC/vernalization locus involving DNA methylation and the Polycomb repression system with the H3K27me3 mark (see Chapter 6). Surprisingly, there also appear to be links between DNA methylation and the active H3K36me3 mark at the exons of transcribed genes, this time suggesting a link between DNA methylation, regulation of transcription initiation and mRNA maturation (Hodges et al. 2009). It is interesting to note that there is a synergy of action between several epigenetic mechanisms to enhance function, transcriptional repression or activation.

2.4. Chromatin topology

Chromosomes are often visualized as X-shaped (see Figure 2.1). This shape does exist, but it is important to know that it only represents a very transitory state of chromosome conformation, namely during mitosis. The rest of the time, in the cell nucleus, the three-dimensional organization of chromosomes is quite different.

2.4.1. *Epigenetic landscapes*

We have just talked about epigenetic modifications without really considering their distribution along chromosomes. Their distribution is studied using chromatin immunoprecipitation, coupled with high-throughput sequencing (ChIP-seq). Antibodies that specifically recognize the various histone modifications enable precise mapping of these marks along the genome. It turns out that this distribution forms truly epigenetic landscapes, with peaks, mountains, valleys and plains (Figure 2.3(d)).

We are talking about an "epigenome" here, with the landscapes varying according to the species and cell type. In general, active H3K4me3 and H3K4me1 marks will form peaks at active gene promoters and enhancer regulatory regions, respectively. Acetylation at the lysine ends of H3 (in particular H3K27ac) and H4 will coincide fairly well with these active methylation marks. The active H3K36me3 and H3K4me2 marks will form domains in the gene body (Lim et al. 2010), with the H3K36me3 mark being more strongly enriched in exons than in introns (Hodges et al. 2009) (Figure 2.3(d)).

The repressive marks, H3K27me3 for repression by the Polycomb system at euchromatin and H3K9me2/3 for repression of heterochromatin via the HP-1[8] system, form larger chromatin domains that often cover the entire genomic region to be repressed (Lim et al. 2010) (Figure 2.3(d)). A third type of repressive chromatin exists, known as Black, Void or Null chromatin, but unlike the first two, it is not characterized by specific epigenetic marks. It is enriched in the H1 linker histone, involved in chromatin compaction. This chromatin is not negligible, however, as it can represent up to 50% of the genome (considering unrepeated sequences) (Filion et al. 2010; Ho et al. 2014).

Chromatin marks thus define true epigenetic landscapes or signatures, which are of course important for genome functionality. They also largely influence the folding and organization of chromosomes within nuclei.

2.4.2. *From nucleosome to chromosome territories*

As mentioned above, the nucleosome is the basic unit of chromatin, representing a first level of organization of the DNA molecule, with the succession of nucleosomes forming the chromatin fiber (Figure 2.4). These nucleosomes associate with one another irregularly, notably via electrostatic interactions, to form nucleosome clutches (named in analogy with the organization of eggs in an incubated nest) (Ricci et al. 2015). Many of these clutches then form chromatin nanodomains or CNDs (Miron et al. 2020; Szabo et al. 2020). Clutches and CNDs, respectively, organize DNA in the range of 10 kilobases (kb, 10,000 bp) to 200 kb (Figure 2.4). The presence of these two levels of organization has been observed very recently thanks to the combination of new chromatin visualization techniques with very high-resolution microscopy. The exact functions of these structures are still poorly understood today, but they do contribute to the higher-order topological organization of the chromatin fiber.

At the beginning of the 21st century, molecular and genomic techniques derived from Chromosome Conformation Capture (3C), and in particular its genome-wide version (Hi-C), revealed topologically associating domains (TADs), which represent areas of the genome that are characterized by much stronger internal interactions than interactions with adjacent chromatin domains (Szabo et al. 2019). TADs are

8. The heterochromatin protein 1 (HP1) family consists of highly conserved proteins with important functions in chromatin structure in all eukaryotes.

separated by border zones, enabling genome partitioning. Depending on the species, TADs range in size from 100 kb to a megabase (Mb, 1 million bp). If we refer to the above scales, CNDs may in some cases correspond to TADs, and in others cases to substructures within TADs. Some of these substructures could be defined by epigenetic determinants (Rao et al. 2014) (Figure 2.4).

On a larger scale, Hi-C and cytological approaches involving hybridization of fluorescent probes on DNA have revealed the segregation of the genome into large chromosomal regions, with transcriptionally active zones grouped together in "compartment A" and inactive zones grouped together in "compartment B". Lastly, at the end of this scale, chromosomes form chromosomal territories, where each chromosome will occupy its own space within nuclei, with some having a preferential radial position (in reference to the periphery of the nucleus) (Szabo et al. 2019) (Figure 2.4).

2.4.3. *Focus on topologically associating domains*

Among all these genomic structures, TADs have received particular attention because they appear at a scale where most nuclear processes take place. With the exception of certain eukaryotes where the genome is small and compact (the yeast *Saccharomyces cerevisiae*, the plant *Arabidopsis thaliana*, etc.), TADs are described in all animal and plant species. In some cases, TADs correspond to the epigenetic landscapes described above, notably in *Drosophila melanogaster* or in plants, including rice, cotton, corn, tomato, etc., where the division of epigenetic domains corresponds perfectly to the partitioning of the genome into TADs (Szabo et al. 2019).

In this case, transcriptionally inactive domains (H3K27me3 and Black chromatin) form TADs, and the border regions between these TADs correspond to areas that are enriched in active epigenetic marks. In other cases, such as mammals, where the coding part of genes only represents approximately 1.5% of the genome and where there are many repeated sequences (see Chapter 3), TADs appear larger, in the range of 1 Mb, mixing several epigenetic states and governed by other mechanisms. These include the loop-extrusion mechanism involving the cohesin complex, which is capable of pulling the chromatin fiber through its molecular "ring" structure until it encounters DNA-bound CTCF factors in an inverted orientation at TAD boundaries (Figure 2.4).

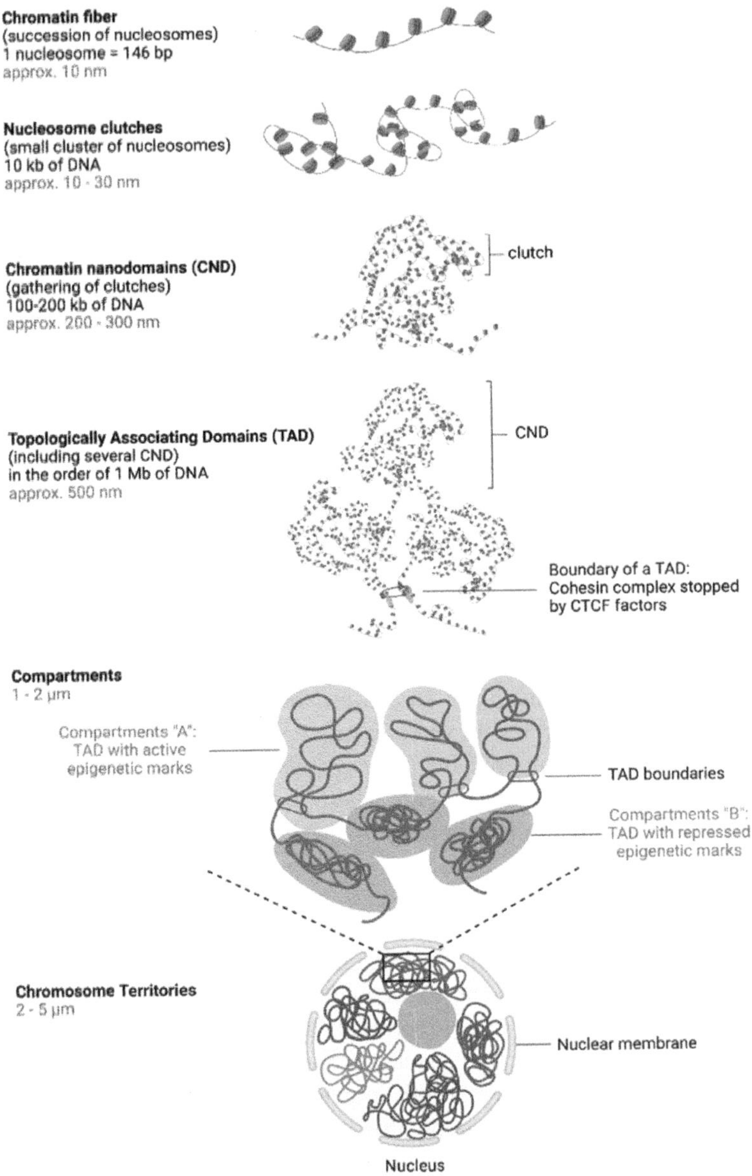

Figure 2.4. *Chromatin topology. The different levels of chromosome folding in the cell nucleus are shown. From nucleosomes and chromatin fibers to chromosome territories, nucleosome clutches, CNDs, TADs and compartments. Created with BioRender.com. For a color version of this figure, see www.iste.co.uk/grunau/epigenetics.zip*

The important aspect of TADs is that they could represent functional units of chromatin, within which connections between regulatory elements, such as interactions between gene promoters and enhancers carrying active epigenetic marks (enhancers can be several tens of kb away from their gene) or interactions between repressor elements of the Polycomb group (carrying the H3K27me3 mark), would be favored (Figure 2.5). These connections could enhance the molecular functions of these players. Disruptions in the separation between TADs can lead to transcriptional deregulations that cause developmental diseases or cancers (Spielmann et al. 2018). TAD organization could also promote a double-strand break repair or sister chromatid replication (Arnould et al. 2021).

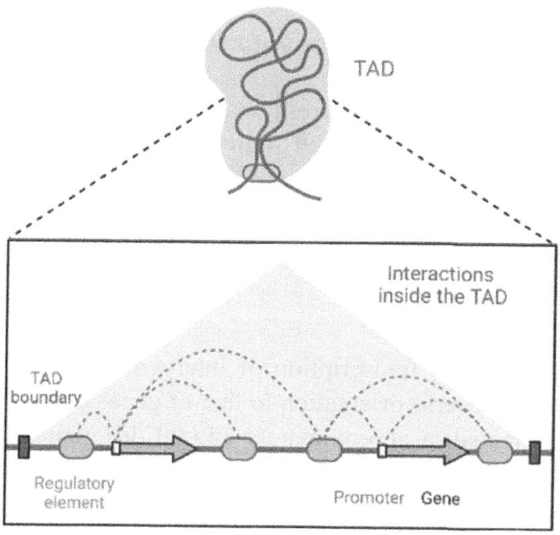

Figure 2.5. *TAD function. Chromatin interactions are favored within TADs, notably between regulatory elements (enhancers or elements with repression marks) and gene promoters for their transcriptional activation or repression. For a color version of this figure, see www.iste.co.uk/grunau/epigenetics.zip*

In short, TADs represent important structural units of genome organization. In particular, they integrate epigenetic information contained at the chromatin fiber level, potentiating their functional relationships. In addition, TAD organization could also play a role in genome structural variation and stabilization (insertions, deletions, inversions, chromosomal rearrangements), and thus influence genome and species evolution (Szabo et al. 2019).

2.5. Regulatory RNAs

Regulatory RNAs may be involved in controlling the nature and position of various epigenetic marks. In genomes, there are two main classes of transcripts: 1) "coding" transcripts, which account for the majority of transcripts and enable protein synthesis after translation; 2) "non-coding" transcripts, such as ribosomal RNAs (rRNAs), transfer RNAs (tRNAs), which are involved in protein synthesis, and regulatory RNAs. Many of the latter are involved in controlling the positioning of epigenetic marks and, depending on the species, different control pathways have been characterized.

2.5.1. *Regulatory RNAs, from discovery to biogenesis*

Regulatory RNAs are synthesized from natural loci in the genome, as well as be produced from transgenes9 integrated into the genome or transcripts from pathogens. A first classification of regulatory RNAs is based on their size, with long non-coding RNAs (lncRNAs) of over 200 nucleotides (nt) and small ones of 20–30 nt.

2.5.1.1. *Long regulatory RNAs*

lncRNAs result from the transcription of intergenic or gene sites, where their transcription may be in reverse orientation to that of genes located at the same locus. Their number is estimated at approximately 15,000 different lncRNAs in humans (Ali and Grote 2020). Some are involved in RNA maturation, splicing, translation regulation or the cellular localization of certain proteins, while others are involved in the control of epigenetic marks (see section 2.5.2.1) (Bardou et al. 2014; Zhang et al. 2019; Roule et al. 2022).

2.5.1.2. *Small regulatory RNAs*

Analysis of the first regulatory RNAs that were identified shows their involvement in various gene regulation pathways. Three main pathways appear to coexist, although the names of the effector proteins and partners may differ, depending on the organism studied. It is important to note that a small regulatory RNA cannot be alone. It is, in fact, loaded onto a protein of the Argonaute family,

9. A transgene is a DNA sequence isolated from a genomic region of interest. It can be introduced into a host genome (the same or another species) and lead to phenotypic changes in the organism.

which contains two subfamilies called AGO and PIWI. The different repression mechanisms in which they act constitute universal regulatory pathways in eukaryotes. These pathways play a key role in a wide range of biological processes such as development, stress response, maintenance of genome stability, chromosome architecture and pathogen repression (viruses, bacteria). These pathways act on gene regulation, either at the post-transcriptional level (Post-Transcriptional Gene Silencing or PTGS) by inducing destabilization and/or degradation of messenger RNAs or decreasing their translational efficiency, or at the transcriptional level (Transcriptional Gene Silencing or TGS), through the deposition and positioning of specific epigenetic marks (Figure 2.6).

The first PTGS pathway is the microRNA pathway described from a genetic screen carried out in the nematode *Caenorhabditis elegans*. Based on the identification of *lin-4* and *lin-14* mutants, this screen established that the functional product of the *lin-4* gene is a small RNA, or more precisely a microRNA, which has the ability to repress the translation of *lin-14* mRNA by pairing at specific complementary sites located in its 3'UTR region (the transcribed but untranslated terminal region of the gene) (Lee et al. 1993; Wightman et al. 1993). MicroRNAs are generated from discrete genomic loci, whose precursor transcripts fold into short hairpins 70 nt long. These hairpin-folded precursors are cleaved at RNA duplexes by DROSHA or DICER RNAs, releasing homogeneous RNAs of 20–22 nt. MicroRNAs then bind to effector proteins belonging to the AGO subfamily and act on their mRNA target by PTGS, either inducing their degradation or inhibiting their translation (Iwakawa and Tomari 2015) (Figure 2.6). Interestingly, some miRNA targets may be transcripts that correspond to writers or erasers of epigenetic marks (Yao et al. 2019).

In the 1990s, another pathway, that of siRNA (small interfering RNA), was described, by artificially and specifically repressing the expression of genes of interest with the so-called "interfering RNA" technique (introduction of transgenes enabling the synthesis of antisense RNA, or RNA complementary to the mRNA of the targeted gene). This discovery was initially made in the nematode *Caenorhabditis elegans* (Guo and Kemphues 1995; Fire et al. 1998). The antisense RNA produced binds to the sense mRNA by complementarity, forming an RNA duplex which is fragmented by a DICER RNAse into precise, sequential populations of 21–23 nt siRNAs. Once produced, siRNAs loaded onto AGO proteins will target complementary RNAs and induce their cleavage by PTGS (Figure 2.6).

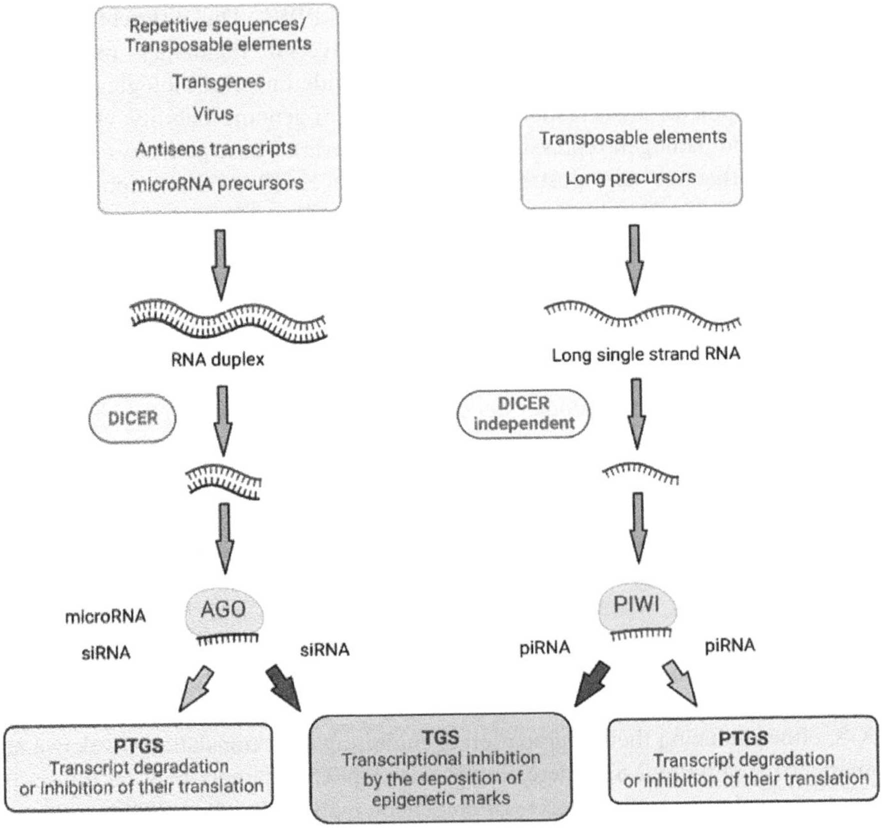

Figure 2.6. *PTGS (Post-Transcriptional Gene Silencing) and TGS (Transcriptional Gene Silencing) initiation pathways. The left DICER-dependent pathway converges on the AGO subfamily, and the right DICER-independent pathway converges on the PIWI subfamily. These pathways can lead to PTGS or TGS. For a color version of this figure, see www.iste.co.uk/grunau/epigenetics.zip*

In certain organisms, such as plants, it has been clearly established that siRNAs can also cause histone modifications and DNA methylation by TGS (Butel et al. 2021). The first observations relate to so-called "co-suppression" phenomena. It has been observed, for example, that following the insertion of two independent transgenes into the tobacco genome, one was methylated and therefore repressed in certain lines (Matzke et al. 1989). At the same time, and in order to produce more intensely colored Petunia flowers, two laboratories generated transgenic plants, which overexpress genes encoding enzymes involved in the flower pigment biosynthesis pathway. Curiously, in several independent transgenic lines, the targeted gene was not overexpressed but, on the contrary, under-expressed, and their flowers, instead of being more colorful, were partially or totally white (Napoli et al. 1990; Van der Krol et al. 1990). Similar observations were also made in tomatoes (Smith et al. 1990). In these situations, the introduction of a transgene leads to the production of siRNAs, which guide and enable the establishment of new epigenetic marks to modulate the expression of targeted genes. In plants, siRNAs can come from a wide variety of sources, including RNA duplexes derived from natural antisense transcripts resulting from the transcription of repeated sequences, often transposable elements (TEs), portions of genes, as in the case of natsiRNAs (natural antisense siRNAs), specific loci such as TAS or PHAS, or transgenes (Hamilton and Baulcombe 1999; Jones et al. 1999; Mette et al. 1999; Matzke et al. 2001; Vaucheret and Fagard 2001; Lahmy et al. 2009; Mosher et al. 2009). In these organisms, as well as in the worm *Caenorhabditis elegans*, the yeast *Schizosaccharomyces pombe* and the ciliophora *Tetrahymena thermophila*10, siRNAs can be amplified by the RDRP enzyme (RNA-Dependent RNA Polymerase), enabling the formation of new RNA duplexes (secondary siRNA) (Holoch and Moazed 2015). This amplification can be crucial for plants and insects, for which siRNAs can play a defensive role against pathogens. In this case, replication or transcription of infectious genomes initiates the production of specific siRNAs (Ratcliff et al. 1997; Mourrain et al. 2000; Morel et al. 2002; Boutet et al. 2003; Van Rij et al. 2006; Raja et al. 2008). The pathogen's transcripts are then degraded (PTGS pathway) or its genome is specifically methylated and thus repressed (TGS pathway) (see Chapter 6) (Figure 2.5). In some organisms, such as the yeast *Schizosaccharomyces pombe*, siRNAs are required for the deposition of epigenetic marks that are essential for the establishment of chromosome centromeres11 (Volpe et al. 2002). When siRNAs impact epigenetic marks, they are generally referred to as hcsiRNA (heterochromatic siRNA) (see section 2.5.2.3).

A final pathway is that of piRNAs (PIWI interacting RNA). It should be noted that piRNAs were originally called rasiRNAs for "repeat-associated siRNAs". They were first described in 2003 in various stages of development in Drosophila (Aravin et al. 2003). However, this name was inappropriate, as piRNA biogenesis is very different from that of the two classes of regulatory RNA described above. In fact, piRNAs are produced independently of DICER, and are loaded onto proteins of the PIWI subfamily that are mainly expressed in gonads in the animal kingdom (Figure 2.5). Between 23 and 30 nt long, they either originate from the degradation

---

10. Thanks to their small size and ease of cultivation, they are all important study models for biologists.

11. Centromeres are important chromosome regions that serve as anchor points for chromosome segregation during mitosis.

of transcripts produced by heterochromatic regions known as "piRNA clusters" (Brennecke et al. 2007), or from 3'UTR regions of genes, or even from intergenic regions, depending on the species and stage of development. The heterochromatic regions that produce them are rich in TEs. In these regions, TEs have accumulated numerous mutations, rendering them unable to mobilize in the genome (see Chapter 3). The sequence of piRNAs is rarely conserved, even between closely related species, and their extraordinary diversity favors species-specific functions. Depending on the species, piRNAs can control certain genes and TEs, either post-transcriptionally (PTGS) or transcriptionally (TGS) (Figure 2.5 and see section 2.5.2.2) (Wu et al. 2020).

2.5.2. *Modes of action of regulatory RNAs on the establishment of epigenetic marks*

Depending on the organism studied, different pathways involving RNA can modulate epigenetic marks. In the following sections, we describe those that have been most extensively studied in recent years.

2.5.2.1. *Regulation by lncRNAs*

Numerous studies have demonstrated that different lncRNAs regulate epigenetic marks and control the expression of certain genes. We have already seen the example of *Xist* involved in the inactivation of the X chromosome in mammals. Other examples include the *ThymoD* lncRNA which, when transcribed, modulates DNA topology and thus prevents the methylation of regulatory sequences located in the region concerned. The gene at the same locus is then active and transcribed. When transcription of the *ThymoD* lncRNA is stopped, the regulatory sequence becomes methylated and the colocalized gene is inactivated (Ali and Grote 2020). A similar example is seen in plants with the *COOLAIR* and *COLDWRAP* lncRNAs, which are involved in controlling flowering by modulating histone modifications (see Chapter 6). Other lncRNAs can act on the epigenetic marks of distantly located genes. In the yeast *Saccharomyces cerevisiae*, in the absence of galactose, the *GAL10* lncRNA is transcribed and inhibits the expression of the *GAL10* and *GAL1* genes, via hypoacetylation of their promoter histones. In the presence of galactose, transcription of the *GAL10,* lncRNA is stopped, and the promoters of the *GAL10* and *GAL1* genes become hyperacetylated, leading to their activation (Ali and Grote 2020). Other lncRNAs, such as *APOLO* and *ELENA* in plants, as well as *Upperhand* in animals, also regulate remotely localized genes (Mach 2017; Seo et al. 2017; Han et al. 2019; Ariel et al. 2020). Although all the mechanisms involving lncRNAs have yet to be fully identified, it appears that they can, at loci that are complementary to them, enable the recruitment of enzymes involved in setting up epigenetic marks.

They can also modulate chromatin conformation, making genomic regions more or less close or accessible (Lleres et al. 2021).

2.5.2.2. *Regulation by piRNAs*

PIWI nuclear proteins, guided by piRNAs, often lead to the transcriptional silencing of TEs (Figure 2.7). This RNA-dependent process is probably accompanied by the deposition of the repressive chromatin mark H3K9me3, as well as the elimination of the active mark H3K4me2. The histone-modifying enzymes involved in the piRNA pathway are KMT SetDB1 for H3K9me3 deposition, KDM Su(var)3-3/LSD1 for H3K4me2 demethylation and HDAC for H3K9ac deacetylation (Mugat et al. 2020). Previous work has also identified GTSF1, Panoramix (Panx) and Maelstrom (Mael) as specific factors in the piRNA pathway (Sienski et al. 2012, 2015; Ohtani et al. 2013). The Panx protein is specific to the Drosophila model, while all other proteins are conserved. Recently, it has been suggested that Su(var)2-10, an E3 ubiquitin ligase, connects Piwi/Panx with KMT SetDB1 (Ninova et al. 2020).

In mice, the pattern of TE transcriptional silencing by H3K9me3 deposition is reinforced by DNA methylation (Ninova et al. 2019). Here, we find a synergy between these two epigenetic mechanisms for repression. Importantly, the deposition of these epigenetic marks, mediated by small regulatory RNAs, can affect the expression of genes that flank TE sequences in the genome.

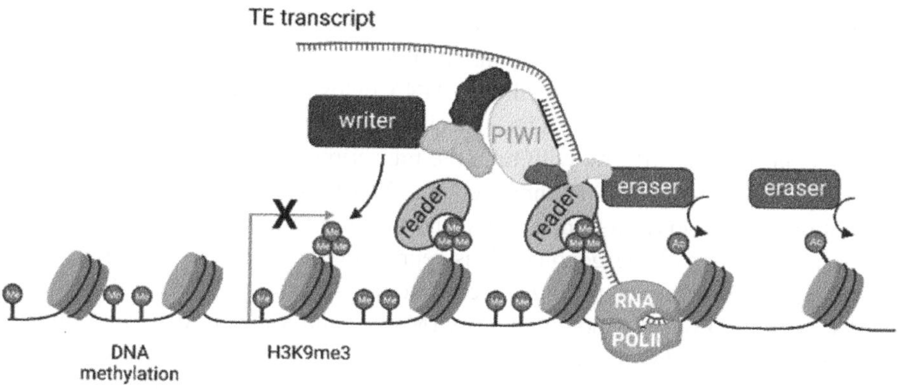

Figure 2.7. *Model of transcriptional repression involving regulatory RNAs in the germ line. Created with BioRender.com. For a color version of this figure, see www.iste.co.uk/grunau/epigenetics.zip*

COMMENT ON FIGURE 2.7. – *Regulatory RNAs guide the PIWI complex towards the emerging transcripts of their complementary transposable elements (TEs). The PIWI complex interacts with numerous cellular factors such as writers, erasers and readers, inducing DNA methylation and/or histone modifications.*

2.5.2.3. *Regulation by hcsiRNAs in plants*

In plants, a particular class of small RNAs, the hcsiRNAs, is specifically involved in controlling de novo DNA methylation through the RdDM (RNA-directed DNA methylation) pathway (Figure 2.8). This class accounts for over 80% of small RNAs in the model plant *Arabidopsis thaliana* (Sanan-Mishra et al. 2021). This pathway is mainly involved in the repression of TEs, as well as, in some cases, in regulating development and responding to certain environmental stresses. The pathway involves two plant-specific RNA polymerase complexes, POLIV and POLV (Lahmy et al. 2009, 2010; Mosher et al. 2009).

POLIV usually initiates hcsiRNA production, producing long transcripts that are then taken over by RDR2 (RDRP enzyme), which enables the synthesis of complementary RNA strands and thus the formation of long RNA duplexes. These are then cut by DCL3 (Dicerlike enzyme), which releases hcsiRNA duplexes of 20–25 nt. The hcsiRNAs will then bind to AGO4, one of the 10 AGO proteins present in *Arabidopsis thaliana*, leading to the degradation of one strand of the duplex. The hcsiRNAs associated with AGO4 will then be recruited by POLV through the "AGO hook" platform of the NRPE1 (subunit of the POLV complex) and SPT5like (KTF1) proteins (Bies-Etheve et al. 2009; Lahmy et al. 2009; Azevedo et al. 2011). This interaction then recruits the DNA methyltransferase DRM2 which, in association with DRM3, will methylate DNA cytosines in a highly specific way (Costa-Nunes et al. 2014). The specificity of action results from the complementarity of hcsiRNAs with both the transcripts produced by POLV and the transcribed DNA sequence (Lahmy et al. 2016) (Figure 2.8). This targeted methylation mainly serves to repress TEs, and thus limit their transpositions, leading to a certain stability of the genome. This repression pathway can also be induced after integration of exogenous DNA pieces into the genome, or after viral infections in order, again, to protect genome integrity (Matzke and Mosher 2014) (see Chapter 6). Importantly, POLIV appears to be recruited by SHH1, a reader of histone modifications, while POLV is recruited by MEthylTransferases (Ebbs et al. 2005; Ebbs and Bender 2006; Law et al. 2011, 2013). Recent observations indicate that the RdDM pathway can also be initiated from POLII transcripts (Sigman et al. 2021).

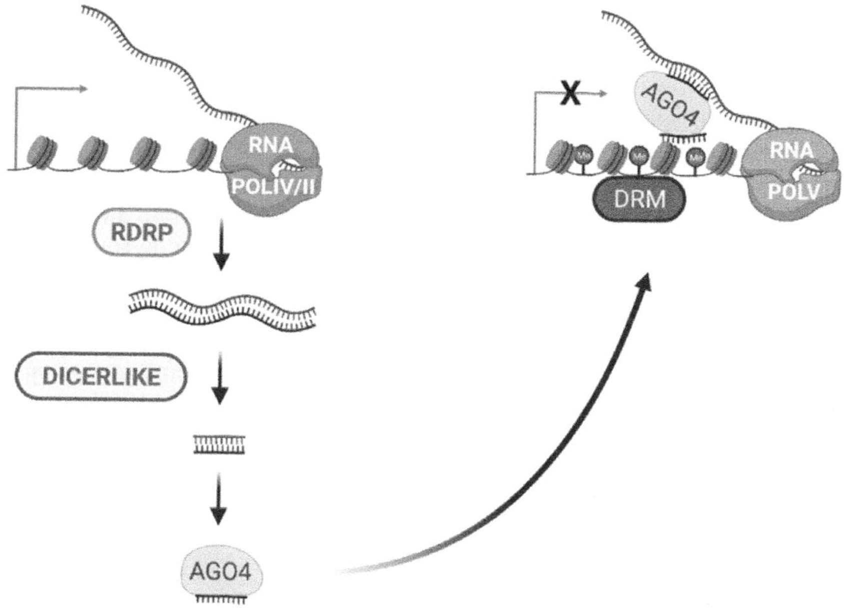

Figure 2.8. *Production and action pathways of hcsiRNAs in plants. The RdDM (RNA-directed DNA methylation) pathway is initiated in plants by the production of POLIV (or POLII) transcripts, which enable the synthesis of hcsiRNAs that, through sequence complementarity with POLV transcripts, will enable de novo methylation by DRM action. Created with BioRender.com. For a color version of this figure, see www.iste.co.uk/grunau/epigenetics.zip*

2.5.2.4. *ScanRNA regulation in ciliates*

In ciliates, the differentiation of the somatic genome from the germ line genome is characterized by the massive and reproducible deletion of TEs. This DNA sequence deletion involves small regulatory RNAs, produced by the germ line in the micronucleus, the scanRNAs. Their production requires proteins related to DICER enzymes. Combined with PIWI proteins, they initiate the elimination of homologous sequences in the developing somatic macronucleus. ScanRNAs appear to give a particular state to chromatin in terms of the TE sequences to be eliminated. In paramecium, scanRNAs are thought to recruit the Ezl1 protein, similar to the catalytic subunit of mammalian PRC2, which deposits the H3K9me3 and H3K27me3 marks. This "heterochromatinization" of TE sequences would lead to the recruitment of the PiggyMac endonuclease for their elimination (Deleris et al. 2021).

2.6. Conclusion

In this chapter, we have seen a panoply of very precise epigenetic mechanisms involving a great specificity of carriers, players and deposited marks. Epigenetic marks are dynamically positioned along chromosomes, influencing chromatin compaction and topology. These epigenetic marks represent signals that appear to be indispensable to various genome functions (gene transcription, TE and virus silencing, replication, DNA repair, chromosome segregation, etc.). Together, these marks control cell identity and their fate during differentiation and development, and contribute to genome integrity. Epigenetic marks also ensure the adaptation of organisms to environmental stress, thus enabling their survival, and they can even be transmitted across generations, thus contributing to a form of epigenetic heredity. In addition to these various epigenetic mechanisms, there are post-transcriptional modifications at the RNA level (known as "epitranscriptomic modifications"), such as methylations observed in tRNAs, rRNAs and even messenger RNAs (Angelova et al. 2020). The identification of all these epigenetic mechanisms demonstrates the extreme complexity of genome expression regulation.

2.7. References

Ali, T. and Grote, P. (2020). Beyond the RNA-dependent function of LncRNA genes. *Elife*, 9, e60583.

Angelova, M., Dimitrova, D., Da silva, B., Marchand, V., Jacquier, C., Achour, C., Brazane, M., Goyenvalle, C., Bourguignon-Igel, V., Shehzada, S. et al. (2020). tRNA 2'-O-methylation by a duo of TRM7/FTSJ1 proteins modulates small RNA silencing in Drosophila. *Nucleic Acids Res.*, 48(4), 2050–2072.

Aravin, A.A., Lagos-Quintana, M., Yalcin, A., Zavolan, M., Marks, D., Snyder, B., Gaasterland, T., Meyer, J., Tuschl, T. (2003). The small RNA profile during *Drosophila melanogaster* development. *Dev Cell*, 5(2), 337–350.

Ariel, F., Lucero, L., Christ, A., Mammarella, M.F., Jegu, T., Veluchamy, A., Mariappan, K., Latrasse, D., Blein, T., Liu, C. et al. (2020). R-loop mediated trans action of the APOLO long noncoding RNA. *Mol Cell*, 77(5), 1055–1065.

Arnould, C., Rocher, V., Finoux, A.L., Clouaire, T., Li, K., Zhou, F., Caron, P., Mangeot, P.E., Ricci, E.P., Mourad, R. et al. (2021). Loop extrusion as a mechanism for formation of DNA damage repair foci. *Nature*, 590(7847), 660–665.

Azevedo, J., Cooke, R., Lagrange, T. (2011). Taking RISCs with Ago hookers. *Curr Opin Plant Biol*, 14(5), 594–600.

Bardou, F., Ariel, F., Simpson, C.G., Romero-Barrios, N., Laporte, P., Balzergue, S., Brown, J.W., Crespi, M. (2014). Long noncoding RNA modulates alternative splicing regulators in Arabidopsis. *Dev Cell*, 30(2), 166–176.

Bartels, A., Han, Q., Nair, P., Stacey, L., Gaynier, H., Mosley, M., Huang, Q.Q., Pearson, J.K., Hsieh, T.F., An, Y.C. et al. (2018). Dynamic DNA methylation in plant growth and development. *Int J Mol Sci*, 19(7), 2144–2160.

Bies-Etheve, N., Pontier, D., Lahmy, S., Picart, C., Vega, D., Cooke, R., Lagrange, T. (2009). RNA-directed DNA methylation requires an AGO4-interacting member of the SPT5 elongation factor family. *EMBO Rep*, 10(6), 649–654.

Bird, A.P. and Southern, E.M. (1978). Use of restriction enzymes to study eukaryotic DNA methylation: I. The methylation pattern in ribosomal DNA from *Xenopus laevis*. *J Mol Biol*, 118(1), 27–47.

Boutet, S., Vazquez, F., Liu, J., Beclin, C., Fagard, M., Gratias, A., Morel, J.B., Crete, P., Chen, X., Vaucheret, H. (2003). Arabidopsis HEN1: A genetic link between endogenous miRNA controlling development and siRNA controlling transgene silencing and virus resistance. *Curr Biol*, 13(10), 843–848.

Brennecke, J., Aravin, A.A., Stark, A., Dus, M., Kellis, M., Sachidanandam, R., Hannon, G.J. (2007). Discrete small RNA-generating loci as master regulators of transposon activity in *Drosophila*. *Cell*, 128(6), 1089–1103.

Butel, N., Yu, A., Le Masson, I., Borges, F., Elmayan, T., Taochy, C., Gursanscky, N.R., Cao, J., Bi, S., Sawyer, A. et al. (2021). Contrasting epigenetic control of transgenes and endogenous genes promotes post-transcriptional transgene silencing in *Arabidopsis*. *Nat Commun*, 12(1), 2787.

Ciabrelli, F., Comoglio, F., Fellous, S., Bonev, B., Ninova, M., Szabo, Q., Xuereb, A., Klopp, C., Aravin, A., Paro, R. et al. (2017). Stable Polycomb-dependent transgenerational inheritance of chromatin states in *Drosophila*. *Nat Genet*, 49(6), 876–886.

Costa-Nunes, P., Kim, J.Y., Hong, E., Pontes, O. (2014). The cytological and molecular role of domains rearranged methyltransferase3 in RNA-dependent DNA methylation of *Arabidopsis thaliana*. *BMC Res Notes*, 7, 721.

Deleris, A., Berger, F., Duharcourt, S. (2021). Role of Polycomb in the control of transposable elements. *Trends Genet*, 37(10), 882–889.

Ebbs, M.L. and Bender, J. (2006). Locus-specific control of DNA methylation by the Arabidopsis SUVH5 histone methyltransferase. *Plant Cell*, 18(5), 1166–1176.

Ebbs, M.L., Bartee, L., Bender, J. (2005). H3 lysine 9 methylation is maintained on a transcribed inverted repeat by combined action of SUVH6 and SUVH4 methyltransferases. *Mol Cell Biol*, 25(23), 10507–10515.

Filion, G.J., Van Bemmel, J.G., Braunschweig, U., Talhout, W., Kind, J., Ward, L.D., Brugman, W., De Castro, I.J., Kerkhoven, R.M., Bussemaker, H.J. et al. (2010). Systematic protein location mapping reveals five principal chromatin types in *Drosophila* cells. *Cell*, 143(2), 212–224.

Fire, A., Xu, S., Montgomery, M.K., Kostas, S.A., Driver, S.E., Mello, C.C. (1998). Potent and specific genetic interference by double-stranded RNA in *Caenorhabditis elegans*. *Nature*, 391(6669), 806–811.

Gardner, K.E., Allis, C.D., Strahl, B.D. (2011). Operating on chromatin, a colorful language where context matters. *J Mol Biol*, 409(1), 36–46.

Guo, S. and Kemphues, K.J. (1995). par-1, a gene required for establishing polarity in *C. elegans* embryos, encodes a putative Ser/Thr kinase that is asymmetrically distributed. *Cell*, 81(4), 611–620.

Hamilton, A.J. and Baulcombe, D.C. (1999). A species of small antisense RNA in posttranscriptional gene silencing in plants. *Science*, 286(5441), 950–952.

Han, X., Zhang, J., Liu, Y., Fan, X., Ai, S., Luo, Y., Li, X., Jin, H., Luo, S., Zheng, H. et al. (2019). The lncRNA Hand2os1/Uph locus orchestrates heart development through regulation of precise expression of Hand2. *Development*, 146(13).

He, X.J., Chen, T., Zhu, J.K. (2011). Regulation and function of DNA methylation in plants and animals. *Cell Res*, 21(3), 442–465.

Henikoff, S., Furuyama, T., Ahmad, K. (2004). Histone variants, nucleosome assembly and epigenetic inheritance. *Trends Genet*, 20(7), 320–326.

Ho, J.W., Jung, Y.L., Liu, T., Alver, B.H., Lee, S., Ikegami, K., Sohn, K.-A., Minoda, A., Tolstorukov, M.Y., Appert, A. et al. (2014). Comparative analysis of metazoan chromatin organization. *Nature*, 512(7515), 449–452.

Hodges, E., Smith, A.D., Kendall, J., Xuan, Z., Ravi, K., Rooks, M., Zhang, M.Q., Ye, K., Bhattacharjee, A., Brizuela, L. et al. (2009). High definition profiling of mammalian DNA methylation by array capture and single molecule bisulfite sequencing. *Genome Res*, 19(9), 1593–1605.

Holoch, D. and Moazed, D. (2015). RNA-mediated epigenetic regulation of gene expression. *Nat Rev Genet*, 16(2), 71–84.

Iwakawa, H.O. and Tomari, Y. (2015). The functions of MicroRNAs: mRNA decay and translational repression. *Trends Cell Biol*, 25(11), 651–665.

Iyer, L.M., Zhang, D., Aravind, L. (2016). Adenine methylation in eukaryotes: Apprehending the complex evolutionary history and functional potential of an epigenetic modification. *Bioessays*, 38(1), 27–40.

Johnson, T.B. and Coghill, R.D. (1925). Researches on pyrimidines. C111. The discovery of 5-methyl-cytosine in tuberculinic acid, the nucleic acid of the tubercle bacillus1. *J Am Chem Soc*, 47(11), 2838–2844.

Jones, L., Hamilton, A.J., Voinnet, O., Thomas, C.L., Maule, A.J., Baulcombe, D.C. (1999). RNA-DNA interactions and DNA methylation in post-transcriptional gene silencing. *Plant Cell*, 11(12), 2291–2301.

Kota, S.K. and Feil, R. (2010). Epigenetic transitions in germ cell development and meiosis. *Dev Cell*, 19(5), 675–686.

Lahmy, S., Pontier, D., Cavel, E., Vega, D., El-Shami, M., Kanno, T., Lagrange, T. (2009). PolV(PolIVb) function in RNA-directed DNA methylation requires the conserved active site and an additional plant-specific subunit. *Proc Natl Acad Sci USA*, 106(3), 941–946.

Lahmy, S., Bies-Etheve, N., Lagrange, T. (2010). Plant-specific multisubunit RNA polymerase in gene silencing. *Epigenetics*, 5(1), 4–8.

Lahmy, S., Pontier, D., Bies-Etheve, N., Laudie, M., Feng, S., Jobet, E., Hale, C.J., Cooke, R., Hakimi, M.A., Angelov, D. et al. (2016). Evidence for ARGONAUTE4-DNA interactions in RNA-directed DNA methylation in plants. *Genes Dev*, 30(23), 2565–2570.

Law, J.A., Vashisht, A.A., Wohlschlegel, J.A., Jacobsen, S.E. (2011). SHH1, a homeodomain protein required for DNA methylation, as well as RDR2, RDM4, and chromatin remodeling factors, associate with RNA polymerase IV. *PLoS Genet*, 7(7), e1002195.

Law, J.A., Du, J., Hale, C.J., Feng, S., Krajewski, K., Palanca, A.M., Strahl, B.D., Patel, D.J., Jacobsen, S.E. (2013). Polymerase IV occupancy at RNA-directed DNA methylation sites requires SHH1. *Nature*, 498(7454), 385–389.

Lee, R.C., Feinbaum, R.L., Ambros, V. (1993). The C. elegans heterochronic gene lin-4 encodes small RNAs with antisense complementarity to lin-14. *Cell*, 75(5), 843–854.

Liang, Z., Geng, Y., Gu, X. (2018). Adenine methylation: New epigenetic marker of DNA and mRNA. *Mol Plant*, 11(10), 1219–1221.

Lim, P.S., Shannon, M.F., Hardy, K. (2010). Epigenetic control of inducible gene expression in the immune system. *Epigenomics*, 2(6), 775–795.

Lleres, D., Imaizumi, Y., Feil, R. (2021). Exploring chromatin structural roles of non-coding RNAs at imprinted domains. *Biochem Soc Trans*, 49(4), 1867–1879.

Luger, K., Mader, A.W., Richmond, R.K., Sargent, D.F., Richmond, T.J. (1997). Crystal structure of the nucleosome core particle at 2.8 A resolution. *Nature*, 389(6648), 251–260.

Mach, J. (2017). The long-noncoding RNA ELENA1 functions in plant immunity. *Plant Cell*, 29(5), 916.

Machida, S., Takizawa, Y., Ishimaru, M., Sugita, Y., Sekine, S., Nakayama, J.I., Wolf, M., Kurumizaka, H. (2018). Structural basis of heterochromatin formation by human HP1. *Mol Cell*, 69(3), 385–397.

Markulin, L., Skiljaica, A., Tokic, M., Jagic, M., Vuk, T., Bauer, N., Leljak Levanic, D. (2021). Taking the wheel – de novo DNA methylation as a driving force of plant embryonic development. *Front Plant Sci,* 12, 764999.

Matzke, M.A. and Mosher, R.A. (2014). RNA-directed DNA methylation: An epigenetic pathway of increasing complexity. *Nat Rev Genet*, 15(6), 394–408.

Matzke, M.A., Primig, M., Trnovsky, J., Matzke, A.J. (1989). Reversible methylation and inactivation of marker genes in sequentially transformed tobacco plants. *EMBO J*, 8(3), 643–649.

Matzke, M., Matzke, A.J., Kooter, J.M. (2001). RNA: Guiding gene silencing. *Science*, 293(5532), 1080–1083.

Meng, H., Cao, Y., Qin, J., Song, X., Zhang, Q., Shi, Y., Cao, L. (2015). DNA methylation, its mediators and genome integrity. *Int J Biol Sci*, 11(5), 604–617.

Mette, M.F., Van Der Winden, J., Matzke, M.A., Matzke, A.J. (1999). Production of aberrant promoter transcripts contributes to methylation and silencing of unlinked homologous promoters in trans. *EMBO J*, 18(1), 241–248.

Miron, E., Oldenkamp, R., Brown, J.M., Pinto, D.M.S., Xu, C.S., Faria, A.R., Shaban, H.A., Rhodes, J.D.P., Innocent, C., de Ornellas, S. et al. (2020). Chromatin arranges in chains of mesoscale domains with nanoscale functional topography independent of cohesin. *Sci Adv*, 6(39), 1–16.

Miyoshi, N., Stel, J.M., Shioda, K., Qu, N., Odajima, J., Mitsunaga, S., Zhang, X., Nagano, M., Hochedlinger, K., Isselbacher, K.J. et al. (2016). Erasure of DNA methylation, genomic imprints, and epimutations in a primordial germ-cell model derived from mouse pluripotent stem cells. *Proc Natl Acad Sci USA*, 113(34), 9545–9550.

Morel, J.B., Godon, C., Mourrain, P., Beclin, C., Boutet, S., Feuerbach, F., Proux, F., Vaucheret, H. (2002). Fertile hypomorphic ARGONAUTE (ago1) mutants impaired in post-transcriptional gene silencing and virus resistance. *Plant Cell*, 14(3), 629–639.

Mosher, R.A., Melnyk, C.W., Kelly, K.A., Dunn, R.M., Studholme, D.J., Baulcombe, D.C. (2009). Uniparental expression of PolIV-dependent siRNAs in developing endosperm of *Arabidopsis*. *Nature*, 460(7252), 283–286.

Mourrain, P., Beclin, C., Elmayan, T., Feuerbach, F., Godon, C., Morel, J.B., Jouette, D., Lacombe, A.M., Nikic, S., Picault, N. et al. (2000). Arabidopsis SGS2 and SGS3 genes are required for posttranscriptional gene silencing and natural virus resistance. *Cell*, 101(5), 533–542.

Mugat, B., Nicot, S., Varela-Chavez, C., Jourdan, C., Sato, K., Basyuk, E., Juge, F., Siomi, M.C., Pelisson, A., Chambeyron, S. (2020). The Mi-2 nucleosome remodeler and the Rpd3 histone deacetylase are involved in piRNA-guided heterochromatin formation. *Nat Commun*, 11(1), 2818.

Napoli, C., Lemieux, C., Jorgensen, R. (1990). Introduction of a chimeric chalcone synthase gene into petunia results in reversible co-suppression of homologous genes in trans. *Plant Cell*, 2(4), 279–289.

Ninova, M., Fejes Toth, K., Aravin, A.A. (2019). The control of gene expression and cell identity by H3K9 trimethylation. *Development*, 146(19).

Ninova, M., Chen, Y.A., Godneeva, B., Rogers, A.K., Luo, Y., Fejes Toth, K., Aravin, A.A. (2020). Su(var)2-10 and the SUMO Pathway Link piRNA-guided target recognition to chromatin silencing. *Mol Cell*, 77(3), 556–570.

Ohtani, H., Iwasaki, Y.W., Shibuya, A., Siomi, H., Siomi, M.C., Saito, K. (2013). DmGTSF1 is necessary for Piwi-piRISC-mediated transcriptional transposon silencing in the Drosophila ovary. *Genes Dev*, 27(15), 1656–1661.

Okamoto, I., Patrat, C., Thepot, D., Peynot, N., Fauque, P., Daniel, N., Diabangouaya, P., Wolf, J.P., Renard, J.P., Duranthon, V. et al. (2011). Eutherian mammals use diverse strategies to initiate X-chromosome inactivation during development. *Nature*, 472(7343), 370–374.

Raja, P., Sanville, B.C., Buchmann, R.C., Bisaro, D.M. (2008). Viral genome methylation as an epigenetic defense against geminiviruses. *J Virol*, 82(18), 8997–9007.

Rao, S.S., Huntley, M.H., Durand, N.C., Stamenova, E.K., Bochkov, I.D., Robinson, J.T., Sanborn, A.L., Machol, I., Omer, A.D., Lander, E.S. et al. (2014). A 3D map of the human genome at kilobase resolution reveals principles of chromatin looping. *Cell*, 159(7), 1665–1680.

Ratcliff, F., Harrison, B.D., Baulcombe, D.C. (1997). A similarity between viral defense and gene silencing in plants. *Science*, 276(5318), 1558–1560.

Ricci, M.A., Manzo, C., Garcia-Parajo, M.F., Lakadamyali, M., Cosma, M.P. (2015). Chromatin fibers are formed by heterogeneous groups of nucleosomes in vivo. *Cell*, 160(6), 1145–1158.

Roule, T., Crespi, M., Blein, T. (2022). Regulatory long non-coding RNAs in root growth and development. *Biochem Soc Trans*, 50(1), 403–412.

Saksouk, N., Simboeck, E., Dejardin, J. (2015). Constitutive heterochromatin formation and transcription in mammals. *Epigenetics Chromatin*, 8, 3.

Sanan-Mishra, N., Abdul Kader Jailani, A., Mandal, B., Mukherjee, S.K. (2021). Secondary siRNAs in plants: Biosynthesis, various functions, and applications in virology. *Front Plant Sci*, 12, 610283.

Schuettengruber, B., Bourbon, H.M., Di Croce, L., Cavalli, G. (2017). Genome regulation by polycomb and trithorax: 70 years and counting. *Cell*, 171(1), 34–57.

Seo, J.S., Sun, H.X., Park, B.S., Huang, C.H., Yeh, S.D., Jung, C., Chua, N.H. (2017). ELF18-INDUCED LONG-NONCODING RNA associates with mediator to enhance expression of innate immune response genes in *Arabidopsis*. *Plant Cell*, 29(5), 1024–1038.

Shayevitch, R., Askayo, D., Keydar, I., Ast, G. (2018). The importance of DNA methylation of exons on alternative splicing. *RNA*, 24(10), 1351–1362.

Shukla, S., Kavak, E., Gregory, M., Imashimizu, M., Shutinoski, B., Kashlev, M., Oberdoerffer, P., Sandberg, R., Oberdoerffer, S. (2011). CTCF-promoted RNA polymerase II pausing links DNA methylation to splicing. *Nature*, 479(7371), 74–79.

Sienski, G., Donertas, D., Brennecke, J. (2012). Transcriptional silencing of transposons by Piwi and maelstrom and its impact on chromatin state and gene expression. *Cell*, 151(5), 964–980.

Sienski, G., Batki, J., Senti, K.A., Donertas, D., Tirian, L., Meixner, K., Brennecke, J. (2015). Silencio/CG9754 connects the Piwi-piRNA complex to the cellular heterochromatin machinery. *Genes Dev*, 29(21), 2258–2271.

Sigman, M.J., Panda, K., Kirchner, R., Mclain, L.L., Payne, H., Peasari, J.R., Husbands, A.Y., Slotkin, R.K., Mccue, A.D. (2021). An siRNA-guided ARGONAUTE protein directs RNA polymerase V to initiate DNA methylation. *Nat Plants*, 7(11), 1461–1474.

Smith, C.J., Watson, C.F., Bird, C.R., Ray, J., Schuch, W., Grierson, D. (1990). Expression of a truncated tomato polygalacturonase gene inhibits expression of the endogenous gene in transgenic plants. *Mol Gen Genet*, 224(3), 477–481.

Spielmann, M., Lupianez, D.G., Mundlos, S. (2018). Structural variation in the 3D genome. *Nat Rev Genet*, 19(7), 453–467.

Stroud, H., Do, T., Du, J., Zhong, X., Feng, S., Johnson, L., Patel, D.J., Jacobsen, S.E. (2014). Non-CG methylation patterns shape the epigenetic landscape in Arabidopsis. *Nat Struct Mol Biol*, 21(1), 64–72.

Szabo, Q., Bantignies, F., Cavalli, G. (2019). Principles of genome folding into topologically associating domains. *Sci Adv*, 5(4), eaaw1668.

Szabo, Q., Donjon, A., Jerkovic, I., Papadopoulos, G.L., Cheutin, T., Bonev, B., Nora, E.P., Bruneau, B.G., Bantignies, F., Cavalli, G. (2020). Regulation of single-cell genome organization into TADs and chromatin nanodomains. *Nat Genet*, 52(11), 1151–1157.

Van Der Krol, A.R., Mur, L.A., Beld, M., Mol, J.N., Stuitje, A.R. (1990). Flavonoid genes in petunia: Addition of a limited number of gene copies may lead to a suppression of gene expression. *Plant Cell*, 2(4), 291–299.

Van Rij, R.P., Saleh, M.C., Berry, B., Foo, C., Houk, A., Antoniewski, C., Andino, R. (2006). The RNA silencing endonuclease Argonaute 2 mediates specific antiviral immunity in *Drosophila melanogaster*. *Genes Dev*, 20(21), 2985–2995.

Vaucheret, H. and Fagard, M. (2001). Transcriptional gene silencing in plants: Targets, inducers and regulators. *Trends Genet*, 17(1), 29–35.

Volpe, T.A., Kidner, C., Hall, I.M., Teng, G., Grewal, S.I., Martienssen, R.A. (2002). Regulation of heterochromatic silencing and histone H3 lysine-9 methylation by RNAi. *Science*, 297(5588), 1833–1837.

Wightman, B., Ha, I., Ruvkun, G. (1993). Posttranscriptional regulation of the heterochronic gene lin-14 by lin-4 mediates temporal pattern formation in *C. elegans*. *Cell*, 75(5), 855–862.

Wu, P.H., Fu, Y., Cecchini, K., Ozata, D.M., Arif, A., Yu, T., Colpan, C., Gainetdinov, I., Weng, Z., Zamore, P.D. (2020). The evolutionarily conserved piRNA-producing locus pi6 is required for male mouse fertility. *Nat Genet*, 52(7), 728–739.

Yao, Q., Chen, Y., Zhou, X. (2019). The roles of microRNAs in epigenetic regulation. *Curr Opin Chem Biol*, 51, 11–17.

Yearim, A., Gelfman, S., Shayevitch, R., Melcer, S., Glaich, O., Mallm, J.P., Nissim-Rafinia, M., Cohen, A.H., Rippe, K., Meshorer, E. et al (2015). HP1 is involved in regulating the global impact of DNA methylation on alternative splicing. *Cell Rep*, 10(7), 1122–1134.

Zenk, F., Loeser, E., Schiavo, R., Kilpert, F., Bogdanovic, O., Iovino, N. (2017). Germ line-inherited H3K27me3 restricts enhancer function during maternal-to-zygotic transition. *Science*, 357(6347), 212–216.

Zhang, H., Lang, Z., Zhu, J.K. (2018). Dynamics and function of DNA methylation in plants. *Nat Rev Mol Cell Biol*, 19(8), 489–506.

Zhang, X., Wang, W., Zhu, W., Dong, J., Cheng, Y., Yin, Z., Shen, F. (2019). Mechanisms and functions of long non-coding RNAs at multiple regulatory levels. *Int J Mol Sci*, 20(22), 5573–5601.

3

Epigenetics and Transposable Elements

Clémentine VITTE[1], Séverine CHAMBEYRON[2] and Cristina VIEIRA[3]

[1] *GQE – Le Moulon, Gif-sur-Yvette, France*

[2] *Institut de Génétique Humaine, CNRS, Université de Montpellier, France*

[3] *LBBE, CNRS, Université Claude Bernard Lyon 1, Villeurbanne, France*

3.1. Introduction

We often think of the genome as a stable entity, essentially containing genes, which are relatively small protein-coding regions underlying the phenotype. However, this is not quite the case.

Genomes contain genes, which generally remain in place. However, these genes only correspond to a small part of the genome, and are separated from each other by large intergenic regions containing mobile DNA elements, whose mobility is controlled by epigenetic mechanisms.

These "transposable elements" (TEs) have five characteristics that give them a significant role in genome evolution:

1) *Ancestrality*: we now suspect that certain TE families were already present in the common ancestor of eukaryotes.

2) *Universality*: they are found in almost all species.

3) *Abundance*: TEs can represent a significant proportion of genomes.

Epigenetics in Ecology and Evolution,
coordinated by Christoph GRUNAU and Stéphane MAURY.

4) *Mobility*: they have the ability to move from one place to another in the genome, as well as from one species to another.

5) *Mutagenicity*: TEs are powerful mutagens, inducing genetic variability.

The development of genetic and epigenetic methods over the last 20 years has led to a better understanding of how these sequences are regulated in eukaryotic genomes, as well as their genetic and epigenetic impact in natural, domesticated or experimental populations.

In this chapter, we will first present how TEs were discovered and classified, and where they are localized in eukaryotic genomes. Next, we will describe the major bases of their epigenetic regulation, and will present the effects that TEs can have on phenotype at the genetic and epigenetic levels. Lastly, we will describe the impact of these elements on population adaptation and species evolution.

3.2. TEs in genomes

3.2.1. *Discovery and classification*

TEs are dispersed repeated sequences, meaning they are in multiple copies located at different genomic positions (as opposed to tandem repeat sequences such as satellites or microsatellites). They were discovered in the 1950s by the scientist Barbara McClintock (1902–1992) through her maize genetics experiments. By analyzing the color of maize kernels, she was able to deduce the existence of TEs and their effects on kernel color long before the molecular support of genetic information was discovered. She also called them "jumping genes", because of their ability to move around the genome, and/or "control elements", as the elements she analyzed affected the expression of maize kernel color.

TEs are highly diverse in terms of size, structure and their mobility mechanism. In order to better study them, a classification was proposed in 2008, and is now the reference in the scientific community (Wicker et al. 2007). TEs are organized into classes, sub-classes, orders, superfamilies and families. This organization is described schematically in Figure 3.1. They are first divided into two classes on the basis of the molecule (RNA or DNA) that mediates transposition. Class I elements, also known as retrotransposons, are mobilized by a "copy-and-paste" mechanism whereby the element's RNA is reverse-transcribed into a complementary DNA copy (cDNA), then the RNA is degraded and complementary DNA is synthesized, forming a double-stranded DNA copy that integrates elsewhere in the genome. Class II elements are mobilized by a DNA intermediate: the copy itself, in its DNA form, is excised and inserted to another part of the genome. Class II elements are separated into two sub-classes according to the number of DNA strands cut at the donor site. These two major sub-classes transpose using rather different mechanisms: TIR transposons, also known as DNA transposons, are mobilized by a "cut-and-paste" mechanism in which double-stranded DNA is excised and integrated at another genomic location. In a number of cases, repair of the excision site restores the copy at the donor site (e.g. by copying onto a sister chromatin). The mechanism of helitron transposition has not yet been molecularly clarified, but a rolling circle or "peel-and-paste" model has been proposed to explain their transposition. This model involves a circular single-stranded DNA intermediate. The order separates TEs on the basis of their insertion mechanism, their overall structure, the sequence of proteins they encode and their organization within the sequence. For example, long terminal repeat (LTR) retrotransposons are reverse-transcribed within a viral particle, and the resulting double-stranded DNA is inserted following an integrase-catalyzed strand cleavage and transfer reaction. Non-LTR retrotransposons, which include long interspersed nuclear elements (LINE), are not reverse-transcribed within a viral particle, and their chromosomal integration is coupled to reverse transcription. Elements are classified into superfamilies on the basis of protein structure, and are subdivided into families on the basis of DNA sequence conservation. To be classified in the same family, two copies must share at least 80% identity over 80% of their nucleotide sequence length.

TEs are divided into autonomous elements, which carry the genes encoding the proteins required for their mobility, and non-autonomous elements, which lack these and use protein factors encoded by autonomous elements for their transposition. These non-autonomous elements do not constitute a group of their own, but are attached to the families from which they derive. Three types of non-autonomous elements are particularly important. Class I includes LARDs (for large retrotransposon derivatives) and TRIMs (for terminal repeat in miniature), which have the same structure as LTR retrotransposons, and SINEs (for short interspersed nuclear elements), which are related to LINEs. LARDs are similar in size to stand-alone LTR retrotransposons, while TRIMs are small (< 4 kb). SINEs are small (a few hundred nucleotides). Class II includes MITEs (miniature inverted terminal elements), which are small (< 600 bp) TIR elements.

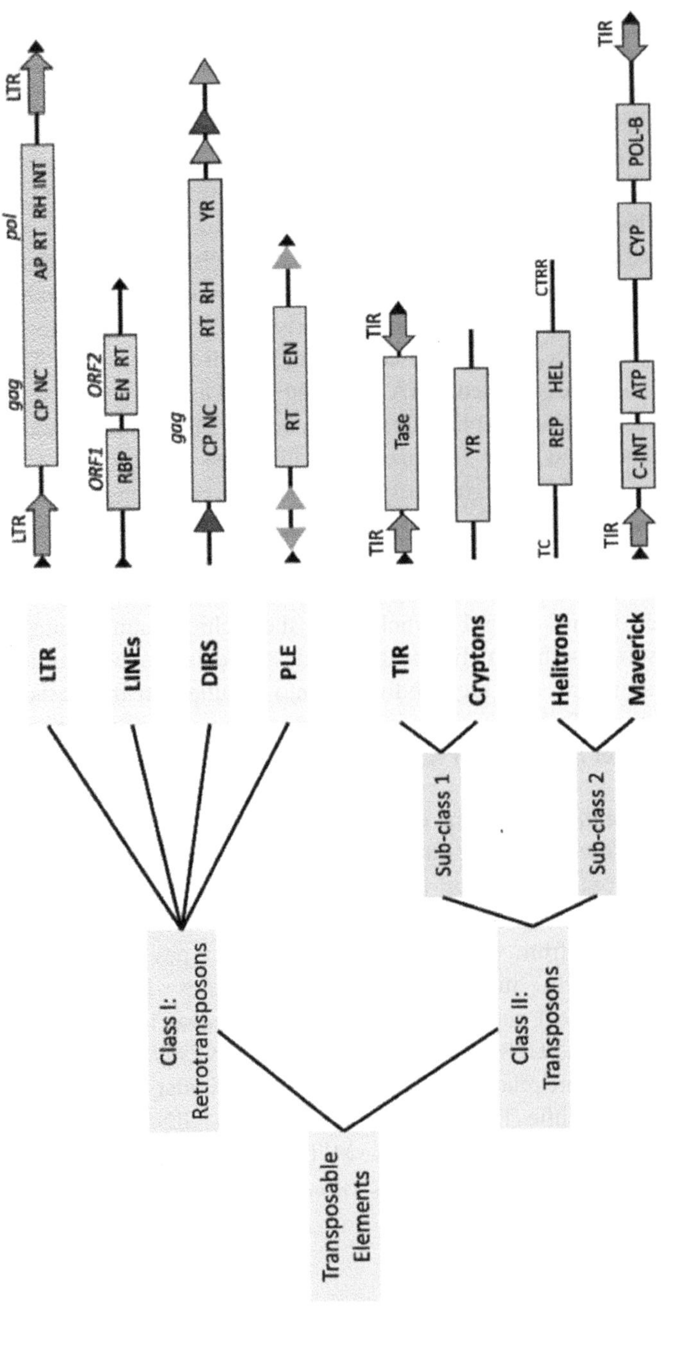

Figure 3.1. *Classification of transposable elements Classification des éléments transposables*

COMMENT ON FIGURE 3.1. – *TEs are classified, from left to right, in terms of class, sub-class and order. For each order, an example of element structure is shown, in order to highlight the structural differences and protein functions encoded by the different elements. These structures are indicative, and variations exist within different elements of the same order. Structural elements: LTR: Long Terminal Repeat; TIR: Terminal Inverted Repeat; protein-coding genes: CP: Caspid Protein; NC: NucleoCapsid; AP: Aspartic Protease; RT: Reverse Transcriptase; RH: RNAseH; INT: Integrase; RBP: Ribo-Binding Protein; EN: Endonuclease; YR: Tyrosine Recombinase; Tase: Transposase; REP: Replication protein; HEL: Helicase; C-INT: C-Integrase; CYP: Cysteine Protease; POLB: DNA Polymerase B.*

3.2.2. *Abundance and location of TEs in genomes*

The sequencing of the human genome in 2001 revealed that TEs account for over 50% of its DNA sequence. Since then, the sequencing of genomes from other species has shown that TEs are major components of most eukaryotic genomes, and that the genomic fraction they occupy varies from species to species (Vitte et al. 2014; Wells and Feschotte 2020). These sequences represent, for example, 15% of the *Arabidopsis thaliana* genome, 25% of that of the *Drosophila melanogaster* fly and over 85% of the genome of certain flowering plants, notably maize, in which they were discovered. Beyond the overall proportion in TEs, the relative proportion of the different elements varies between species. Figure 3.2 illustrates these variations for different species of the eukaryotic tree. For example, the human genome contains predominantly LINE-type elements, while plant genomes are characterized by a high abundance of LTR retrotransposons. Some fish, such as the zebrafish, have genomes rich in class II elements. The yeast *Saccharomyces cerevisiae* contains few TEs, and these are all LTR retrotransposons. Thus, the different classes of TEs appear to have experienced varying degrees of evolutionary success in different species over the course of evolution.

Analysis of TE density along chromosomes in different species, including model organisms such as *A. thaliana* and *D. melanogaster*, has revealed a high abundance of these elements in pericentromeric regions. This is generally true for small genomes. In contrast, the sequencing of larger genomes, such as rice (390 Mb) or maize (2.1 Gb), revealed a very high abundance of TEs all along the chromosome arms (Figure 3.3). In maize, TEs account for over 50% of chromosome arms, making the structure of these regions similar to that of *A. thaliana* pericentromeres.

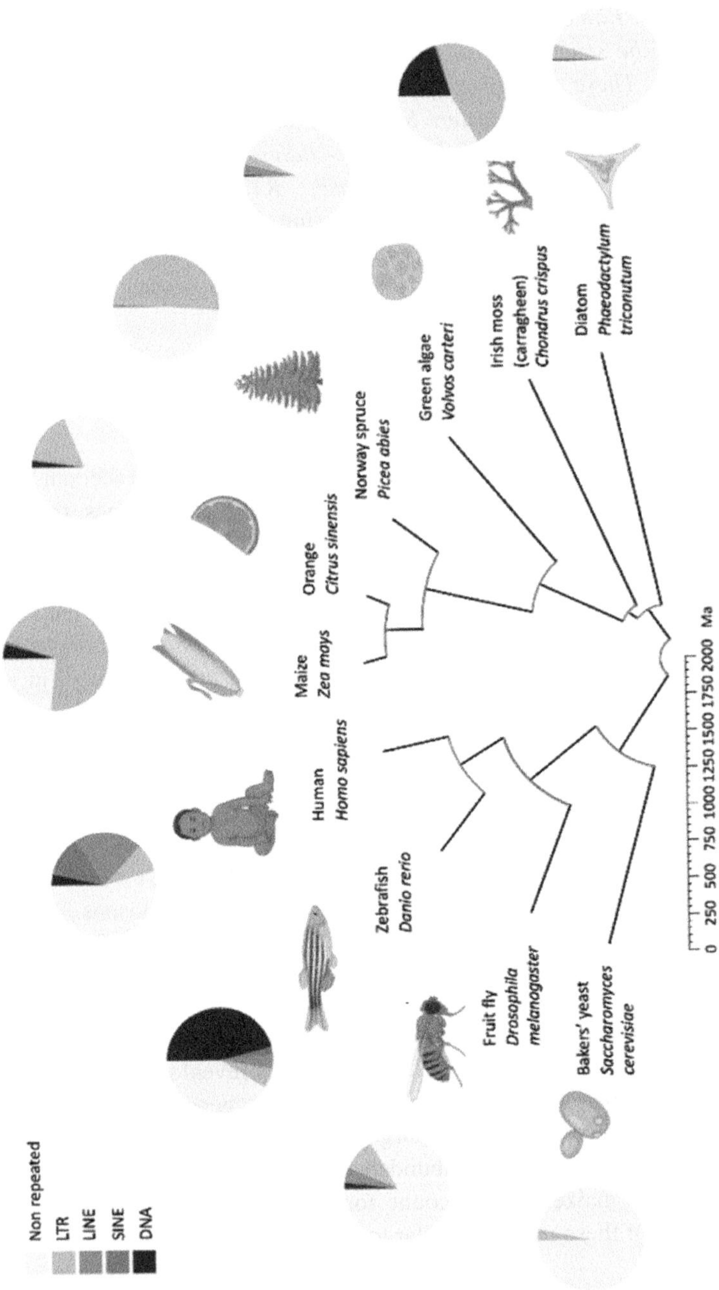

Figure 3.2. *Ratio of the most common types of transposable elements in eukaryotes. Adapted from Wells and Feschotte (2020). Created with BioRender.com*

COMMENT ON FIGURE 3.2. – *Estimates of proportions come from Wells and Feschotte (2020), except for orange, for which the data come from Xu et al. (2013). They are derived from analyses based on sequence identity with TE databases, and are somewhat underestimated. However, they are carried out using the same methodology, which allows comparison between species. Latin species names are shown in italics. The tree indicates the separations between species, and dates are given in millions of years (My).*

Furthermore, the different TE superfamilies are not homogeneously localized along chromosomes. For example, MITEs are often localized within gene regions, while retrotransposons tend to be localized outside these regions. However, in large genomes such as maize, retrotransposons are so abundant that over 70% of genes are located within 2 kb of a TE. Within LTR retrotransposons, there is a difference between the *Gypsy* superfamily elements, which tend to be localized in non-genic regions, and *Copia*, which tend to be localized in genic regions. This variation in density along the chromosomes can be explained by insertion bias, selection bias and variation in recombination rates along the chromosomes. For example, certain families preferentially insert into heterochromatic regions, due to their association with SWI-SNF-type chromatin remodeling factors. Short elements, such as MITEs, are probably better tolerated close to genes, and therefore less counter-selected.

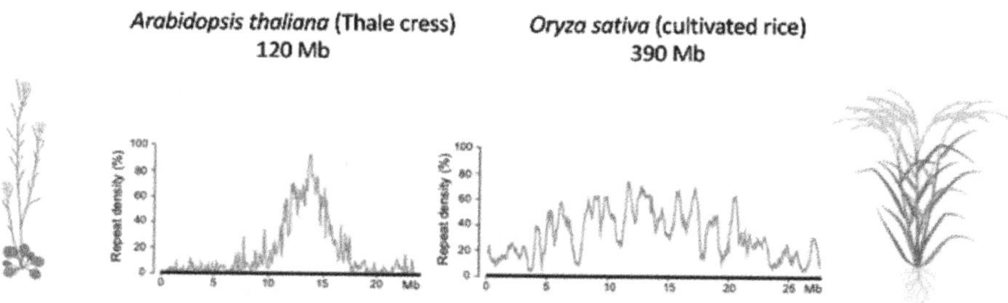

Figure 3.3. *Comparison of chromosome repeat density between* Arabidopsis thaliana *and cultivated rice* Oryza sativa. *Repeat densities are calculated along chromosomes over 100 kb windows. Data from chromosomes 3 and 12 for* A. thaliana *and* O. sativa, *respectively. Adapted from Mirouze and Vitte (2014). Created with BioRender.com*

3.2.3. *Transposition control*

As a result of their ability to multiply and insert themselves into the genome, TEs are mutagenic. In particular, a high rate of transposition can have harmful effects, since the increase in copy number multiplies the probability of generating insertional mutations in genes that contribute to the selective value of individuals. As a result, mechanisms to control transposition activity have been developed over the course of evolution. Control of transposition can be achieved either by preventing transcription of TEs through transcriptional gene silencing (TGS), or by destroying TE mRNAs through post-transcriptional gene silencing (PTGS). Both mechanisms are mediated by small regulatory RNAs.

The players involved in this control obviously depend on the epigenetic machinery of the species under consideration (Mirouze and Vitte 2014) (see Chapter 2), and a good knowledge of these pathways is therefore essential for understanding TE regulation in a given species. For example, DNA methylation plays a major role in TE regulation in plants (Law and Jacobsen 2010), whereas in Drosophila, where methylation is virtually non-existent, the control of TE transcription is achieved by regulatory RNAs that induce the deposition of repressive histone marks (see Chapter 2) (Czech et al. 2013).

Generally speaking, any new copy of a TE that inserts into a genome will be recognized by small regulatory RNAs, triggering the establishment of repressive chromatin marks (DNA methylation and histone marks). These regulatory RNAs can originate either from copies of the same element located elsewhere in the genome, or from the new copy itself (Shpiz et al. 2014). The region of insertion has a major impact on the chromatin marks involved in TE control. In plants, insertions located in non-genic regions are targeted by repressive histone marks allowing chromatin compaction, while insertions located within a few kilobases of genes are only methylated by the regulatory RNA pathway, allowing TEs to be transcriptionally switched off while maintaining an open chromatin state, compatible with transcriptional gene regulation (Sigman and Slotkin 2016). In Drosophila, the production of regulatory RNAs is ensured by specific genomic regions, known as piRNA clusters (Khurana and Theurkauf 2010). Any new insertion belonging to a TE family, inserting itself into these regions or already possessing a copy located in these regions, will therefore be controlled and unable to transpose.

The epigenetic control of TEs is fairly reliable, notably because of the multiplicity of players (and therefore pathways) involved in its implementation. For example, in *A. thaliana*, mutants in DNA methylation pathways lead to the remobilization of a subset of just 10 families, and over 90% of the mobilized copies belong to just three families (one retrotransposon and two DNA transposons) (Quadrana et al. 2019), suggesting that other epigenetic marks are involved in TE repression. This is particularly the case for histone methylation (H3K27me3 and H3K9me). Some mutant alleles of epigenetic control pathway genes are, however, associated with higher transposition rates than wild-type alleles (Baduel et al. 2021), suggesting that even partial modification of these pathways can induce transposition in nature.

Environmental changes can also lead to TE mobilization, and the interaction between genetic variants (of the elements themselves and of epigenetic control pathways) and the environment seems to be one of the major determinants of TE mobility.

Beyond the control of transposition, epigenetic marks also play a role in regulating the negative effects that TEs can have on gene expression. For example, mutants of DNA methylation pathways reveal thousands of cryptic transcription sites located in TE sequences (Le et al. 2020), which do not produce RNA in wild plants, indicating that methylation reduces the phenotypic impact of TEs.

3.3. Impact of TEs on phenotype

3.3.1. *Genetic impact*

Several types of mutations are responsible for genetic diversity. Point mutations, or SNPs (single nucleotide polymorphisms), correspond to the change of a nucleotide in the DNA sequence. In addition, "indels" (insertion/deletion) can be generated if several successive nucleotides are inserted or deleted. These mutations are small in size, and generally carry no specific genetic information. The movement of TEs can also induce mutations. These can be various types of mutations and are a source of significant genetic variability, which are sometimes overlooked. The presence of several copies of the same DNA sequence in the genome also confers special properties to these elements. Furthermore, during transposition, TEs mobilize a large number of DNA bases (from a few hundred base pairs to tens of kilobases), which contain relevant genetic information, including genes and sequences that regulate their expression. A new insertion is therefore a ready-to-use molecular source that can interact with the new genomic environment in which it has been inserted. Below, we describe the different types of genetic mutation that TEs can induce.

3.3.1.1. *Mutations due to recombination*

Homologous recombination, as the name suggests, takes place between homologous regions of two chromosomes. However, there is also "non-homologous" recombination, which takes place between genomic regions sharing the same sequence, but located at different locations in the genome. As repeated sequences, TEs generate copies with high sequence-identity at different locations in the genome, and can therefore be used as templates for this type of event. This type of recombination induces chromosomal rearrangements that modify the composition of the genome, either by eliminating the genomic region located between the two copies if they are in direct orientation, or by inverting this region if they are in an opposing orientation (head-to-tail). These inversions can lead to changes in the gene/regulatory region pair, resulting in significant phenotypic changes. This is the case with the *antennapedia* mutation in Drosophila, where antennae are replaced by legs. In tomato, a complex rearrangement of the retrotransposon-linked region carrying the *SUN* gene induces the overexpression of the gene and an elongated fruit phenotype (Xiao et al. 2008).

3.3.1.2. *Gene sequence mutations*

Since TEs are large sequences (at least several hundred base pairs), their insertion into a gene will generate a change in the gene sequence. Similarly, a poorly repaired excision will induce a modification of the gene sequence. If this modification affects the coding phase, it will induce a change in the protein produced, which is often too strongly modified to be functional. However, since TEs carry their own promoter and terminator sequences, as well as signals for the addition of poly-A tails to the transcript, TE insertions can also have a post-transcriptional effect. Depending on whether it is inserted upstream, downstream or in an intron, an insertion can induce a change in the start or end position of the transcript, or the formation of new transcripts with different splicings. Lastly, TEs also contain their own splicing signals, and their insertion downstream of a gene can generate longer transcripts, which end in the TE sequence. These different effects are illustrated in Figure 3.4.

3.3.1.3. *Mutations by modification of regulatory sequences for expression*

Insertion of a TE in the promoter of a gene or in the proximal sequences that regulate it can destabilize the binding of transcription factors, and thus alter the level of gene expression. However, TEs also possess their own regulatory sequences, and their insertion in the vicinity of a gene's promoter can provide the latter with new proximal *cis*-regulatory sequences. This type of modification is richer than the previous one, as it can provide a wide range of variation, depending on the sequence of the element and the regulatory sequences it carries. Similarly, the expression level and spatio-temporal nature of a gene can be modified by the insertion of a TE in or near distal *cis*-regulatory sequences (located more than 2 kb from the gene promoter). A particular aspect of TEs is that their regulatory sequences are sometimes sensitive to the environment. As a result, their insertion in the vicinity of a gene can induce sensitivity of gene expression to the environment. For example, the red color of blood oranges is linked to the insertion of a retrotransposon upstream of the *Ruby* gene, a key gene in the anthocyanin pathway, making its expression sensitive to temperature. As a result, only oranges carrying the retrotransposon, when exposed to cold, can express anthocyanins, which are responsible for the red color of the fruit pulp (Butelli et al. 2012). Remarkably, blood oranges from Sicily and China have different insertions upstream of the *Ruby* gene, but which provide similar regulation.

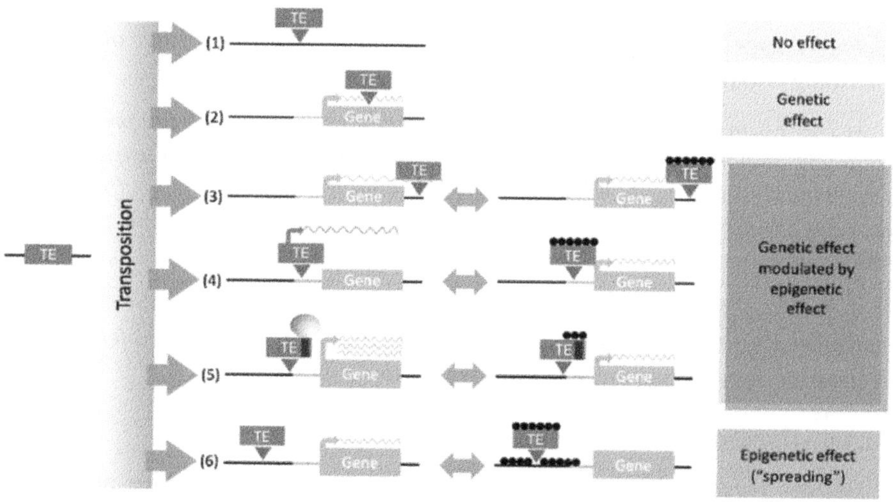

Figure 3.4. *Effects of TEs by insertion site*

COMMENT ON FIGURE 3.4. – *A TE can be inserted at different locations in the genome. Each arrow represents a different type of insertion, located at greater or lesser distances from the gene, with the consequence on transcription shown schematically on the right. The case of an insertion in an exon, generating an aberrant protein, is not shown. The gene is represented by a red square, with the arrow indicating the start of transcription. Transcripts are indicated by wavy lines. The number of lines represents the quantity of transcripts generated. Balls represent chromatin marks (DNA methylation, histone modification). The large ball represents a transcription factor. The box corresponds to the transcription factor binding site carried by the ET. The effect of the TE at the time of insertion is shown to the left of the double arrow, without epigenetic regulation. To the right of the double arrow, it shows the TE effect with epigenetic regulation. Photo from Cropley et al. (2006):*

1) The TE inserts at a great distance from genes, and has no phenotypic effect.

2) The TE inserts into an exon, and induces a transcript including the TE, generating an aberrant protein (not shown).

3) The TE inserts into the 3' region of a gene, and generates a longer transcript, producing a longer, aberrant protein (not shown). Transcript modification can also occur if the TE inserts into the intron, generating either a truncated transcript or a differently spliced transcript (not shown).

4) The TE inserts into the promoter region of a gene. The transcription machinery uses the transcription start site of the TE instead of that of the gene, generating a longer, aberrant transcript.

5) The TE inserts beyond the promoter region of the gene, at the proximal cis-*regulatory region level. As the TE has a regulatory region bearing a transcription factor binding site (purple box), the corresponding transcription factor (purple ball) can regulate the level of gene transcription. Here, a factor increasing the level of transcription is shown.*

6) The TE inserts into the 5' region of the gene. The chromatin state implemented at the TE diffuses locally around it, reaching the promoter region of the gene, and preventing the generation of a transcript.

3.3.1.4. *The emergence of new functions*

Besides regulatory information, autonomous TEs encode proteins required for transcription and transposition. These protein sequences, which are not essential to the life of the host individual since they are only needed for the mobility of the element, are a rich molecular source enabling the emergence of new functions, notably through the positive selection of the protein domains they contain. For example, the protein domain of transposases from TIR DNA elements has been selected several times in the course of evolution, enabling the generation of proteins with new functions through fusion with different proteins. In mammals, for example, the molecular domestication of endogenous retrovirus envelope glycoproteins, which are capable of inducing cell fusion, led to the emergence of syncytin, a key player in the formation of the placental structure. Capture of these genes has occurred several times in different mammalian evolutionary lineages, and would have enabled the repeated emergence of placental structure in different species such as the mouse or the human (Mallet and Prudhomme 2004; Dupressoir and Heidmann 2011).

3.3.1.5. *The emergence of new regulatory networks*

As a result of their ability to regulate their own transcription, TEs often carry transcription factor binding sites (TFBS). Transposition of a TFBS enables amplification of this TFBS to multiple genomic regions. TEs are thus able to bring the same regulatory sequence into the vicinity of different genes, potentially generating new regulatory networks. As some of the regulatory regions contained in TEs are sensitive to the environment, the networks thus generated are also capable of producing a concerted transcriptional reaction in response to the environment. For example, the recent amplification of the *mPing* element in rice, and its propensity to insert itself in the 5' region of genes, has generated a set of genes whose expression is dependent on cold and salinity (Naito et al. 2009). In addition, some proteins encoded by TEs are able to bind to their own DNA sequences. This is the case of Class II transposases. In the course of evolution, new proteins have emerged from the fusion of a gene and a transposase. In addition to their original function, these proteins are able to recognize the DNA sequences of the corresponding TEs, which are numerous and scattered throughout the genome. This molecular assembly enables the unified regulation of a large number of genomic regions. If some or all of these regions are located close to genes, they can generate new networks regulating their expression. This has recently been demonstrated in bats, where the transposase-derived protein KRABINER binds to *cis*-regulatory elements derived from related transposons and controls the unified regulation of a set of genes (Cosby et al. 2021).

3.3.2. *Epigenetic impact*

Most of the genetic effects mentioned above can be modulated by the presence of epigenetic marks on the TE sequence. In particular, the effect of an insertion can be limited by the addition of epigenetic marks on the TE sequence. For example, DNA methylation of a TE generating an aberrant version of the transcript is capable of abolishing the generation of this transcript (Figure 3.4). As this type of mark is typically deposited within a few generations, the transcriptional effects of a new insertion can be rapidly counterbalanced.

A particular feature of epigenetic-mediated transcriptional regulation is the metastability of expression: the gene is expressed variably within the same individual among the same cell type. Thus, most individuals whose TE insertion is regulated by an epigenetic mechanism show mosaic phenotypes, which are due to methylation variability at early stages of embryogenesis, followed by stable somatic transmission within cell lines. The early methylation profile also determines the number of final cells affected, and is responsible for the phenotypic variations observed in these mutants.

For example, the expression of the *SPB* gene in tomato is controlled by the presence of a *Copia* superfamily LTR retrotransposon upstream of the gene. This element provides the gene with regulatory sequences enabling its transcription, which gives the tomato its red color. In *cnr* (colorless non-ripening) mutants, methylation of the TE induces the repression of gene transcription, giving the tomato an orange color (Manning et al. 2006). As a result of the variation in methylation in different cells, we can observe tomatoes that are completely red, completely orange, and tomatoes with red and orange sectors. Similarly, in mice, the normal expression of the *agouti* gene encodes a dark brown coat, resembling the color of the agouti coat. The insertion of IAP (Intracisternal A Particle) LTR retrotransposons into the promoter region of this gene provides an alternative promoter, which is responsible for the generation of a modified transcript that is ectopically expressed and induces the formation of a yellow-colored coat. Methylation of the IAP element prevents the expression of this alternative transcript, allowing normal gene expression and the agouti coat color. Depending on the point in development at which methylation takes place, a greater or lesser number of cells are affected, inducing a greater or lesser number of spots (Figure 3.5).

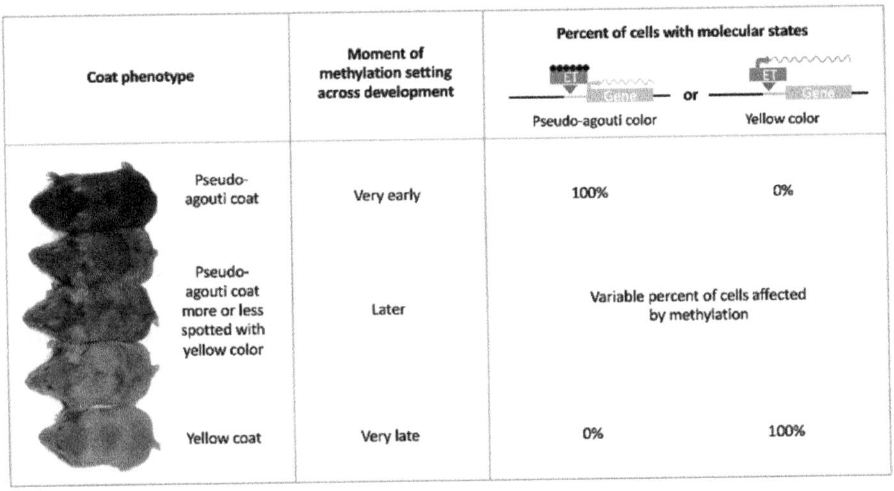

Figure 3.5. *Genetic and epigenetic effect of a TE on gene transcription: the example of the agouti gene. Photo adapted from Cropley et al. (2006)*

COMMENT ON FIGURE 3.5. – *Mice with a TE inserted upstream of the* agouti *gene show a series of phenotypes ranging from agouti-colored coats, to yellow coats, to spotted phenotypes with varying numbers of spots. These phenotypes are linked to epigenetic regulation of the IAP TE located upstream of the gene, which takes place more or less early during development; and a molecular explanation of the percentages of cells affected.*

It is sometimes the inactive state of the element that corresponds to the "wild type" state, as in the case of the *FWA* (Flowering WAgenigen) locus in *Arabidopsis thaliana* (Kinoshita et al. 2007). This gene encodes a transcription factor controlling flowering, and its transcription begins in a partially duplicated SINE. In wild type plants, this promoter region is methylated, leading to transcriptional silencing in vegetative tissues. The presence of the SINE is necessary and sufficient for epigenetic gene silencing (Fujimoto et al. 2008). This methylation state is widely conserved across *A. thaliana* accessions (Vaughn et al. 2007). However, the alteration of small RNA production or DNA methylation at the SINE duplication allows the ectopic expression of the *FWA* gene, resulting in a late-flowering phenotype (Soppe et al. 2000).

3.3.3. *Local spreading of epigenetic marks*

Since TE recognition is sequence-dependent, epigenetic regulation of these sequences is localized and limited to the TE sequence. However, the deposition of epigenetic marks sometimes extends beyond the extremities of the targeted TE. This is known as local diffusion, or "spreading" of epigenetic marks. Insertion of a TE in the vicinity of a gene can thus induce the propagation of these marks at the gene's promoter, inducing its epigenetic repression, and leading to transcriptional silencing (Figure 3.4).

In Drosophila somatic ovary cells, the *expanded* gene has a *Gypsy* TE insertion in its first intron, 1.2 kb downstream of the gene transcription site. The presence of this insertion induces propagation of the H3K9me3 epigenetic mark over more than 10 kb, covering the gene's transcription initiation site and preventing occupation of the site by Polymerase II and transcription of the gene. If the TE is not present or the TE repression pathway is inactive, the H3K9me3 mark does not cover the transcription initiation site and the gene is highly transcribed (Sienski et al. 2015; Mugat et al. 2020; Andreev et al. 2022). In mice, the analysis of the chromatin states of two cell lines with polymorphic retrotransposons revealed that almost all retrovirus-like TE copies are capable of inducing repressive chromatin around their insertion sites and extend to at least 1 kb. This heterochromatin induction has not been observed in other TE families, such as LINE. One of the insertions induces the repression of a nearby gene (Rebollo et al. 2011). The chromatin marks of genes can also extend to adjacent regions, and alter the chromatin status of TEs that are located nearby (Rebollo et al. 2012). Large-scale studies on this issue are rare and difficult, given that they require comparative analyses between lines with and without TE insertion. In Drosophila, such studies show that the presence of insertions alters the chromatin structure (Lee and Karpen 2017), as well as that gene expression levels are negatively correlated with the number of insertions in genes (Fablet et al. 2022).

Together, these mutations and epimutations generate transcript and protein variability. As we have seen from a few examples, these new molecular variants can have an effect on the individual's terminal phenotype. This phenotype is then subject to evolutionary forces (notably drift and selection). In the next section, we will see what effects TEs can have in terms of population adaptation and species evolution.

3.4. The effect of TE in adaptation and evolution

As with any other mutation (SNP, small indels), TE insertions can be considered as neutral, in other words, without any consequences on the phenotype of individuals, or harmful or adaptive when they reduce or increase their fitness.

The vast majority of TE insertions take place in non-coding regions of the genome, or within other TE sequences. These insertions are said to be neutral, as they have no effect on the phenotype of the individuals carrying them. These mutations exist in low frequency in a population, and are therefore very likely to be lost over successive generations. Mutations that cause significant harmful effects, such as gene disruption, are eliminated even more rapidly through natural selection. Some insertions can also reach high frequencies in populations as a result of genetic drift, particularly in populations with a small effective size. Over time, these insertions accumulate mutations affecting their mobility, rendering them incapable of transposing autonomously. In rarer cases, the TE insertion has positive effects on the fitness of individuals, and is positively selected. In such cases, we observe an increase in the frequency of this insertion in the population. Although these cases are rare, they are very interesting, as they are the driving force behind typically significant phenotypic changes, and can enable individuals to adapt locally. In the longer term, they can play a role in the evolution of species.

A classic example is the presence of TEs close to genes providing resistance to insecticides. These adaptations are found recurrently in several populations of several species. This is notably the case of *Accord* TEs in *Drosophila melanogaster* and *Doc* TE insertions in *Drosophila simulans* (Rostant et al. 2012).

Another remarkable example is the peppered moth. This is a classic historical example of directional selection. In England, during the Industrial Revolution, the frequency of black morphs of these moths increased, to the detriment of white morphs, which faced greater predation. Several decades after these observations, we have now understood the underlying molecular mechanisms. It involves a TE insertion mutation in the promoter of the *cortex* gene, which is responsible for the moths' dark phenotype. This mutation has been dated to around the same period as the start of the industrial revolution, and has seen its frequency increase in populations in the regions involved (Hof et al. 2016).

Lastly, during domestication and crop improvement, humans have also selected TE insertions located close to genes whose expression is involved in traits of interest, such as, for plants, fruit color and size or plant architecture and, for animals, morphology or coat color. For example, as we saw earlier, the color of the blood orange is linked to the insertion of a TE upstream of the *Ruby* gene. In this case, and spectacularly, two different insertions were selected twice independently by humans, once in Sicily and once in China. In grapevines, grape berry color is caused by the insertion of a *Gret1* LTR retrotransposon upstream of the *VvbyIA* gene (Kobayashi et al. 2004). The absence of the insertion leads to a black berry, the presence of the insertion leads to a white berry and a recombinant modified insertion leads to a pink berry. Long-distance gene insertions were also selected. For example, the insertion of a retrotransposon 60 kb upstream of the *Tb1* gene is a key event in the domestication of maize, enabling the transition from a bushy architecture to a single-stemmed plant (Studer et al. 2017).

In these examples, the genetic effect of the TE plays a major role, and the high-frequency or increasing-frequency phenotype is linked to the allele carrying the TE insertion. The relative role of the genetic and epigenetic effects of TE insertions on phenotypes in natural or domesticated populations is still relatively poorly established. In most cases, epigenetic marks have not been analyzed, so the epigenetic effect could not be tested. In some cases, however, it is the allele with the methylated TE that is in high frequency in the population. This is the case of the *FWA* (Flowering WAgenigen) locus in *A. thaliana*, which is involved in flowering. Its transcription begins in a partially duplicated SINE. In wild plants, this promoter region is methylated, leading to transcriptional silencing in vegetative tissues. The presence of the SINE is necessary and sufficient for epigenetic gene silencing (Fujimoto et al. 2008). Thus, what has been selected is not the allele without the SINE, but the allele with a methylated SINE. Moreover, this methylation state is widely conserved across *Arabidopsis thaliana* accessions (Vaughn et al. 2007). In contrast, the alteration of small RNA production or DNA methylation at the SINE duplication allows ectopic expression of the *FWA* gene, resulting in a late-flowering phenotype (Soppe et al. 2000). Selection of the allele carrying the methylated TE could therefore enable more plastic regulation of the gene.

For *FLC*, another flowering regulator gene, several alleles carrying TEs have been characterized. It therefore seems that this gene has been the repeated target of TE insertion, several of which have been selected for. This is also true at the Brassicaceae level, the family containing *Arabidopsis thaliana*, suggesting that TE-induced mutations could be a significant source enabling plants to respond to the environment.

Interactions between TE insertions, host genome and chromatin state can also be modulated by the environment (Fablet and Vieira 2011; Lanciano and Mirouze 2018). Indeed, as epigenetic marks are sensitive to biotic and abiotic changes (Blake and Watson 2016; Nätt and Thorsell 2016; Alonso et al. 2019), it is easy to assume that changes in phenotypes and the epigenome are linked to changes in the environment, and that this can have a long-term impact on the evolution of organisms and their adaptation (Marin et al. 2019). Environmental induction of these mechanisms can be placed within a broader conceptual framework of phenotypic plasticity, which may facilitate evolutionary change (see Chapters 7 and 11).

3.5. Conclusion

As a result of their repeated structure, the proteins they encode, and the regulatory elements they contain (TFBS, transcription start and stop signals), transposable elements move around and multiply in genome sequences that carry biological information. This makes them very special mutagens, which, if inserted close to genes, can have an effect on transcription. Beyond this genetic regulation of transcripts, the presence of an epigenetic control of these elements, which can be altered spontaneously, in connection with environmental variations, or following mutations in epigenetic control pathways, gives them plasticity in terms of regulation. These particular characteristics make them major players in genome dynamics and evolution. The relative proportions of genetic and epigenetic effects of their polymorphisms on phenotypes remain to be clarified, as does their involvement in the networks underlying complex polygenic traits. These questions are at the heart of current TE research.

The recent advent of high-throughput sequencing techniques enabling the generation of long reads (Nanopore, Pacific Biosciences) opens up some very interesting prospects for answering these questions. First, the availability of long reads allows us to not only characterize the presence of an insertion at a given location in a genome, but also and above all to fully decrypt the sequence of the inserted element, thus giving access to the entire content of the regulatory repertoire it contains. Second, technologies such as Nanopore provide access to DNA methylation profile over the entire read length (and therefore of the element) at the base level, thus offering the possibility of jointly analyzing the genetic and epigenetic effects of a TE insertion.

The use of such technologies on natural populations or panels of domesticated individuals, crossed with climatic and/or phenotypic data, should enable major advances in understanding the contribution of TEs to local adaptation and evolution.

3.6. References

Alonso, C., Ramos-Cruz, D., Becker, C. (2019). The role of plant epigenetics in biotic interactions. *New Phytolo*, 221, 731–737. doi: 10.1111/nph.15408.

Andreev, V.I., Yu, C., Wang, J., Schnabl, J., Tirian, L., Gehre, M., Handler, D., Duchek, P., Novatchkova, M., Baumgartner, L. et al. (2022). Panoramix SUMOylation on chromatin connects the piRNA pathway to the cellular heterochromatin machinery. *Nat Struct Mol Biol*, 29, 130–142. doi: 10.1038/s41594-022-00721-x.

Baduel, P., Leduque, B., Ignace, A., Gy, I., Gil, J., Loudet, O., Colot, V., Quadrana, L. (2021). Genetic and environmental modulation of transposition shapes the evolutionary potential of *Arabidopsis thaliana*. *Genome Biology*, 22, 138. doi: 10.1186/s13059-021-02348-5.

Blake, G.E. and Watson, E.D. (2016). Unravelling the complex mechanisms of transgenerational epigenetic inheritance. *Curr Opin Chem Biol*, 33, 101–107. doi: 10.1016/j.cbpa.2016.06.008.

Butelli, E., Licciardello, C., Zhang, Y., Liu, J., Mackay, S., Bailey, P., Reforgiato-Recupero, G., Martin, C. (2012). Retrotransposons control fruit-specific, cold-dependent accumulation of anthocyanins in blood oranges. *Plant Cell*, 24, 1242–1255. doi: 10.1105/tpc.111.095232.

Cosby, R.L., Judd, J., Zhang, R., Zhong, A., Garry, N., Pritham, E.J., Feschotte, C. (2021). Recurrent evolution of vertebrate transcription factors by transposase capture. *Science*, 371, eabc6405. doi: 10.1126/science.abc6405.

Cropley, J.E., Suter, C.M., Beckman, K.B., Martin, D.I.K. (2006). Germ-line epigenetic modification of the murine Avy allele by nutritional supplementation. *Proc Nat/Acad Sci*, 103, 17308–17312. doi: 10.1073/pnas.0607090103.

Czech, A., Wende, S., Mörl, M., Pan, T., Ignatova, Z. (2013). Reversible and rapid transfer-RNA deactivation as a mechanism of translational repression in stress. *PLOS Genet*, 9, e1003767. doi: 10.1371/journal.pgen.1003767.

Dupressoir, A. and Heidmann, T. (2011). Les syncytines – Des protéines d'enveloppe rétrovirales capturées au profit du développement placentaire. *Med Sci (Paris)*, 27, 163–169. doi: 10.1051/medsci/2011272163.

Fablet, M. and Vieira, C. (2011). Evolvability, epigenetics and transposable elements. *Biomol Concepts*, 2, 333–341. doi: 10.1515/BMC.2011.035.

Fablet, M., Salcez-Ortiz, J., Jacquet, A., Menezes, B.F., Dechaud, C., Veber, P., Noûs, C., Rebollo, R., Vieira, C. (2022). A quantitative, genome-wide analysis in *Drosophila* reveals transposable elements' influence on gene expression is species-specific. *Genome Biol Evol*, 15(9), 477049. doi: 10.1101/2022.01.20.477049.

Fujimoto, R., Kinoshita, Y., Kawabe, A., Kinoshita, T., Takashima, K., Nordborg, M., Nasrallah, M.E., Shimizu, K.K., Kudoh, H., Kakutani, T. (2008). Evolution and control of imprinted FWA genes in the genus *Arabidopsis*. *PLOS Genet*, 4, e1000048. doi: 10.1371/journal.pgen.1000048.

Hof, A.E. van't, Campagne, P., Rigden, D.J., Yung, C.J., Lingley, J., Quail, M.A., Hall, N., Darby, A.C., Saccheri, I.J. (2016). The industrial melanism mutation in British peppered moths is a transposable element. *Nature*, 534, 102–105. doi: 10.1038/nature17951.

Khurana, J.S. and Theurkauf, W. (2010). piRNAs, transposon silencing, and Drosophila germline development. *J Cell Biol*, 191, 905–913. doi: 10.1083/jcb.201006034.

Kinoshita, Y., Saze, H., Kinoshita, T., Miura, A., Soppe, W.J.J., Koornneef, M., Kakutani, T. (2007). Control of FWA gene silencing in *Arabidopsis thaliana* by SINE-related direct repeats. *Plant Journal*, 49, 38–45. doi: 10.1111/j. 1365-313X.2006.02936.x.

Kobayashi, S., Goto-Yamamoto, N., Hirochika, H. (2004). Retrotransposon-induced mutations in grape skin color. *Science*, 304, 982. doi: 10.1126/science.1095011.

Lanciano, S. and Mirouze, M. (2018). Transposable elements: All mobile, all different, some stress responsive, some adaptive? *Curr Opin Genet Dev*, 49, 106–114. doi: 10.1016/j.gde.2018.04.002.

Law, J.A. and Jacobsen, S.E. (2010). Establishing, maintaining and modifying DNA methylation patterns in plants and animals. *Nat Rev Genet*, 11, 204–220. doi: 10.1038/nrg2719.

Le, N.T., Harukawa, Y., Miura, S., Boer, D., Kawabe, A., Saze, H. (2020). Epigenetic regulation of spurious transcription initiation in *Arabidopsis*. *Nat Commun*, 11, 3224. doi: 10.1038/s41467-020-16951-w.

Lee, Y.C.G. and Karpen, G.H. (2017). Pervasive epigenetic effects of Drosophila euchromatic transposable elements impact their evolution. *eLife*, 6, e25762. doi: 10.7554/eLife.25762.

Mallet, F. and Prudhomme, S. (2004). Notre patrimoine rétroviral. *J Soc Biol*, 198, 399–412. doi: 10.1051/jbio/2004198040399.

Manning, K., Tör, M., Poole, M., Hong, Y., Thompson, A.J., King, G.J., Giovannoni, J.J., Seymour, G.B. (2006). A naturally occurring epigenetic mutation in a gene encoding an SBP-box transcription factor inhibits tomato fruit ripening. *Nat Genet*, 38, 948–952. doi: 10.1038/ng1841.

Marin, P., Genitoni, J., Barloy, D., Maury, S., Gibert, P., Ghalambor, C.K., Vieira, C. (2019). Biological invasion: The influence of the hidden side of the (epi)genome. *Funct Ecol*, 34(2), 385–400. doi: 10.1111/1365-2435.13317.

Mirouze, M. and Vitte, C. (2014). Transposable elements, a treasure trove to decipher epigenetic variation: Insights from Arabidopsis and crop epigenomes. *J Exp Botany*, 65, 2801–2812. doi: 10.1093/jxb/eru120.

Mugat, B., Nicot, S., Varela-Chavez, C., Jourdan, C., Sato, K., Basyuk, E., Juge, F., Siomi, M.C., Pélisson, A., Chambeyron, S. (2020). The Mi-2 nucleosome remodeler and the Rpd3 histone deacetylase are involved in piRNA-guided heterochromatin formation. *Nat Commun*, 11, 2818. doi: 10.1038/s41467-020-16635-5.

Naito, K., Zhang, F., Tsukiyama, T., Saito, H., Hancock, C.N., Richardson, A.O., Okumoto, Y., Tanisaka, T., Wessler, S.R. (2009). Unexpected consequences of a sudden and massive transposon amplification on rice gene expression. *Nature*, 461, 1130–1134. doi: 10.1038/nature08479.

Nätt, D. and Thorsell, A. (2016). Stress-induced transposon reactivation: A mediator or an estimator of allostatic load? *Environ Epigenet*, 2(3), dvw015. doi: 10.1093/eep/dvw015.

Quadrana, L., Etcheverry, M., Gilly, A., Caillieux, E., Madoui, M.-A., Guy, J., Bortolini Silveira, A., Engelen, S., Baillet, V., Wincker, P. et al. (2019). Transposition favors the generation of large effect mutations that may facilitate rapid adaption. *Nat Commun*, 10, 3421. doi: 10.1038/s41467-019-11385-5.

Rebollo, R., Karimi, M.M., Bilenky, M., Gagnier, L., Miceli-Royer, K., Zhang, Y., Goyal, P., Keane, T.M., Jones, S., Hirst, M. et al. (2011). Retrotransposon-induced heterochromatin spreading in the mouse revealed by insertional polymorphisms. *PLOS Genetics*, 7, e1002301. doi: 10.1371/journal.pgen.1002301.

Rebollo, R., Miceli-Royer, K., Zhang, Y., Farivar, S., Gagnier, L., Mager, D.L. (2012). Epigenetic interplay between mouse endogenous retro-viruses and host genes. *Genome Biol*, 13, R89. doi: 10.1186/gb-2012-13-10-r89.

Rostant, W.G., Wedell, N., Hosken, D.J. (2012). Transposable elements and insecticide resistance. In *Advances in Genetics*, Goodwin, S.F., Friedmann, T., Dunlap, J.C. (eds). Academic Press, Cambridge, 169–201. doi: 10.1016/B978-0-12-394394-1.00002-X.

Shpiz, S., Ryazansky, S., Olovnikov, I., Abramov, Y., Kalmykova, A. (2014). Euchromatic transposon insertions trigger production of novel Pi- and endo-siRNAs at the target sites in the drosophila germline. *PLoS Genet.*, 10, e1004138. doi: 10.1371/journal.pgen.1004138.

Sienski, G., Batki, J., Senti, K.-A., Dönertas, D., Tirian, L., Meixner, K., Brennecke, J. (2015). Silencio/CG9754 connects the Piwi-piRNA complex to the cellular heterochromatin machinery. *Genes Dev.*, 29, 2258–2271. doi: 10.1101/gad.271908.115.

Sigman, M.J. and Slotkin, R.K. (2016). The first rule of plant transposable element silencing: Location, location, location. *Plant Cell*, 28, 304–313. doi: 10.1105/tpc.15.00869.

Soppe, W.J., Jacobsen, S.E., Alonso-Blanco, C., Jackson, J.P., Kakutani, T., Koornneef, M., Peeters, A.J. (2000). The late flowering phenotype of fwa mutants is caused by gain-of-function epigenetic alleles of a homeodomain gene. *Mol Cell*, 6, 791–802. doi: 10.1016/s1097-2765(05)00090-0.

Studer, A.J., Wang, H., Doebley, J.F. (2017). Selection during maize domestication targeted a gene network controlling plant and inflorescence architecture. *Genetics*, 207, 755–765. doi: 10.1534/genetics.117.300071.

Vaughn, M.W., Tanurdžić, M., Lippman, Z., Jiang, H., Carrasquillo, R., Rabinowicz, P.D., Dedhia, N., McCombie, W.R., Agier, N., Bulski, A. et al. (2007). Epigenetic natural variation in *Arabidopsis thaliana*. *PLOS Biol*, 5, e174. doi: 10.1371/journal.pbio.0050174.

Vitte, C., Fustier, M.-A., Alix, K., Tenaillon, M.I. (2014). The bright side of transposons in crop evolution. *Brief Funct. Genomics*, 13, 276–295. doi: 10.1093/bfgp/elu002.

Wells, J.N. and Feschotte, C. (2020). A field guide to eukaryotic transposable elements. *Annu Rev Genet*, 54, 539–561. doi: 10.1146/annurev-genet-040620-022145.

Wicker, T., Sabot, F., Hua-Van, A., Bennetzen, J.L., Capy, P., Chalhoub, B., Flavell, A., Leroy, P., Morgante, M., Panaud, O. et al. (2007). A unified classification system for eukaryotic transposable elements. *Nat Rev Genet*, 8, 973–982. doi: 10.1038/nrg2165.

Xiao, H., Jiang, N., Schaffner, E., Stockinger, E.J., van der Knaap, E. (2008). A retrotransposon-mediated gene duplication underlies morphological variation of tomato fruit. *Science*, 319, 1527–1530. doi: 10.1126/science.1153040.

Xu, Q., Chen, L.-L., Ruan, X., Chen, D., Zhu, A., Chen, C., Bertrand, D., Jiao, W.-B., Hao, B.-H., Lyon, M.P. et al. (2013). The draft genome of sweet orange (*Citrus sinensis*). *Nat Genet*, 45, 59–66. doi: 10.1038/ng.2472.

4

Epigenetics: The Same for all Species?

Clémentine VITTE[1] and Nicolas NÈGRE[2]
[1] *GQE – Le Moulon, Gif-sur-Yvette, France*
[2] *DGIMI, Université de Montpellier, France*

In this chapter, we present the origin and evolution of epigenetic mechanisms in the living world, and illustrate the value of cross-species comparative epigenetic approaches for identifying new molecular mechanisms of epigenetic regulation.

4.1. Universal epigenetic mechanisms?

All living organisms on our planet share a common medium for transmitting information: DNA. This robust system was probably only set up once, at the very beginning of the history of life, and links all organisms, past and present, in a long chain of information transmission and evolution. The same cannot be said of epigenetic regulation. First, the mechanisms of epigenetic regulation emerged at different periods in the history of life. Some groups of organisms therefore do not possess the same epigenetic mechanisms as others. We will explore the origins of these systems in section 4.2.

Furthermore, once these mechanisms were implemented, they too began to evolve. Through the acquisition or loss of certain genes, these mechanisms may have varied from one group to another. We will see how DNA methylation, for

Epigenetics in Ecology and Evolution,
coordinated by Christoph GRUNAU, Stéphane MAURY.

example, has different characteristics in plants, fungi and animals. Lastly, while certain groups share the same enzymatic repertoire for the establishment of epigenetic marks, there is nevertheless great diversity in the distribution of these marks along the genome, as well as in their functions. We will see that DNA methylation either occurs on gene bodies or transposons, or on both, or even on neither. We will see that, depending on the organism, this apposition can have a repressive or activating effect, or no effect at all. According to the principles of evolution, it is indeed possible for a mechanism to become vestigial.

If this diversity can be observed for a well-studied mechanism such as DNA methylation, it probably exists for other carriers of epigenetic information such as small RNAs, histone modifications or chromosome organization. Then, which organism should we choose as a model for studying epigenetic regulation? By observing the epigenetic characteristics of different organisms and their diversity, it is easy to see how the pioneering use of a few models may have biased the first biological conclusions, and how the choice of a model has an impact on the lines of research that can be developed from it. Drosophila, for example, has a different methylation profile from other insects, and therefore cannot be representative of them. The same applies to the nematode, the mouse or the Arabidopsis plant models. So, are there any model organisms in epigenetics? At the end of this chapter, we will explore the directions and approaches that could enable us to understand epigenetics in the future, not just as a universal mechanism, but as a set of evolving machineries.

4.2. Origin of the various chromatin components

DNA is the iconic image of genetics, and chromatin that of epigenetics. Chromatin can be defined as the association of the DNA molecule with other nucleic acids and proteins, whose action modulates gene expression. However, not all the components of this system have appeared at the same time throughout the history of life. This history began shortly after the formation of the Earth, 4.5 billion years ago.

4.2.1. *Origin of DNA*

Since its recognition in 1953 as the carrier of genetic information – the molecule of life theorized by Schrödinger – DNA – has been found in all forms of cellular life. This immediately suggests that DNA was already present at the time of the

LUCA[1]. Yet, most studies date the emergence of the LUCA to 3.8 billion years ago, perhaps even 4.5 billion years ago according to recent studies incorporating molecular clock models (Betts et al. 2018). These distant emergence times, close to the age of the planet itself, beg the following question: How could the DNA-based information transmission system have formed so soon after Earth's formation? Classical experiments carried out in the 1950s by the Americans Miller and Urey, under reducing conditions close to those of the original Earth, showed that spontaneous nucleotide synthesis was possible (Miller and Urey 1959). This work was corroborated more recently by Teichert et al. (2019) from the University of Munich. However, other clues tend to suggest that it is RNA that is at the origin of life and not DNA. Indeed, in our cells, dNTPs are synthesized enzymatically from ribonucleotides. Several enzymes are required to replicate the DNA molecule, whereas RNA is theoretically capable of self-replication. In France, Patrick Forterre's work on the origin of life has produced several syntheses on the subject (Forterre et al. 2000; Forterre 2002). Was the LUCA then DNA-based? The hypothesis of an RNA-based original biome seems more likely, with the possible presence of DNA replicated by reverse transcription. The hypothesis of the LUCA possessing a DNA genome is possible but requires the replacement of the ancestral proteins necessary for its functioning (helicases, polymerases, primases) by proteins that differ according to the evolutionary lineage, notably through the contribution of DNA by viruses (Forterre et al. 2005).

4.2.2. *Origin of small RNAs*

Just as DNA is the common medium for transmitting information, the molecular mechanisms for reading DNA – transcription and translation – are conserved in both eukaryotes and prokaryotes. Transcription is carried out by DNA-dependent RNA polymerases, in other words, using one strand of DNA to synthesize a single-stranded RNA molecule of a complementary sequence. These RNAs are the true carriers of genetic information, as they are taken over by ribosomes and decoded by transfer RNAs to produce an amino acid polymer corresponding to the RNA sequence. However, even in the process of gene transcription in prokaryotes, several portions of the RNA molecule are in fact non-coding, in the sense that they are not read by ribosomes and do not carry an open reading frame. Several RNA sequence determinants are in fact important motifs for the biology of the RNA itself: ribosome attachment, transcription termination, etc. Furthermore, the majority of RNAs produced by both eukaryotic and prokaryotic cells are non-coding: these are

1. Last Universal Common Ancestor.

ribosomal RNAs and transfer RNAs, the production of which can account for up to 85% of the cellular metabolism. As you can see, non-coding RNAs are an integral part of the molecular biology system, some components of which may even predate the establishment of DNA as an information carrier (see section 4.2.1). In these conditions, it is not surprising to find that eukaryotes, as well as bacteria (Waters and Storz 2009), possess numerous mechanisms for producing small regulatory non-coding RNAs. These small RNAs are either produced by the transcription of independent genes or by cleavage of the 3' end of messenger RNAs. These small RNAs have regulatory effects, either by binding to target proteins, which modifies their function, or by hybridization to target mRNAs, which can cause their degradation by the recognition of RNA:RNA duplexes. They can also mask or unmask regulatory sites on the mRNA, such as the ribosome recognition site. The observation of such diversification in bacteria indicates that small RNA regulatory mechanisms were implemented very early in evolution, probably at the same time as the development of transcription and translation mechanisms, which they are both the product and the target of.

4.2.3. *Origin of DNA methylation*

Since DNA methylation is based on the presence of nitrogenous bases modified by the presence of a methyl group, it seems logical that methylated bases could have been used by living organisms at a time when a system for synthesizing and assembling nitrogenous bases was being set up. Indeed, DNA methylation systems are found in eukaryotes (Aliaga et al. 2019), archaea (Blum and Payne 2019) and bacteria (Blow et al. 2016), indicating that this system emerged in their common ancestor. In bacteria, there are several types of DNA modification: 6-methyladenosine (6mA), 4-methylcytosine (4mC) and the predominant one found also in eukaryotes, 5-methylcytosine (5mC). As in eukaryotes, these modifications are carried out by DNA methyltransferases (MTases). However, in bacteria, they form part of a defense system against restriction enzymes (REs). Indeed, by producing REs capable of degrading DNA from viral or other exogenous sources, bacteria are exposed to their own endonuclease activity, which could damage their own DNA. They protect themselves by producing an MTase with the same sequence specificity as the RE. These methylated sites are thus inaccessible to the bacterial RE. Beyond this system, "orphan" MTases – with no associated RE – have been identified in bacterial genomes, some of which have functions in genome regulation. *Dam* methylase, for example, Marinus and Casadesus (2009), which methylates adenosines in *E. coli*, does not just play a role in the initiation of replication, as well as in the promoter accessibility of certain genes, and thus in their transcriptional regulation. DNA modifications in bacteria could therefore represent the establishment of one of the first epigenetic systems (Sánchez-Romero and Casadesús 2020).

4.2.4. *Origin of chromatin*

The arrangement of chromosomal DNA in eukaryotes is classically associated with a structure called chromatin, the basic unit of which is the wrapping of about 147 base pairs of DNA around an octamer of histones called the nucleosome. These structures are not found in bacteria, but the bacterial chromosome is nevertheless organized into a similar structure called the nucleoid. Indeed, nucleoid-associated proteins (NAPs) are able to organize the bacterial chromosome into more or less condensed three-dimensional structures (Dame and Tark-Dame 2016), which resemble the three-dimensional organization of eukaryotic genomes. Archaea also possess a circular chromosome organized by proteins of the NAP class, but in a phylum of archaea called *Euryarcheota*, atypical histone proteins can also be found (Williams and Embley 2014; Peeters et al. 2015). It has been postulated that this proto-chromatin structure is necessary for these extremophilic organisms to protect DNA from denaturation by high temperatures. Eukaryotic histones thus appear to derive from a common ancestor with archaea (Hajheidari et al. 2019). However, diversification of histone-modifying enzymes actually took place in early eukaryotes, as did chromatin remodeling systems.

4.3. Evolution of epigenetic systems

We have just seen that most epigenetic mechanisms are found in bacteria today, indicating a common origin between eukaryotes and prokaryotes. The age of the first prokaryotes is estimated at –3.5 billion years. We can assume that DNA, the mechanisms of molecular biology, the production of non-coding RNAs, DNA methylation and the organization of DNA into proto-chromatin structures, was already present by then. The first eukaryotes, resulting from a cellular symbiosis between archaea and bacteria, are thought to have appeared between 1.8 and 2.3 billion years ago (Hedges et al. 2004). In these early cells, chromosomes may already have been organized into chromatin by proto-histones. It is also likely that most of the molecular interactions between methylation, small RNAs, chromatin and transcription already took place in these early eukaryotic cells. In order to clarify the timeline of the emergence of epigenetic mechanisms, we need to turn to evolutionary approaches, for example, by establishing the phylogeny of the various chromatin players. Indeed, we are currently able to sequence, assemble and annotate a very large number of genomes. At the time of writing, some 390,000 prokaryotic genomes and over 21,000 eukaryotic genomes are available on the NCBI website[2].

---

2. See: www.ncbi.nlm.nih.gov/genome/browse#!/overview/.

This large number makes it possible to compare gene content, not just on a coarse scale (bacteria vs. archaea vs. eukaryotes), but on a much finer scale, such as within a phylogenetic group such as insects, mammals or angiosperms. These two scales enable different observations to be made. Between very distant clades, we can identify conserved mechanisms. Within a clade, we can identify their diversification and evolution. This approach that combines epigenetics and chromatin was recently conceptualized during the EvoChromo workshop (Drinnenberg et al. 2019). Indeed, due to the ancient origin of epigenetic mechanisms, these systems have had ample time to evolve. When we compare these mechanisms between different species, we can therefore find different biogenesis mechanisms, players and functions. To illustrate this concept, let us look at an example of epigenetic evolution, that of DNA methylation in eukaryotes.

4.4. Example of the evolution of DNA methylation in different groups of organisms

Since DNA methylation originated in bacteria and archaea, eukaryotes (plants, animals, fungi) all have a DNA methylation system. Unlike in bacteria, it is mainly cytosines that bear the methylation marks in eukaryotes, in the form of 5-methylcytosines (5mC). As DNA methylation is a conserved process within eukaryotes, it offers a good model for exploring the evolution of this system's function within this group. To address this question, we can first look at the presence and/or absence of methylation players in different groups of organisms: DNA methyltransferases (DNMTs) (Laisné et al. 2018). In mammals, including humans, three DNMTs have been described: DNMT1, DNMT3A and DNMT3B (see Chapter 2). As a reminder, DNMT1 is considered to be a maintenance enzyme, capable of methylating the neosynthesized strand by recognizing the presence of methylation on the complementary strand. Thus, after mitosis, each daughter cell possesses the methylation marks of the mother cell. DNMT1 is therefore crucial for propagating methylation to offspring and making it an epigenetic mark. DNMT3A and B are enzymes capable of methylating DNA de novo, i.e. in the absence of existing methylation on the DNA strand. This activity is not spontaneous in the cell and is triggered by a recruitment signal: transcription factors, chromatin factors or small RNAs. Both enzymes are found in plants and animals, but not in fungi (Zemach and Zilberman 2010). As fungi and animals are more closely related to each other than they are to plants, this suggests that these proteins were present in the common ancestor of all three groups, and were lost in fungi during evolution. This type of loss is also observed over shorter evolutionary scales. For example,

DNMT1 is absent from the model yeasts *Saccharomyces cerevisiae* and *Schizosaccharomyces pombe*, and neither DNMT1 nor DNMT3 is found in the model organisms *Drosophila melanogaster* and *Caenorhabditis elegans*. These variations correspond to independent evolutionary events, and suggest that these enzymes have been lost several times during evolution.

In plants, four methyltransferases have been identified: MET1, CMT2, CMT3 and DRM2. MET1 is the ortholog of DNMT1 and, such as DNMT1, is responsible for maintaining the methylation state in the CpG context (He et al. 2011). CMT2 and CMT3 are plant-specific methyltransferases, and enable the maintenance of cytosine methylation in CpHpH and CpHpG contexts (where H is A, T or C), also known as "non-CpG" contexts, which are not methylation targets in animals. The DRM1 and DRM2 proteins enable de novo methylation, and target cytosines in CpG, CpHpG and CpHpH contexts. Interestingly, these proteins are degraded homologues of animal DNMT3s, and are not the "true" orthologs of these proteins. The study of mutants of these genes in the moss *Physchomytrella patens* revealed an uncoupling of de novo methylation as a function of chromatin compaction – DNMTs affecting heterochromatic regions, DRMs affecting euchromatic regions – and suggests that the emergence of the DRM function in plants would be linked to euchromatin targeting (Yaari et al. 2019). However, what impact do these variations have on DNA methylation function?

For example, we can look at the distribution of methylation marks in genomes. In mammals, it is well known that hypomethylated areas correspond to CpG islands – CpG-rich regions – located close to gene promoters. Methylation of these regions has an inhibitory regulatory impact on gene transcription. This behavior is different in insects, for example, Bewick et al. (2017), where very few cytosines (between 1 and 2%) are methylated, mainly in a CpG context. However, CpG islands are not as easily identifiable in insects as they are in mammals. Far from being randomly arranged, methylated cytosines are, in fact, localized in gene exons, particularly within those that are transcriptionally active. This example illustrates that DNA methylation can also have a conserved role: the regulation of gene expression, and a different type of regulation, adapted to the structure of the genomes of these species. In plants, the highly demethylated CpG and CpHpG regions correspond to *cis*-regulatory regions of gene expression. However, activation of these regions is associated with increased chromatin accessibility, and the effect of increasing or decreasing target gene expression is mediated by histone marks. This regulation, which is also linked to the presence of transcription factors able to bind to sites located in low methylation regions, is dependent on tissue, cell-type and environmental condition.

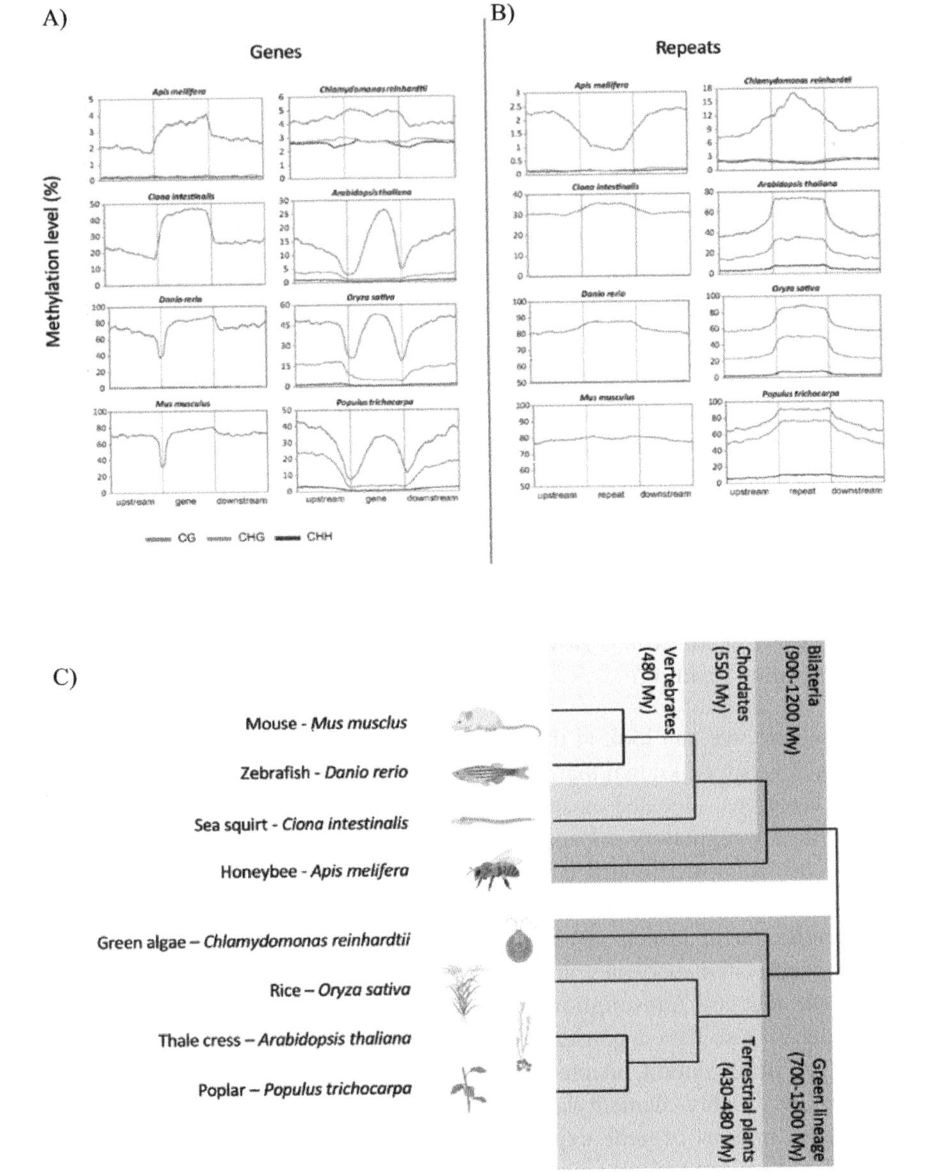

Figure 4.1. *Distribution of methylation in genes (A) and repeat regions (B), for different species of animal (A) and the green lineage (B). Adapted from Feng et al. (2010). Created with BioRender.com. C) Phylogeny of the species shown in (A), (B). The major phylogenetic groups are framed in different shades of gray, and the approximate age of the group is indicated*

As an epigenetic mark, another role of DNA methylation involves enabling molecular distinction between genes and transposable elements (Figure 4.1), which are regulated in very different ways in a genome: the expression of TEs must be avoided, while that of genes must be regulated during development and in response to the environment. This function is well conserved between species, except in vertebrates, where the modalities of molecular distinction widely vary. In vertebrates (zebrafish, mouse), the distinction takes place at the gene promoter regions (Figure 4.1), which are demethylated (Feng et al. 2010). In bees (Figure 4.1), which have a very low basal methylation level (2%), genes are hypermethylated (4%) and repeated elements are hypomethylated (1%). Within the green lineage (Figure 4.1), there is a variation in relative methylation levels between genes and repeat elements, but the cytosines involved are not the same: in *Chlamydomonas reinhardtii*, the CpHpG and CpHpH contexts are equally methylated (approximately 2.5%) in genes and repeated elements, and it is the level of methylation in the CpG context, which shows a peak in repeat regions, that distinguishes the two types of regions. In contrast, in angiosperms, genes are only moderately methylated in the CpG context, whereas repeated elements are highly methylated in all three contexts. Taken together, these examples illustrate the wide variation of mechanisms in place in different evolutionary lineages, as well as their co-adaptation with genome structure and gene expression regulation.

4.5. Which model organisms for epigenetics?

As we have just seen, the various epigenetic mechanisms appeared at different times in evolution, probably in prokaryotic life forms, similar to today's bacteria, existing for several hundred million years before the emergence of eukaryotic life forms. However, these epigenetic systems are not found in today's bacteria in a form similar to that found in eukaryotes. Their origins, which are both ancient relative to present-day life forms and late relative to the original life forms, explain why we find these mechanisms more diversified and less conserved between large groups of organisms, unlike the most fundamental mechanisms of molecular genetics, such as replication, transcription or translation.

Given that these mechanisms are less conserved and more diversified between groups of organisms, it is difficult to establish study models that are representative of all organisms. The most obvious example, in the case of DNA methylation, is the fundamental difference between insects and mammals. They do not have the same enzyme sets, the same methylation distribution or the same regulatory function. Between a Drosophila and a mouse, these mechanisms are present, but are not comparable. However, differences are also present within the same group. In insects, for instance, Drosophila is not a model of its own group: when we look at the group of MTases found in Diptera and Hymenoptera, it is different. DNMT3, for example, is present in Hymenoptera, among others, but not in Diptera (Bewick et al. 2017). In bees, DNA methylation plays a role in determining caste (worker or queen), depending on the larval diet. A diet containing royal jelly activates developmental genes that produce functional gonads. A functional methylome is thus found in social Hymenoptera (Yan et al. 2015). How many different mechanisms and functions are present in other insect orders? How about within each order?

In conclusion, there is no single model in epigenetics. There is a great diversity of models. Evolution has exploited and hijacked these regulatory mechanisms in a multitude of ways, and they have co-evolved with the gene composition of epigenetic regulation, and with genome structure. In these times of massive sequencing, when the ability to analyze genomes and epigenomes is no longer limited to a small number of species, further exploration of the living world will undoubtedly reveal even greater diversity and variation in the role of epigenetics in the physiology and ecology of organisms.

4.6. References

Aliaga, B., Bulla, I., Mouahid, G., Duval, D., Grunau, C. (2019). Universality of the DNA methylation codes in eucaryotes. *Scientific Reports*, 9(1), 173. doi: 10.1038/s41598-018-37407-8.

Betts, H.C., Puttick, M.N., Clark, J.W., Williams, T.A., Donoghue, P.C.J., Pisani, D. (2018). Integrated genomic and fossil evidence illuminates life's early evolution and eukaryote origin. *Nature Ecology & Evolution*, 2(10), 1556–62. doi: 10.1038/s41559-018-0644-x.

Bewick, A.J., Vogel, K.J., Moore, A.J., Schmitz, R.J. (2017). Evolution of DNA methylation across insects. *Molecular Biology and Evolution*, 34(3), 654–65. doi: 10.1093/molbev/msw264.

Blow, M.J., Clark, T.A., Daum, C.G., Deutschbauer, A.M., Fomenkov, A., Fries, R., Froula, J., Kang, D.D., Malmstrom, R.R., Morgan, R.D. et al. (2016). The epigenomic landscape of prokaryotes. *PLOS Genetics*, 12(2), e1005854. doi: 10.1371/journal.pgen.1005854.

Blum, P. and Payne, S. (2019). Evidence of an epigenetics system in archaea. *Epigenetics Insights*, 12, 2516865719865280. doi: 10.1177/2516865719865280.

Dame, R.T. and Tark-Dame, M. (2016). Bacterial chromatin: Converging views at different scales. *Cell Nucleus*, 40, 60–65. doi: 10.1016/j.ceb.2016.02.015.

Drinnenberg, I.A., Berger, F., Elsässer, S.J., Andersen, P.R., Ausió, J., Bickmore, W.A., Blackwell, A.R. et al. (2019). EvoChromo: Towards a synthesis of chromatin biology and evolution. *Development (Cambridge, England)*, 146(19). doi: 10.1242/dev.178962.

Feng, S., Cokus, S.J., Zhang, X., Chen, P.-Y., Bostick, M., Goll, M.G., Hetzel, J., Jain, J., Strauss, S.H., Halpern, M.E. et al. (2010). Conservation and divergence of methylation patterning in plants and animals. *Proceedings of the National Academy of Sciences*, 107(19), 8689–94. doi: 10.1073/pnas.1002720107.

Forterre, P. (2002). The origin of DNA genomes and DNA replication proteins. *Current Opinion in Microbiology*, 5(5), 525–532. doi: 10.1016/S1369-5274(02)00360-0.

Forterre, P., Filée, J., Myllykallio, H. (2000). Origin and evolution of DNA and DNA replication machineries. *Madame Curie Bioscience Database* [Online]. Available at: https://www.ncbi.nlm.nih.gov/books/NBK6360/.

Forterre, P., Gribaldo, S., Brochier, C. (2005). Luca: The last universal common ancestor. *Medecine Sciences: M/S*, 21(10), 860–865. doi: 10.1051/medsci/20052110860.

Hajheidari, M., Koncz, C., Bucher, M. (2019). Chromatin evolution-key innovations underpinning morphological complexity. *Frontiers in Plant Science*, 10 [Online]. Available at: https://www.frontiersin.org/article/10.3389/fpls.2019.00454.

He, X.-J., Chen, T., Zhu, J.-K. (2011). Regulation and function of DNA methylation in plants and animals. *Cell Research*, 21(3), 442–465. doi: 10.1038/cr.2011.23.

Hedges, S.B., Blair, J.E., Venturi, M.L., Shoe, J.L. (2004). A molecular timescale of eukaryote evolution and the rise of complex multicellular life. *BMC Evolutionary Biology*, 4, 2. doi: 10.1186/1471-2148-4-2.

Laisné, M., Gupta, N., Kirsh, O., Pradhan, S., Defossez, P.-A. (2018). Mechanisms of DNA methyltransferase recruitment in mammals. *Genes*, 9(12), 617. doi: 10.3390/genes9120617.

Marinus, M.G. and Casadesus, J. (2009). Roles of DNA adenine methylation in host-pathogen interactions: Mismatch repair, transcriptional regulation, and more. *FEMS Microbiology Reviews*, 33(3), 488–503. doi: 10.1111/j.1574-6976.2008.00159.x.

Miller, S.L. and Urey, H.C. (1959). Organic compound synthesis on the primitive earth. *Science*, 130(3370), 245–251. doi: 10.1126/science.130.3370.245.

Peeters, E., Driessen, R.P.C., Werner, F., Dame, R.T. (2015). The interplay between nucleoid organization and transcription in archaeal genomes. *Nature Reviews Microbiology*, 13(6), 333–341. doi: 10.1038/nrmicro3467.

Sánchez-Romero, M.A. and Casadesús, J. (2020). The bacterial epigenome. *Nature Reviews Microbiology*, 18(1), 7–20. doi: 10.1038/s41579-019-0286-2.

Teichert, J.S., Kruse, F.M., Trapp, O. (2019). Direct prebiotic pathway to DNA nucleosides. *Angewandte Chemie (International Ed. in English)*, 58(29), 9944–9947. doi: 10.1002/|anie.201903400.

Waters, L.S. and Storz, G. (2009). Regulatory RNAs in bacteria. *Cell*, 136(4), 615–628. doi: 10.1016/j.cell.2009.01.043.

Williams, T.A. and Embley, T.M. (2014). Archaeal 'dark matter' and the origin of eukaryotes. *Genome Biology and Evolution*, 6(3), 474–481. doi: 10.1093/gbe/evu031.

Yaari, R., Katz, A., Domb, K., Harris, K.D., Zemach, A., Ohad, N. (2019). RdDM-Independent de novo and heterochromatin DNA methylation by plant CMT and DNMT3 orthologs. *Nature Communications*, 10(1), 1613. doi: 10.1038/s41467-019-09496-0.

Yan, H., Bonasio, R., Simola, D.F., Liebig, J., Berger, S.L., Reinberg, D. (2015). DNA methylation in social insects: How epigenetics can control behavior and longevity. *Annual Review of Entomology*, 60, 435–452. doi: 10.1146/annurev-ento-010814-020803.

Zemach, A. and Zilberman, D. (2010). Evolution of eukaryotic DNA methylation and the pursuit of safer sex. *Current Biology*, 20(17), R780–85. doi: 10.1016/j.cub.2010.07.007.

5

Epigenome Modifications as a Therapeutic and Research Tool

Nelia LUVIANO[1], Francesco CALZAFERRI[2] and Marie LOPEZ[2]

[1] *IHPE, Université de Perpignan Via Domitia, France*

[2] *IBMM, CNRS, Montpellier, France*

5.1. Introduction

As outlined in Chapter 1, the epigenome consists of different epigenetic marks that are reversible and can be transmitted through mitosis and sometimes by meiosis being inherited to subsequent generations. The epigenome can present alterations that are called epimutations. Three types of epimutations can be defined according to their origin (Roquis et al. 2016):

1) The environmentally induced changes that can appear right after an environmental stimulus, such as the presence of chemical compounds, temperature changes, stressing stimuli, etc.

2) The genotypically conditioned changes that consist of the epimutations, determined by the genotypic characteristics of a locus making an allele to fall under epigenetic control.

3) The random or spontaneous changes that consist of the epimutations resulting from errors in the establishment and maintenance of the epigenotype.

Epimutations appear more frequently and more rapidly in nature than the genetic mutations. The resulting environmental epimutations are associated with changes in

Epigenetics in Ecology and Evolution,
coordinated by Christoph GRUNAU and Stéphane MAURY.

the gene transcription and determine phenotypic variations similar to genetic mutations (Horsthemke 2006).

It was discovered that an excellent approach to study the function and role of the epigenetic marks was to intentionally modify these epigenetic marks and then study the effects of these modifications in the gene transcription and the phenotype. Different approaches can be used to experimentally modify epigenetic marks. Among them, two are widely used: 1) exposure to chemical compounds that interfere with the function of epigenetic enzymes such as epigenetic inhibitors; and 2) the use of epigenome engineering tools based on adapted genomic edition techniques.

5.2. Epigenetic modification strategies

5.2.1. *Epigenetic ligands*

Generally speaking, protein ligands are chemical entities that modify the activity of a target protein. Two classes of protein ligands can be defined depending on the nature of their targets. On the one hand, enzyme inhibitors modify the catalytic activity of a target enzyme. On the other hand, non-catalytically active protein ligands interact with a target protein to modify its recognition and binding properties. Therefore, by modification of protein catalytic activity or recognition properties, the use of such agents in appropriate models, i.e. in cells, in vivo, ex vivo, etc., leads to epigenetic modifications.

5.2.1.1. *Enzyme inhibitors*

A simple way to apprehend enzyme inhibitors is the lock–key concept where a key enters a specific lock to open or close a door. Similarly, enzyme inhibitors interact with their target enzyme to inhibit its activity. As a very simplified approach, the lock–key concept has several limits since enzyme inhibitors are not always specific to one enzyme, they can also interact with several sites on the same target and more complex approaches and chemical entities can be used to modulate enzyme activity; nevertheless, this concept helps us to understand the main principles of enzyme inhibition.

Enzyme inhibitors differ in their mode of action (Figure 5.1). The inhibitor can locate and interact in the catalytic site, catalytic inhibitor or outside the catalytic pocket, allosteric inhibitor. These interactions between the inhibitor, catalytic or allosteric, lead to a loss of enzyme activity by preventing the substrate to reach the catalytic pocket. For example, penicillins are antibiotics that are catalytic inhibitors of PBP (penicillin-binding protein) enzymes and lead to bacterial cell wall disruption (Kelly Judith et al. 1982; Pares et al. 1996). As allosteric inhibitors, Rifapentine is a clinically used anti-infectious against *Mycobacterium tuberculosis* targeting RNA polymerase (Artsimovitch et al. 2005; Sheik Amamuddy et al. 2020).

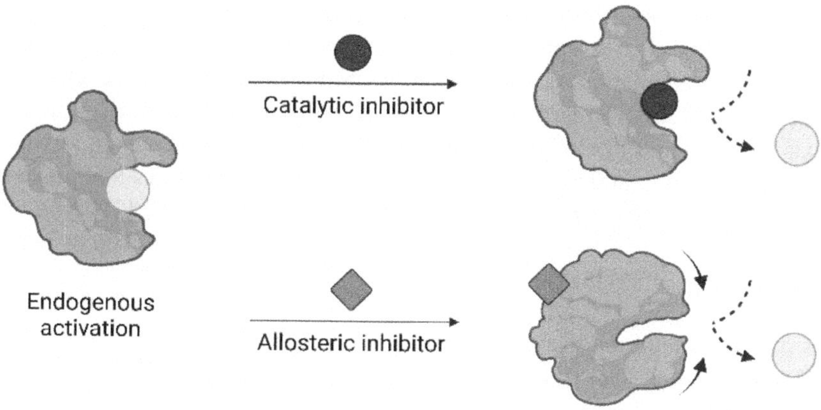

Figure 5.1. *Allosteric versus catalytic enzyme inhibitor*

Another important feature of enzyme inhibitors is their type of interaction. A non-covalent inhibitor interacts via weak bonds with the enzyme, i.e. hydrogen, electrostatic, ionic or hydrophobic bonds, in this case, interactions are reversible. On the opposite, the interaction of a covalent inhibitor with the target enzyme will result in a chemical reaction and a subsequent covalent bond between the inhibitor and the enzyme. With very few exceptions, covalent interactions lead to irreversible inhibitors. For instance, decitabine (Dacogene™), a DNA methyltransferase inhibitor, leads to the formation of a covalent bond between the sulfur of the catalytic cysteine and the carbon at position 6 of the cytosine of the inhibitor (Santi et al. 1984). On the other side, SAHA (Vorinostat®), a histone deacetylase inhibitor, interacts in a non-covalent manner with the enzyme where the hydroxamic acid moiety chelates the zinc ion present in the catalytic pocket (Finnin et al. 1999).

5.2.1.2. *Non-enzyme protein ligands*

There are several types of protein ligands targeting non-enzyme proteins, including receptor protein ligands and protein–protein interaction (PPI) disrupter.

Receptor ligands lead to an activation or a deactivation of the receptor activity. For example, Clonazepam, an antiepileptic drug, activates the γ-aminobutyric acid (GABA) receptor, whereas Ticagrelor, used against stroke and acute coronary syndrome, deactivates G protein-coupled receptor. In all cases, as for enzyme inhibitors, receptor ligands can locate at the recognition site of the receptor or in other part of the protein as allosteric ligands, such as for Clonazepam and Ticagrelor (Sheik Amamuddy et al. 2020).

Interaction between proteins is of great importance since numerous biological processes are governed by protein complexes, for example, cell differentiation or apoptosis, plant adaptation or virus infection (Struk et al. 2019; Luck et al. 2020; Gordon et al. 2020). Deregulation of PPIs is implicated in various pathological contexts; that is why PPI modulations are of increasing interest (Lu et al. 2020) as an alternative to enzyme inhibitors. PPI disrupters are chemical entities that prevent the interaction between target proteins. For example, disrupting the Gp120/ type 5 C–C chemokine receptor (CCR5) complex Maraviroc was approved in 2006 as anti-HIV treatment (Dorr et al. 2005).

As for enzyme inhibitors, non-enzyme protein ligands can interact directly with the substrate binding site, such as DRI-C23041, preventing ACE2-spike interaction in the context of SRAS-CoV2 infection (Bojadzic et al. 2021), or via an allosteric mechanism by interaction with other parts on the protein, such as 6,6-dihydropyrazolo[1,5-a]pyrazin-4(5H)-one, an allosteric inhibitor of mGluR2 receptors mainly studied in CNS disorders (Conde-Ceide et al. 2016) (Figure 5.2).

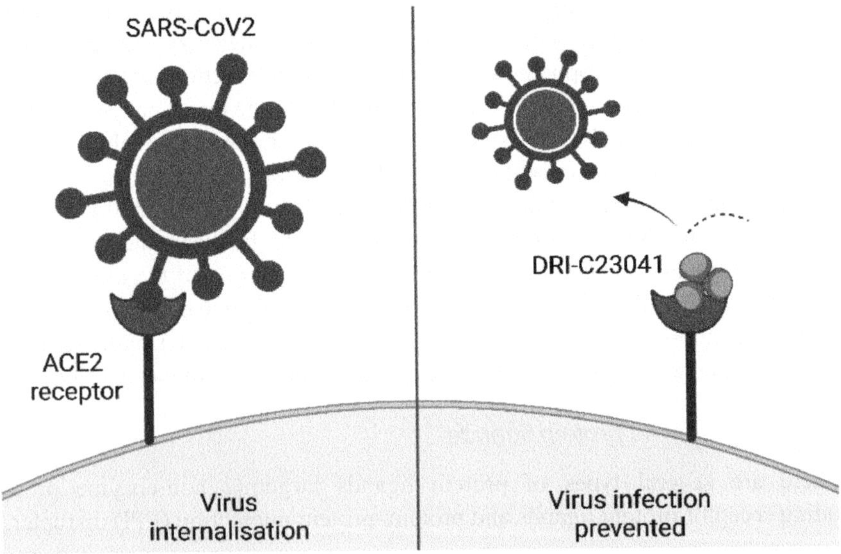

Figure 5.2. *Disruption of the SARS-CoV2-ACE2 receptor interaction by the PPI disrupter*

Similarly, to enzyme inhibitors, non-enzyme protein ligands can interact via covalent or non-covalent bonds.

5.2.1.3. *Interest of epigenetic protein ligands*

As described previously, the chromatin adopts different structures (hetero- or euchromatin) depending on epigenetic marks. Since, by definition, epigenetic marks are reversible, they can be reverted and epigenetic pattern can be modified. Epigenetic ligands are chemical agents that target epigenetic writers, erasers or readers, to modify epigenetic marks and modulated associated gene expression. The use of chemical epigenetic inhibitors can be used to study the impact of epigenetic changes on the phenotype and to counteract epigenetic deregulation in pathological contexts, for example. However, since the epigenetic control is crucial in every living organism, most of the epigenetic targeting molecules are toxic, elicit untargeted secondary effects and lack selectivity causing global changes (Chiappinelli et al. 2016), highlighting the importance of a fine tune when modulating epigenetic processes.

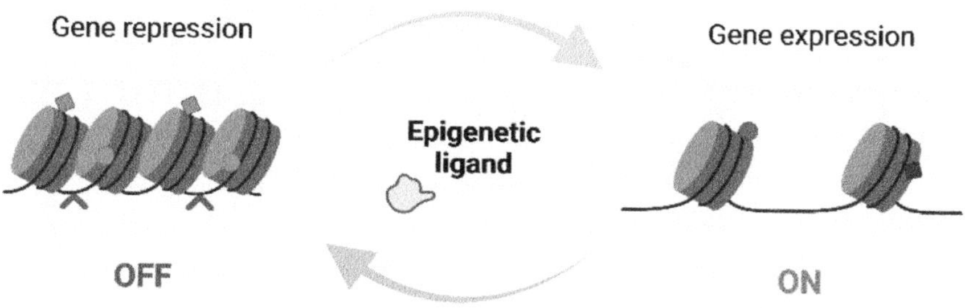

Figure 5.3. *Epigenetic ligands modulate gene expression*

5.2.2. *Epigenetic engineering*

Epigenome editing allows the modification of the epigenetic marks in target genes using engineered enzymes fused to one or more epigenetic enzymes. The programmable enzymes used are the zinc finger proteins (ZFPs), the transcription activator-like effectors (TALENs) and the CRISPR-Cas systems (Clustered Regularly Interspaced Short Palindromic Repeats) (Brocken, Tark-Dame and Dame 2018). ZPFs and TALENs are DNA-binding proteins that can be engineered to form interactions between the amino acid sequence of the DNA-binding domain (DBD) and the nucleotides of target DNA sequences. Nevertheless, ZFPs and TALENs have been surpassed by the CRISPR-Cas9 system that targets specific DNA sequences by using RNA: DNA complementarity (Laufer and Singh 2015).

CRISPR-Cas modules are adaptive immune response systems of many archaea and bacteria. It can identify and memorize foreign nucleic acids by capturing a small nucleotide sequence that is then recognized in a subsequent encounter and destroyed by the Cas enzymes, protecting prokaryotes from invading viruses (Koonin et al. 2016). In this system, a guide RNA (gRNA) directs the Cas nuclease to cut DNA at a specific site by the recognition of a short sequence of 2–6 base pairs called PAM (protospacer adjacent motif). There are different PAMs for different Cas nucleases. The most used Cas endonuclease is the Cas9 from the bacteria *Streptococcus pyogenes* whose PAM sequence is -NGG (Figure 5.4) (Cong et al. 2013).

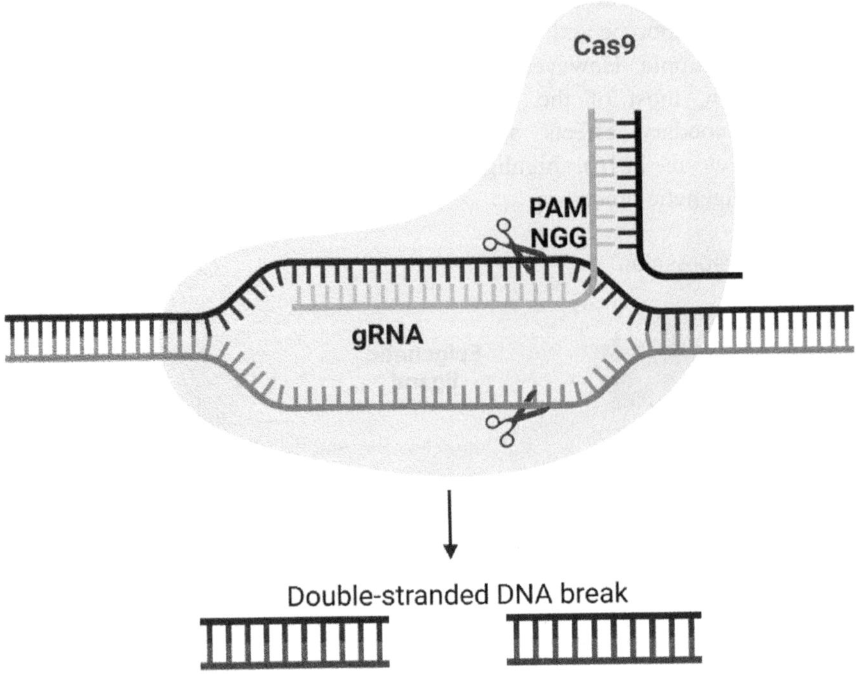

Figure 5.4. *Cas 9 enzyme from Streptococcus pyogenes that cuts three nucleotides upstream the PAM sequence in a target DNA sequence guided by a gRNA*

Cas9 protein has two domains (RuvC and HNH) that target the DNA strand. Thus, both domains generate double-stranded breaks (DSBs). The introduction of silencing mutations to each domain, namely D10A and H840A, stops the nucleolytic activity of Cas9 but does not impact its targeted binding specificity (Qi et al. 2013). Therefore, the deactivated version of Cas9 called dead Cas9 or dCas9 is still able to target a DNA sequence thanks to a gRNA without cutting it; this modified enzyme is used to deliver enzymes that modify epigenetic marks to specific DNA sites (Figure 5.5).

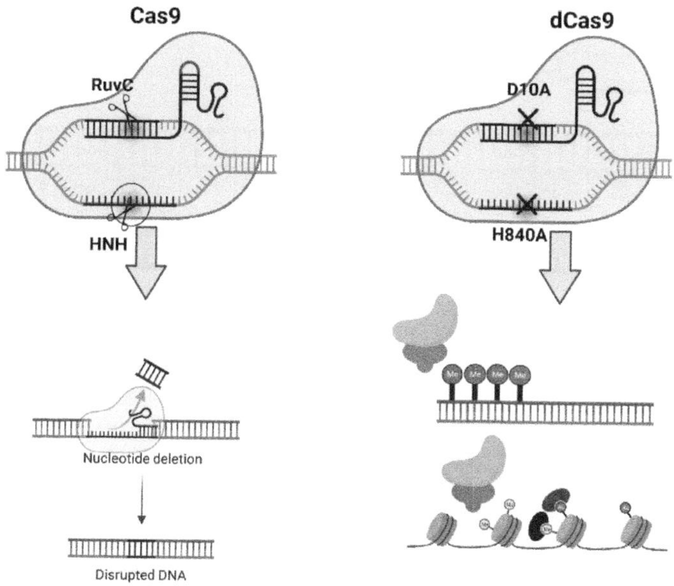

Figure 5.5. *Cas9 enzyme is engineered by two mutations in the RuvC and NHN domains to inactivate its nuclease activity, converting the enzyme to a dCas9 or deadCas9 that can no longer cut DNA but is still being able to be directed to a target sequence guided by a gRNA and it can bring an epigenetic enzyme to write or erase epigenetic marks*

The dCas9 protein can be fused to a multitude of enzymes, including epigenetic "writers" and "erasers" that deposit or remove, respectively, specific marks on the DNA and histone sites. The fusion of dCas9 to epigenetic writers or erasers has been principally applied to modify the most studied epigenetic marks, i.e. DNA methylation and histone modifications. However, more recently, this technology has also been applied to nuclear architecture, RNA modifications, transcription factors and chromatin remodelers (Brocken et al. 2018).

The profile of epigenetic marks in cells principally impacts gene expression. A gene is a DNA sequence that contains the instructions to produce a protein. Gene sequences contain several subdomains: 1) a sequence called promoter where RNA polymerase binds and transcribes DNA into RNA, 2) introns and 3) exons. Usually, only exon sequences are kept during the splicing process that converts RNA into messenger RNA (mRNA). The resulting mRNA is translated into a protein in the ribosomes. This process where a gene sequence is finally converted into a functional protein is called gene expression (Figure 5.6).

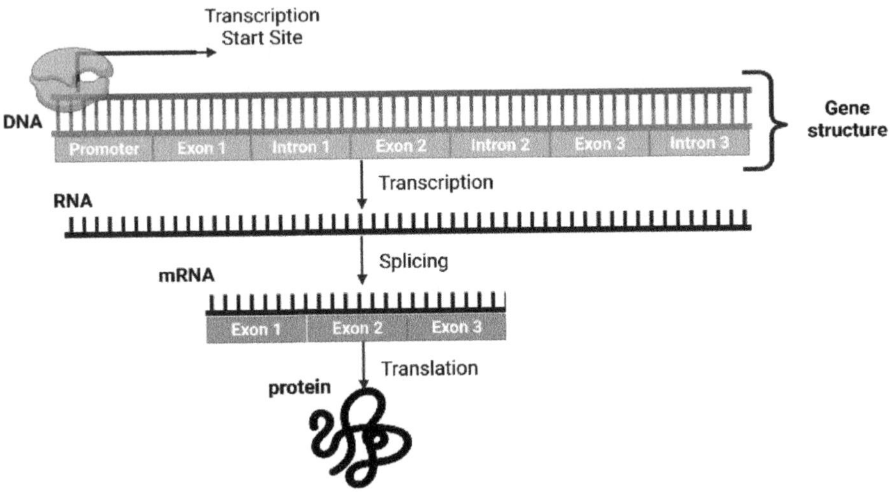

Figure 5.6. *The gene expression process starts with DNA transcription into RNA, its subsequent conversion to mRNA by the splicing step and finally mRNA translation into a functional protein*

Significant improvements were achieved in epigenome editing technologies in the last few years. New tools were developed to recruit multiple epigenetic enzyme domains (also called epigenetic effectors) in a target DNA sequence. Moreover, it was shown that the action of multiple epigenetic effectors in a targeted gene is more effective to produce changes in its expression level than the use of a single effector. One of these tools consists of recruiting multiple protein copies to a polypeptide scaffold protein called "SunTag" that consist of multiple fusion proteins called single-chain variable fragments (scvF) that bind the variable regions of an antibody (Figure 5.7(A)). The SunTag system was proved to be very effective in increasing epigenetic editing levels and is also used to amplify fluorescence signals by adding fluorescent proteins to the SunTag scaffold (Tanenbaum et al. 2014).

A different strategy to recruit multiple epigenetic effectors is the use of RNA aptamers included in the gRNA (Zalatan et al. 2015). These aptamers are RNA structures that are specifically recognized by RNA-binding proteins forming a scaffolding RNA (scRNA). The most used RNA aptamers are MS2, PP6 and PUF binding site (PBS) that, fused to multiple epigenetic effectors, are directed to the target sequence by indirectly being associated with the dCas9 through the gRNA (Figure 5.7(B)).

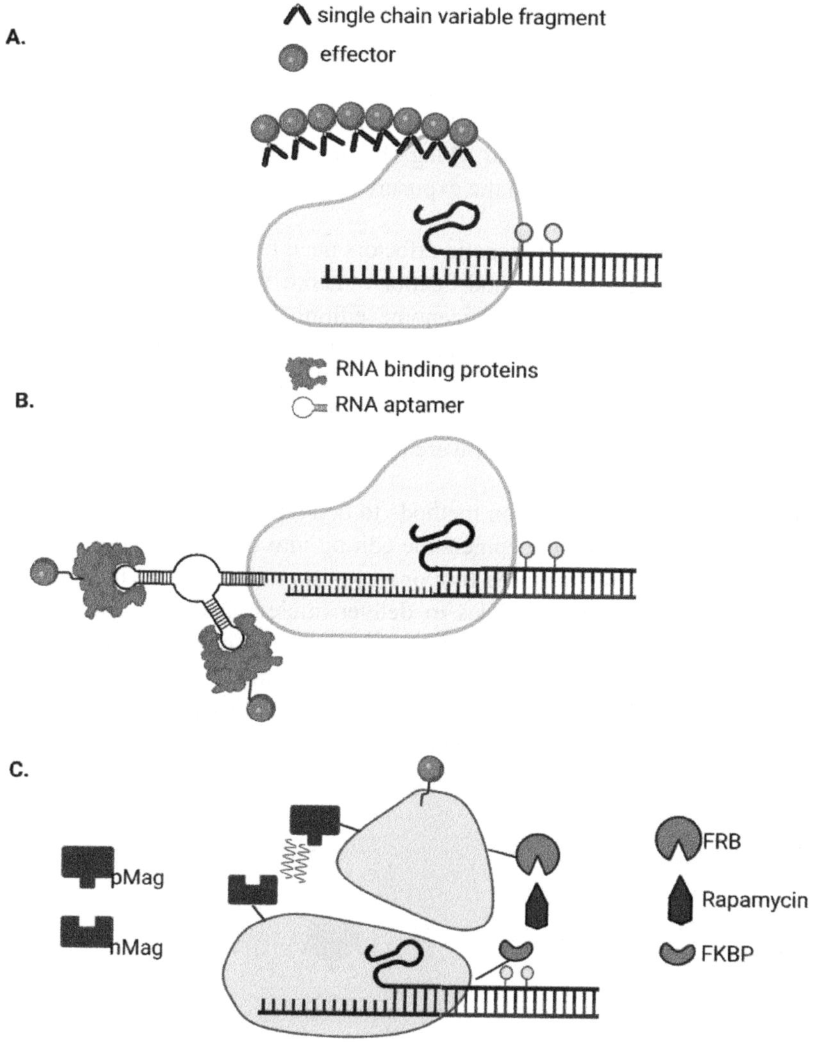

Figure 5.7. *Different systems to attach epigenetic effectors to a targeted sequence*

There also exists a strategy to control dCas9 DNA binding activity in space and time using split dCas9 that consist of the Cas9 enzyme divided into two fragments, i.e. *C*- and *N*-terminal fragments, that can bind together into a full functional complex only in the presence of a full-length single guide RNA (gRNA). Therefore, the use of a truncated sgRNA prevents the complex formation, which allows the division in two fragments (Wright et al. 2015). The assembly of both fragments can also occur in the presence of a chemical or light inductor (Zetsche et al. 2015). For chemical induction, *C*- and *N*-terminal dCas9 fragments are fused with FKBP and FRB domains, respectively, that bind together only in the presence of rapamycin (Banaszynski et al. 2005). Split dCas9 fragments can also be reassembled by using photoinducible magnets pMag and nMag (Figure 5.7(C)); each one is fused to a fragment and binds together upon the exposure to blue light (Kawano et al. 2015).

Engineered enzymes and epigenetic effectors need to be delivered into cells to be able to produce the epigenetic modifications. Three types of models are used to deliver nucleic acids expressing epigenome editing tools into the cells: in vitro, in vivo and ex vivo. In vitro means that the nucleic acids coding for the dCas9-epigenetic effector and sgRNA system are delivered into cell cultures, in vivo that they are delivered in a living animal targeting some of its tissues and ex vivo that they are delivered in tissues that were taken from a living animal (Figure 5.8).

There are different transfection methods to deliver the DNA, mRNA or proteins responsible for effectuating the epigenome editing into the cells. There are viral and non-viral transfection methods; the first one consists of the use of viral vectors such as lentiviral and adenoviral particles to deliver nucleic acids into cells. The viral transfection methods are highly efficient but require a lot of monitoring of biosafety rules that are available only in biosafety levels 3 and 4 (BSL3-4) laboratories; besides, they are very expensive. On the contrary, non-viral methods, including physical and chemical techniques such as electroporation, and lipofection (Yin et al. 2014), are less laborious and are cost-effective (Figure 5.8).

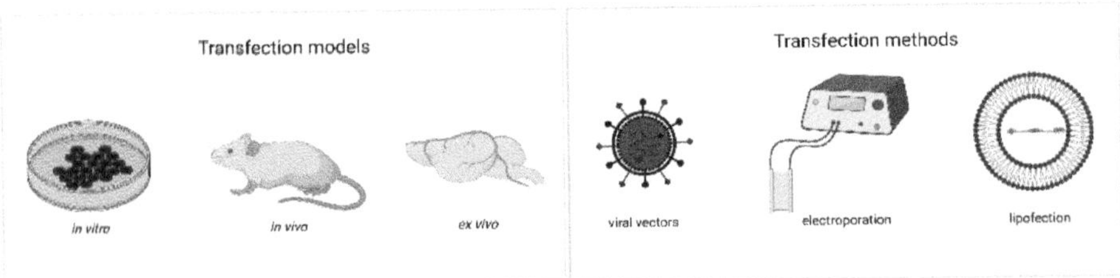

Figure 5.8. *Test environments and transfection methods to deliver nucleic acids coding for epigenome editing proteins*

The principal applications of epigenome editing are for basic and biomedical research (Goell and Hilton 2021). Epigenome editing tools open the possibility to understand causal relationships between the epigenetic marks, chromatin compaction, genome regulation and phenotype of organisms. Changing epigenetic marks can also help to understand the role of these modified epigenetic marks in the non-coding genomic regions, their functional interaction, their contribution to cellular memory, the way they shape the nuclear architecture and their influence on chromatin accessibility (Policarpi et al. 2021). Here, we describe the dCas9 epigenome editing techniques according to the epigenetic mark they targeted, and we show some application examples.

5.3. Epigenetic modification targeting DNA methylation

5.3.1. *DNMT inhibitors*

Since DNA methylation of gene promoters leads to gene silencing, increasing or lowering DNA methylation level in these regions will lead to lower or higher gene expression level, respectively.

As described earlier, enzymes responsible for DNA methylation marks are DNA methyltransferases (DNMTs) and ten–eleven translocation (TET) methylcytosine dioxygenases that act as DNA methylation writers and erasers, respectively. Inhibition of DNMTs and TETs can therefore control the DNA methylation level and gene expression.

Several DNMT inhibitors (DNMTi) have been described (Figure 5.9). First, nucleoside DNMTi are non-natural analogues of cytidine. Examples of such nucleoside DNMTi are 5-azacytidine (5aza, Vidaza®), 5-aza-2'-deoxycytidine (5azadC, Dacogen™) (Jones and Taylor 1980), both FDA-approved for anticancer treatment, and guadecitabine, a 5azadC prodrug. For these cytidine analogues, carbon atom C5 is replaced by a nitrogen atom which results in a covalent interaction with the catalytic cysteine of DNMTs. As nucleoside analogues, 5aza and 5azadC incorporate in the genome without distinction of site or cell distinction, which induces unselective activity and side effects. Zebularin, another nucleoside analogue, is not a covalent inhibitor and forms a reversible complex with DNMT; however, its weaker activity requires the use of higher toxic doses.

Considering the interest of these first nucleoside DNMTi as well as their drawbacks, a lot of research has been carried out on non-nucleoside DNMTi. In this chapter, some examples will be reported, but the description of non-nucleoside DNMTi will not be exhaustive. First, *in silico* studies identified RG108 as a DNMTi

(Brueckner et al. 2005). However, whereas an inhibition of in-cell methylation was observed, enzyme assays on purified DNMT (Suzuki et al. 2010; Halby et al. 2012) showed the absence of activity which illustrates a probable indirect effect. SGI-1027 is a DNA ligand with DNMTi activity (Yoo et al. 2013) that induces DNMT degradation and gene reexpression but does not show a genomic DNA methylation inhibition. Virtual screening also led very recently to the identification of GSK3484862 (Pappalardi et al. 2021) that showed submicromolar DNMT inhibition activity, in-cell efficacy with moderate toxicity (Azevedo Portilho et al. 2021) and is, to date, one of the most promising DNMTi. Applying another strategy, Halby et al. (2016) identified bisubstrate-analogue DNMTi with potent but poor pharmacological properties. In vitro high-throughput screening (Ceccaldi et al. 2011) and structure optimization (Pechalrieu et al. 2020) identified flavonoid derivatives as potent in vitro DNMTi with in-cell activities. However, the low stability of these derivatives does not allow them to enter clinical trials. This illustrates the difficulties to identify non-nucleoside DNMTi, and to date no such inhibitor is in clinical trials.

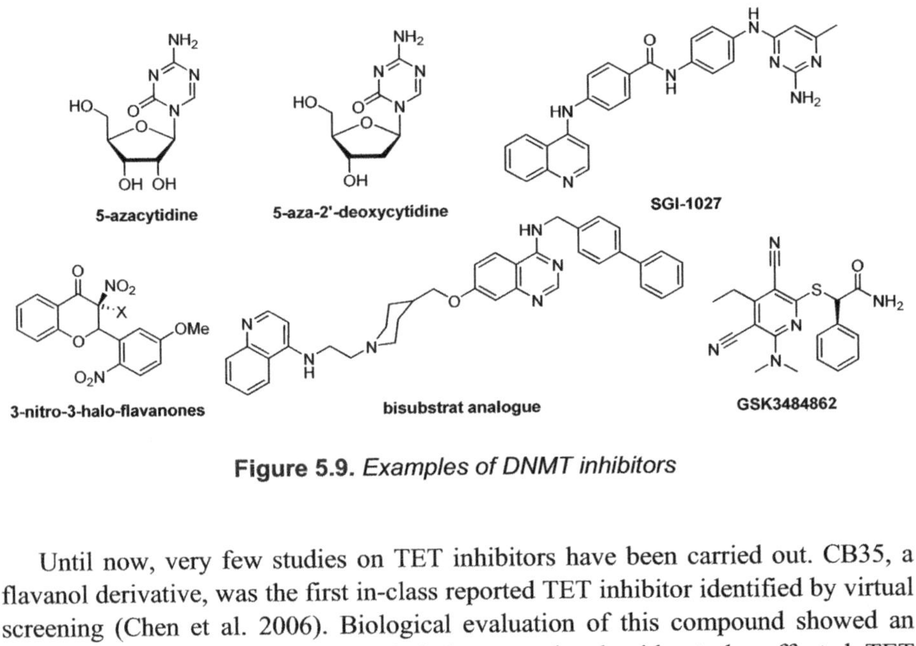

Figure 5.9. *Examples of DNMT inhibitors*

Until now, very few studies on TET inhibitors have been carried out. CB35, a flavanol derivative, was the first in-class reported TET inhibitor identified by virtual screening (Chen et al. 2006). Biological evaluation of this compound showed an inhibition of TET-mediated demethylation associated without the affected TET protein complex. Recently, BC339, a cytosine analogue, showed mi-micromolar TET inhibition activity and in-cell increase of DNA methylation (Chua et al. 2019) with limited effects on cell viability. Importantly, BC339 does not show DNMT inhibition activity. Based on substrate similarity, TETi76, an *N*-oxalylglycine analogue, was described with submicromolar TET inhibition activity within cell and in vivo ability to decrease TET activity (Guan et al. 2021). Two other enzyme inhibitors, namely DMOG, a substrate analogue (Zhang et al. 2016), and pyrimidine-based derivative SC1 (Pluripotin) (Chen et al. 2006) showed TET inhibition activity but as unselective inhibitor further optimizations are required (Figure 5.10).

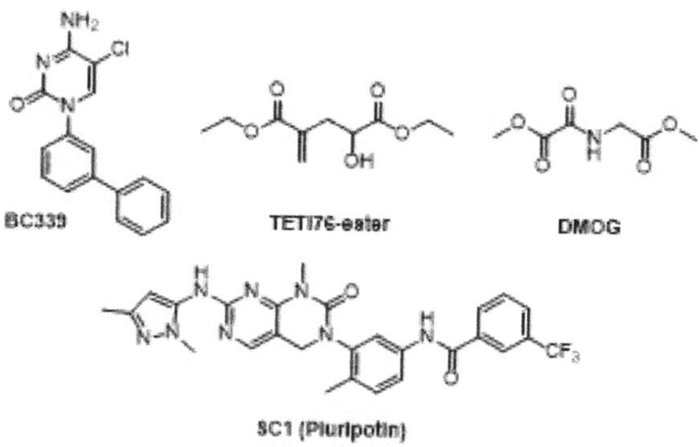

Figure 5.10. *Examples of TET inhibitors*

5.3.2. *dCas9-based technology to edit DNA methylation*

The dCas9 coupled to an enzyme implicated in DNA methylation or demethylation process can be directed to modify methylation marks of a targeted gene, enabling researchers to study the role of this epigenetic mark in chromatin structure, gene expression and phenotypic variation at a particular gene. The principal function that has been explored by this method is the role of DNA methylation, principally in gene promoters, on gene expression, and if the DNA methylation level modification can repress or activate genes.

The function of DNA methylation in the promoter of genes has been largely studied in mice and human cancer cell lines using dCas9 coupled to the catalytic domain of the DNA methyltransferase DNMT3A (an enzyme that add methyl groups de novo to the cytosines) or to the catalytic domain of the DNA demethylase ten–eleven translocation dioxygenase 1 TET1 that catalyzes the demethylation of 5-methylcytosine (Choudhury et al. 2016; Liu et al. 2016). In addition to DNMTs and TETs, prokaryotic CpG methyltransferase derived from *Mollicutes siroplasma* (*M.SssI*), strain MQ1 is used to elicit target DNA methylation in human cells and in mouse embryos, introducing high levels of 5-methylcytosine (5mC) (Lei et al. 2016). Nevertheless, the function of DNA methylation in other species, such as the invertebrates that includes many taxa (insects, arachnids, mollusks, crustaceans, corals, worms, echinoderms, sponges, cnidarians, etc.) has been poorly studied at present. Furthermore, the type of DNA methylation is different in invertebrates since it is found principally in the gene bodies (introns and exons) and not in the gene promoters as in the vertebrates. Therefore, methylome editing techniques are promising tools to study the function of this type of DNA methylation in those understudied species.

Furthermore, the tool dCas9+DNMT3A, TET1 or MQ1 can be improved using the system SunTag, which allows the recruitment of multiple copies of an antibody-fusion protein. Therefore, many enzymes DNMT3A, TET1 or MQ1 can be recruited at the same time and directed to a targeted DNA sequence, demonstrating that they generated a greater impact in the DNA methylation level of targeted gene expression (Huang et al. 2016).

The use of methylome editing techniques allowed the finding of DNA methylation function in the transcription activity of gene promoters principally in humans and mammals. It was found in mammals that the presence of high levels of DNA methylation in gene promoters led to gene silencing and that a low promoter methylation level activates gene expression. Therefore, the addition of methyl groups to the DNA sequence of a gene promoter with dCas9-DNMT3A represses that gene, and vice versa, the edition with dCas9-TET1 to erase methylation groups from a promoter of a specific gene activates its gene expression at a certain level. Nevertheless, these methylome editing techniques induced a certain level of methylation and demethylation that depends on genetic and chromatin context and on the delivery efficiency of the nucleic acids coding for dCas9-DNMT3A or dCas9-TET (Siegfried and Simon 2010).

Since methylation marks in gene promoters can modify gene expression, methylome editing techniques are applied to disease-related genes that are silenced by high levels of DNA methylation in their promoters, opening the possibility of therapeutic options by editing their DNA methylation levels. Multiple methylome editing tools have been tested in vitro in different cancer cell lines, and in vivo in mice models. Dysregulations in DNA methylation patterns are linked with several diseases, including cancer, developmental and neurological disorders (Robertson 2005).

5.3.3. *Examples of DNA methylation modifications*

The in vivo lentiviral delivery of dCas9-TET3CD in a kidney fibrosis mouse model resulted in targeted promoter demethylation and the consequent reactivation of antifibrotic genes that resulted in the attenuation of the kidney fibrosis and the stabilization of the kidney function (Xu et al. 2018).

Interestingly, TET inhibitors have a great interest in neurological disfunctions often related to *BDNF* gene overexpression; for example, SC1 TETi enabled to revert cognitive disfunction in mice (Zhang et al. 2020). Besides, TETi BC339 also showed a promising result in osteoporosis models, where genes implicated with bone mineral density, for example, *Sost*, *Dlx5*, were shown to be hypomethylated (Dusadeemeelap et al. 2022). CB35 TETi was shown to reprogram somatic cells modifying TET-mediated DNA methylation level.

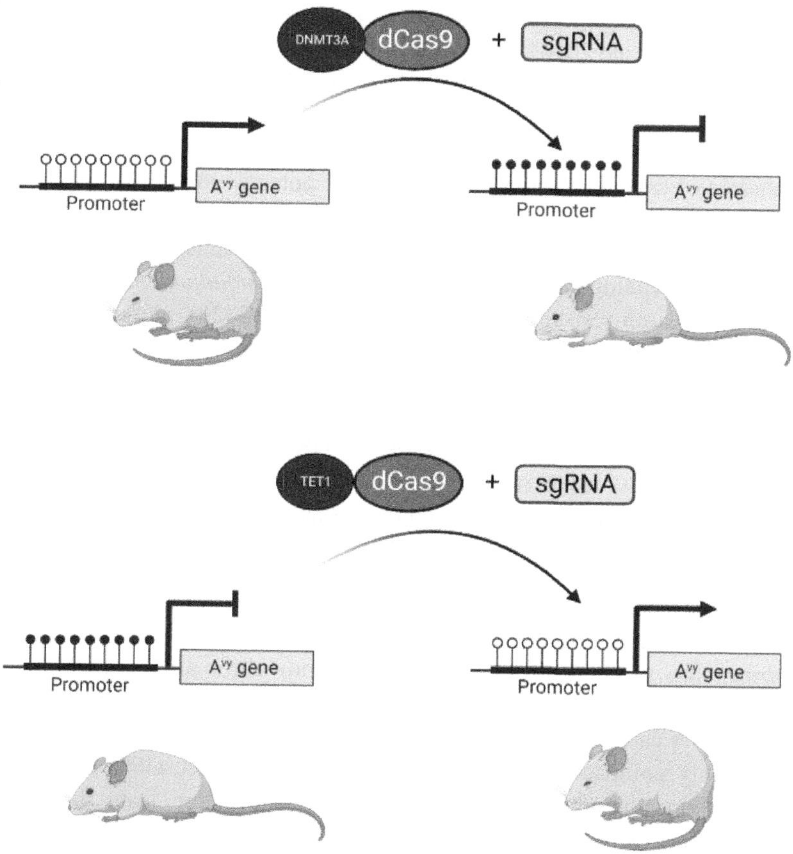

Figure 5.11. *Agouti viable yellow (A^{vy}) coat color in mice is associated with obesity, diabetes and cancer. This A^{vy} phenotype is provoked by the expression of the A^{vy} gene; therefore, the use of dCas9-DNMT3A to methylate the promoter region of this gene can silence the gene and change the phenotype to brown color and healthy mice (Wei et al. 2019); vice versa, it is possible to induce the A^{vy} phenotype by using dCas9-TET1 to demethylate the A^{vy} gene promoter*

An interesting example is the study of the Agouti gene (Duhl et al. 1994). The Avy-related coat color gene can be silenced by editing with dCas9-DNMT3A that successfully methylates the promoter region of the Avy locus causing the phenotype of a brown coat color (Figure 5.11). The inverse is also possible with dCas9-TET1 to activate the gene and induce the Avy coat color that is associated with obesity, diabetes and cancer (Sharma and Aazmi 2019; Wei et al. 2019).

DNA methylation was also modulated using DNMTi. Since cancer-related use of DNMTi is widely documented in many reviews and book chapters (Lopez et al. 2016; Pechalrieu et al. 2016), other applications will be briefly described here to give some examples of field diversity related to DNMTi. DNMTi were studied in neurological diseases, and synaptic plasticity was shown to be modulated by the 5azadC treatment in hippocampus slices (Nelson et al. 2008). In the context of schizophrenia, 5aza and zebularine show good potency to restore *RELN* gene, a memory suppressor gene, expression in cells but present low ability to cross the blood–brain barrier (BBB) (Kundakovic et al. 2006). In parasite infection, treatment with bisubstrate-analogue DNMTi resulted in *Plasmodium falciparum* death, including artemisinin-resistant parasite (Nardella et al. 2020), and an important role of flavonoid DNMTi was reported in Biomphalaria glabrata, a vector mollusc of the parasite Schistosoma mansoni (Luviano et al. 2021). Applications of DNMTi in plants are also important since several phenotypic characteristics observed depending on growth conditions are related to DNA methylation. For instance, the expression of Xa21G gene in *Japonica* rice (*Oryza sativa* L.) confers bacterial resistance and treatment with 5azadC led to a resistant line (Ronald 1996).

5.4. Epigenetic modification of histone epigenetic marks

Histones are proteins that package DNA into chromatin and chromosomes. Chromatin states control DNA accessibility to transcription factors, to put it simply: accessible DNA can be transcribed to proteins, and this state is called euchromatin, whereas highly compacted chromatin leads to DNA inaccessible to transcription factors, leading to gene silencing, and this chromatin state is known as heterochromatin (Murakami 2013) (see Chapter 2). Certain covalent histone modifications are associated with an active transcription of genes and some others to gene repression. However, recent findings demonstrated that post-translational modifications (PTMs) of histones interact with each other and with distinctive combinations of proteins, resulting in different transcriptional activities (Li and Zhang 2014; Li et al. 2014).

As we have seen in Chapter 2, epigenetic enzymes that write or erase histone modifications can directly affect gene transcription (Figure 5.12). Histone writer enzymes associated with active transcription are histone acetyltransferases p300 and CREB-binding protein (CBP) and histone methyltransferases SMYD3, MLL3, PRDM9 and DOT1L (Gomez et al. 2019). On the opposite, histone writer and eraser enzymes associated with gene repression are histone deacetylases HDACs that remove an acetyl group from and the histone lysine methyltransferases G9A, SUV39H1 and EZH2 (Lawrence et al. 2016).

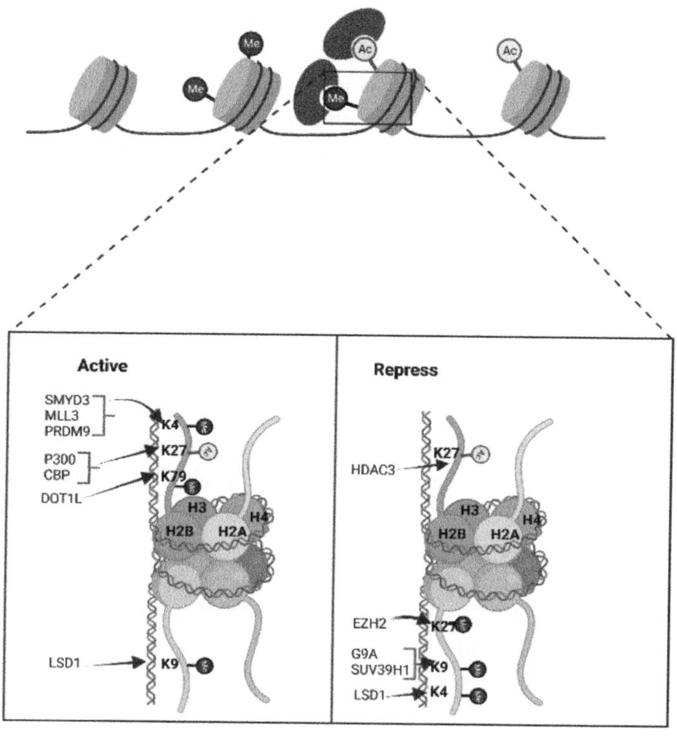

Figure 5.12. *Epigenetic enzymes that modify histone modifications to activate or repress gene expression*

Therefore, considering the implication of histone epigenetic modifications in gene expression and their implication in environment susceptibility, biological processes and disease development, their modulations using enzyme inhibition or gene editing strategies have been widely used.

5.4.1. *Inhibitors of histone-targeting epigenetic marks*

Enzyme inhibitors targeting histone acetylation or methylation for anticancer therapies have been widely documented (examples of review: Falkenberg and Johnstone 2014; McCabe et al. 2016; Lopez et al. 2016a; Michalak et al. 2019). Therefore, in this chapter, examples in other fields of application will be described.

Histone acetylation can be modulated targeting writers, i.e. histone acetyltransferases (HATs) or its corresponding erasers, i.e. histone deacetylases (HDACs). A widely studied HAT inhibitor family is the p300 inhibitors. Interestingly, complementary to HAT inhibition, HAT activator TTK21, conjugated on nanosphere (CSP), was reported to induce histone acetylation in hippocampus and frontal cortex and to improved brain functions in mouse model (Chatterjee et al. 2013). Besides, in heart failure model, curcumin, a HAT inhibitor targeting p300, was reported to prevent ventricular hypertrophy and maintain systolic function, decreasing the risk of heart failure (Shi et al. 2022). Finally, in the context of bone mineralization, HDAC inhibitors such as Trichostatin A (TSA) and SAHA were shown to promote osteoblast differentiation attenuating bone loss (McGee-Lawrence et al. 2011).

Histone methylation is controlled by writers, i.e. histone methyltransferases (HMTs), and erasers, i.e. histone demethylases (HDMs). Although histone methylation has a crucial role in gene expression, its modulation has not been as widely studied as histone acetylation, due to the complexity of histone methylation impact and to the multiplicity of enzyme involved. Nevertheless, HMT or HDM inhibitors were shown to be relevant in several contexts. Besides their application on cancers, HMT and HDM inhibitors were for instance shown to have an impact in diabetic kidney disease model where the inhibition of a HDM, i.e. JMJDA2A, by 2,4-PDCA, a pyridine-based inhibitor reverse migration, proliferation and inflammation caused by hyperglycemia (Sun et al. 2017).

5.4.2. *dCas9-based technology to edit post-translational histone modifications*

The dCas9-p300 system was transfected in a human kidney cell line, targeting the promoters of the genes *MYOD1* and *OCT4* that are important for the mammalian embryonic development. It was found that dCas9-p300 produced robust transcriptional activation of the human *MYOD* and *OCT4* genes from promoters (Hilton et al. 2015). *OCT4* expression is associated with the pluripotency of stem cells, which is the capacity of a single cell type to differentiate into multiple other types of cells (Obokata and Vacanti 2014; Zeineddine et al. 2014) and *MYOD* plays a key role in regulating muscle differentiation (Tajbakhsh et al. 1996).

dCas9 fused to LSD1 is used for targeted histone demethylation that can repress gene expression. This system was delivered as viral vectors in mouse embryonic stem cells where it targeted the *OCT4* and *SOX2* genes, inducing its repression and triggering morphological changes in the embryonic stem cells (Kearns et al. 2015).

Histone phosphorylation plays crucial roles in chromatin remodeling related to other nuclear processes (Rossetto et al. 2012); the dCas9-dMSK1 system helped to find a positive relationship between phosphorylation of the histones and human gene activation. Furthermore, three genes that are implicated in melanoma drug resistance and seven genes related to melanoma resistance were found thanks to this tool (Li et al. 2021).

Different combinations of histone and DNA methyltransferases fused to the dCa9 enzyme can be done to maximize silencing of target genes. For example, dCas9 fused to the histone methyltransferase Ezh2 co-transfected with DNMT3A-dCas9 and DNMT3L, directed to the human *HER2* promoter, induced the long-term repression of the gene *HER2* in a cancer cell line and change the chromatin state to heterochromatin, the highly condensed state (O'Geen et al. 2019).

5.5. dCas9-based technology to edit nuclear architecture

The 3D organization of the genome within the cell nucleus regulates gene expression (Bickmore and van Steensel 2013), genes located at the nuclear periphery are associated with repressed or very low expression and genes located at the nuclear interior are associated with active gene expression (van Steensel and Belmont 2016).

The CRISPR-genome organizer (CRISPR-GO) consists of the CRISPR-dCas9 system coupled to nuclear-compartment-specific proteins via ligand-mediated interaction (Figure 5.13). It allows repositioning of specific genomic loci to the nuclear periphery and nuclear bodies, offering a tool to study the relationship between spatial genome organization, gene expression and cellular function (Wang et al. 2018). Relocation of genomic loci to the nuclear periphery with CRISPR-GO silenced gene expression. Moreover, this system was used to study how telomere reorganization in the nuclear compartments affected cellular phenotypes; the telomeres were relocated to the nuclear periphery causing a decrease of the cell viability (Wang et al. 2018).

A similar system called CRISPR-positioning in the nucleus (CRISPR-PIN) enables the non-invasive control of chromosome organization by specific re-localization of a genomic locus to the nuclear periphery directed by a gRNA. This system uses the dockerin (Doc) and cohesin (Coh) interactions to link together a dCas9 protein fused to a cohesin domain with a nuclear membrane protein ligated to a dockerin domain (Lin et al. 2019). The cohesin and dockerin domains have a calcium-dependent interaction and come from cellulolytic bacteria that produce free cellulases such as *Clostridium thermocellum* (Brown et al. 2015). One of the advantages of CRISPR-PIN over other systems is that there is no need for the addition of a chemical intermediary to fusion the proteins associated with the nuclear periphery and to the nuclear interior, respectively (Lin et al. 2016). This system can help to study the role of gene positioning in gene expression and further questions in chromosome biology.

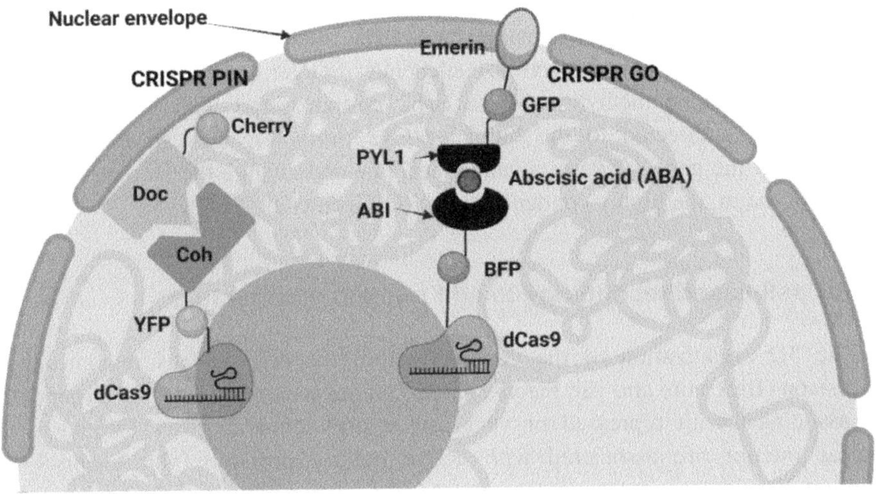

Figure 5.13. *CRISPR-PIN and CRISPR-GO systems for targeting genomic loci to the nuclear periphery. CRISPR-GO targets genomic loci to the nuclear periphery through the co-expression of ABI-BFP-dCas9 and PYL1-GFP-Emerin. ABI and PYL1 are assembled in the presence of ABA and cause relocalization of the targeted genomic loci to the nuclear periphery. CRISPR-PIN relocates genomic locus to the nuclear periphery directed by a gRNA associated with a dCas9 protein fused to a cohesin (Coh) domain that interacts with a dockering domain (Doc) ligated to a nuclear membrane protein*

One more nuclear architecture editing tool is the CRISPR-dCas9-CLOuD9 system that is used to establish selectively and reversibly chromatin loops. This system consists of two dCas9 enzymes, one from *Staphylococcus aureus* (CSA) and one from *S. pyogenes* (CSP); each dCas9 is directed towards a target locus using gRNAs, and the connection of both chromosomal loci is induced with the addition of the plant hormone ABA which facilitates reversible association of the chemical-induced proximity proteins (Figure 5.14). The edition of chromatin loops by the dCas9-CLOuD9 system has demonstrated that modifications of loop structure can reversibly alter gene expression (Morgan et al. 2016). In erythroleukemia cell lines, β-globin gene is silenced and, in its place, there is an aberrant high expression of the fetal γ-globin gene. The dCas9-CLOuD9 tool was used to re-establish β-globin gene expression through the formation of a chromatin loop between this gene and a locus control region (LCR) (Morgan et al. 2016).

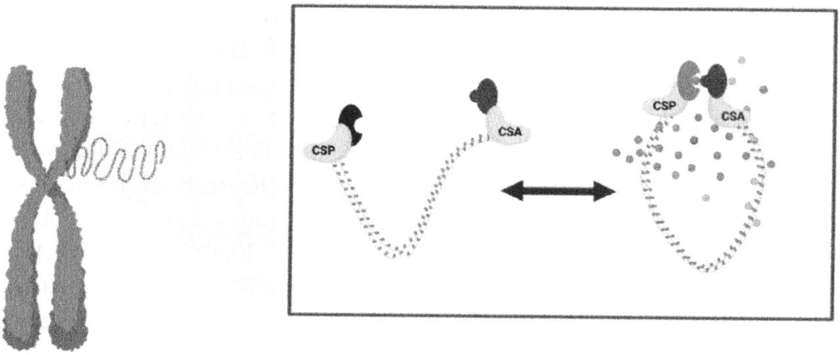

Figure 5.14. *Addition of plant hormone abscisic acid (ABA) represented by the green circles brings two complementary CLOuD9 constructs CSP from* S. pyogenes *and CSA from* S. aureus, *into proximity forming chromatin loops. Removal of ABA restores the initial chromatin conformation*

5.6. dCas9 fused to transcription factors

Transcription factors that are repressors and activators of gene expression are fused to the dCas9 enzyme to repress or activate targeted genes. When dCas9 is fused to transcription activators, it is called CRISPR activator (CRISPRa). The VP64 is a transcription activator domain that recruits transcriptional machinery to induce gene expression. VP64 can be fused to dCas9 to specifically increase the gene expression of a targeted gene; dCas9-VP64 is a tool for experiments that require modest gene expression activation. Other transcriptional activators that can be fused to dCas9 to activate the expression of a targeted gene are VP48, VP120, p65 and P65-HSF1 (Chavez et al. 2016). dCas9-VP64 was used to activate the expression of the *Sim1* gene and reverse obesity resulting from *Sim1* allelic insufficiency (Gemberling et al. 2021). CRISPRa has been applied to treat genetic or acquired diseases such as epilepsy, muscular dystrophy, diabetes and retinitis (Zhang et al. 2019; Becirovic 2022).

When dCas9 is associated with a transcription repressor, it is called CRISPR interference (CRISPRi). The transcriptional repressor domain Kruppel-associated box (KRAB) induces the silencing of gene expression through the recruitment of corepressor proteins. KRAB can be fused to dCas9 to specifically decrease the gene expression of a targeted locus. Alternative transcriptional repressors that can also be fused to dCas9 to repress a targeted gene are SID4x, Mxi1 and the transcription factor "Friend of GATA-1" (FOG1) (Gilbert et al. 2013; Konermann et al. 2013). CRISPRi is used to induce gene knockdowns and study gene function by be screening of phenotypic effects but they differ from the genetic knockdowns because the DNA sequence is not modified; instead, it is silenced by transcriptional repressors that recruit multiple effectors that modify the chromatin configuration. CRISPRi has been demonstrated to be transferable by bacterial conjugations repressing target genes in recipient bacteria; therefore, this system could be used to control bacterial populations (Ji et al. 2014). CRISPRi with dCas12a has been applied to metabolic engineering in cyanobacteria to build cell factories that convert CO_2 into value-added chemicals (Choi and Woo 2020). dCas9-KRAB was transfected in vivo in mouse targeting Nrl, a rod photoreceptor gene inducing the knockdown of this gene whose mutations are associated with retinitis pigmentosa and retinal degenerative diseases (Moreno et al. 2018).

The transcriptional regulators used in CRISPRa/i techniques are naturally abundant in cells affecting gene expression, as well as having indirect and multilateral effects on the states of the epigenome since they recruit multiple effector partners.

5.7. dCas13-based technology to edit RNA modifications

Finally, it is also possible to modify the RNA modifications in a targeted and specific manner with CRSPR Cas enzymes. The enzyme dCas9 that targets DNA is replaced by the enzyme dCas13 that targets single-stranded RNA for targeted RNA modifications. dCas13 is the deactivated form of Cas13 that comes from the bacterium *Leptotrichia shahii* and there are two members dCas13a and dCas13b (Cox et al. 2016). The targeted RNA modifications are a better alternative to transcriptional repressor techniques such as CRISPRi since this last may remove all splice variants of the target gene leading to unpredictable results. Therefore, it is very important to target and control the activity of the transcripts that are the cellular RNAs after the transcription (Hutvagner 2005).

The N^6-methyladenosine (m6A) is the most abundant mRNA modification in eukaryotes inducing alteration in the chromatin accessibility. The nucleus-localized dCas13 fused to a fragment of the METTL3 methyltransferase domain and cytoplasm-localized fusions with a methyltransferase complex MTTL3:METTL14 can modify targeted RNA methylation by the incorporation of m6A in a specific sequence. This targeted RNA methylation changes the transcript abundance and the gene splicing (Wilson et al. 2020). On the other hand, the fusion of CRISPR-Cas9 with ALKBH5 or FTO allows site-specific demethylation of RNAs (Liu et al. 2019). The m6A demethylation is applied to improve crop performance, FTO leads to m6A demethylation of mRNAs and repeat RNAs in rice nuclei, causing an open chromatin state and transcriptional activation. Additionally, numerous metabolic pathways related to photosynthesis and regulation of nitrogen are upregulated in FTO-expressing rice promoting root cell proliferation (Shen and Yu 2021).

Targeted demethylation with dCas13-FTO in mRNA of lipogenic genes induced the knockdown on these genes validating the role of m6A on sterol regulatory element transcription and showing that the knockdown of lipogenic genes alleviated fatty liver disease in mice (Hu et al. 2020). Nevertheless, the function of m6A remains largely unknown and the use of the RNA editing tools will help to elucidate its function.

RNA editing holds promise for treating genetic disease, particularly at the mRNA level, where disease-relevant sequences can be rescued to produce functional protein products. dCas13 fused to ADAR to modify adenosine to inosine deaminase activity in target transcripts allows us to edit transcripts and induce RNA knockdowns. The dCas13-ADAR2 system is also called RNA Editing for Programmable A to I Replacement (REPAIR) and has been successfully proved to edit full-length transcripts containing pathogenic mutations. It was recently reported that REPAIR was applied against two disease relevant mutations, one linked to urinary diabetes (nephrogenic diabetes insipidus) and another linked to Fanconi anemia showing an efficiency of 35% and 23%, respectively (Cox et al. 2016). Many diseases originated from variations during the splicing step (Angarola and Anczuków 2021). For example, the accumulation of repeat transcripts into the nucleus is a characteristic of neurological and neuromuscular diseases such as Huntington's disease and myotonic dystrophy and the use of targeted RNA editing modifications can eliminate these toxic RNAs (Tang et al. 2021).

5.8. Conclusion

There exist multiple epigenetic effectors that can be fused to the enzyme dCas9 to be directed to a target DNA sequence to edit the epigenome in that region. There are also multiple strategies to improve the epigenome edition such as the use of recruiters of multiple epigenetic effectors to induce a higher activation or deactivation of a targeted gene than the use of a single effector. However, there are limitations in the use of these epigenetic editing tools, one of them is the delivery efficiency, which is very low mainly in vivo and in species that are not model organisms. Additional limitations are the off-target effects that refer to the effects produced in sequences that are very similar to the targeted sequences as well as to the effects produced in the adjacent sequences to the targeted ones (Uddin et al. 2020). One more limitation is that the dCas9 enzyme and its orthologs produce an immune response; it has been found that more than 50% of patients subjected to gene therapy by Cas9 presented anti-Cas9 antibodies (Charlesworth et al. 2019). Moreover, viral vectors are a highly efficient delivery method frequently used in epigenome editing; however, it was found that adenoviral vectors could induce very strong immune responses when used for gene therapy in humans, meaning that the use of the dCas9-epigenetic effector system as a therapeutic tool is not ready to be delivered into humans without considering undesirable effects.

5.9. References

Angarola, B.L. and Anczuków, O. (2021). Splicing alterations in healthy aging and disease. *Wiley Interdiscip Rev RNA*, 12(4), e1643.

Artsimovitch, I., Vassylyeva, M.N., Svetlov, D., Svetlov, V., Perederina, A., Igarashi, N., Matsugaki, N., Wakatsuki, S., Tahirov, T.H., Vassylyev, D.G. (2005). Allosteric modulation of the RNA polymerase catalytic reaction is an essential component of transcription control by rifamycins. *Cell*, 122(3), 351–363.

Azevedo Portilho, N., Saini, D., Hossain, I., Sirois, J., Moraes, C., Pastor, W.A. (2021). The DNMT1 inhibitor GSK-3484862 mediates global demethylation in murine embryonic stem cells. *Epigenet Chromatin*, 14(1), 56.

Banaszynski, L.A., Liu, C.W., Wandless, T.J. (2005). Characterization of the FKBP.rapamycin. FRB ternary complex. *Journal of the American Chemical Society*, 126(13), 4615–4621.

Bashtrykov, P. and Jeltsch, A. (2016). Epigenome editing in the brain. In *Neuroepigenomics in Aging and Disease*, Delgado-Morales, R. (ed.). Springer, Berlin, 409–424.

Becirovic, E. (2022). Maybe you can turn me on: CRISPRa-based strategies for therapeutic applications. *Cellular and Molecular Life Sciences*, 69(2), 130.

Bickmore, W.A. and van Steensel, B. (2013). Genome architecture: Domain organization of interphase chromosomes. *Cell*, 152(6), 1260–1284.

Bojadzic, D., Alcazar, O., Chen, J., Chuang, S.-T., Condor Capcha, J.M., Shehadeh, L.A., Buchwald, P. (2021). Small-molecule inhibitors of the coronavirus spike: ACE2 protein–protein interaction as blockers of viral attachment and entry for SARS-CoV-2. *ACS Infectious Diseases*, 6(6), 1519–1534.

Brocken, D.J.W., Tark-Dame, M., Dame, R.T. (2018). dCas9: A versatile tool for epigenome editing. *Current Issues in Molecular Biology*, 26, 15–32.

Brown, S.D., Sander, K.B., Wu, C.-W., Guss, A.M. (2015). *Clostridium thermocellum*: Engineered for the production of bioethanol. In *Direct Microbial Conversion of Biomass to Advanced Biofuels*, Himmel, M.E. (ed.). Elsevier, London/New York, 321–333.

Brueckner, B., Garcia Boy, R., Siedlecki, P., Musch, T., Kliem, H.C., Zielenkiewicz, P., Suhai, S., Wiessler, M., Lyko, F. (2005). Epigenetic reactivation of tumor suppressor genes by a novel small-molecule inhibitor of human DNA methyltransferases. *Cancer Research*, 65(14), 6305–6311.

Ceccaldi, A., Rajavelu, A., Champion, C., Rampon, C., Jurkowska, R., Jankevicius, G., Sénamaud-Beaufort, C., Ponger, L., Gagey, N., Dali Ali, H. et al. (2011). C5-DNA methyltransferase inhibitors: from screening to effects on zebrafish embryo development. *ChemBioChem*, 12(9), 1336–1345.

Charlesworth, C.T., Deshpande, P.S., Dever, D.P., Camarena, J., Lemgart, V.T., Cromer, M.K., Vakulskas, C.A., Collingwood, M.A., Zhang, L., Bode, N.M. et al. (2019). Identification of preexisting adaptive immunity to Cas9 proteins in humans. *Nature Medicine*, 25(2), 249–254.

Chatterjee, S., Mizar, P., Cassel, R., Neidl, R., Selvi, B.R., Mohankrishna, D.V., Vedamurthy, B.M., Schneider, A., Bousiges, O. and Mathis, C. et al. (2013). A novel activator of CBP/p300 acetyltransferases promotes neurogenesis and extends memory duration in adult mice. *Journal of Neuroscience*, 33(26), 10698.

Chavez, A., Tuttle, M., Pruitt, B.W., Ewen-Campen, B., Chari, R., Ter-Ovanesyan, D., Haque, S.J., Cecchi, R.J., Kowal, E.J., Buchthal, J. et al. (2016). Comparison of Cas9 activators in multiple species. *Nature Methods*, 13(6), 563–566.

Chen, S., Do Jeong, T., Zhang, Q., Yao, S., Yan, F., Peters Eric, C., Schöler Hans, R., Schultz Peter, G., Ding, S. (2006). Self-renewal of embryonic stem cells by a small molecule. *Proceedings of the National Academy of Sciences*, 103(46), 16266–16261.

Chiappinelli, K.B., Strissel, P.L., Desrichard, A., Li, H., Henke, C., Akman, B., Hein, A., Rote, N.S., Cope, L.M. and Snyder, A. et al. (2016). Inhibiting DNA methylation causes an interferon response in cancer via dsRNA including endogenous retroviruses. *Cell*, 162(5), 974–986.

Choi, S.Y. and Woo, H.M. (2020). CRISPRi-dCas12a: A dCas12a-Mediated CRISPR interference for repression of multiple genes and metabolic engineering in cyanobacteria. *ACS Synthetic Biology*, 9(9), 2351–2361.

Choudhury, S.R., Cui, Y., Lubecka, K., Stefanska, B., Irudayaraj, J. (2016). CRISPR-dCas9 mediated TET1 targeting for selective DNA demethylation at BRCA1 promoter. *Oncotarget*, 6(29), 46545–46556.

Chua, G.N.L., Wassarman, K.L., Sun, H., Alp, J.A., Jarczyk, E.I., Kuzio, N.J., Bennett, M.J., Malachowsky, B.G., Kruse, M., Kennedy, A.J. (2019). Cytosine-based TET enzyme inhibitors. *ACS Medicinal Chemistry Letters*, 10(2), 180–185.

Conde-Ceide, S., Alcázar, J., de Diego, S.A.A., López, S., Martín-Martín, M.L., Martínez-Viturro, C.M., Pena, M.A., Tong, H.M., Lavreysen, H. and Bridges, T.M. et al. (2016). Preliminary investigation of 6,6-dihydropyrazolo[1,5-a]pyrazin-4-one derivatives as a novel series of mGlu5 receptor positive allosteric modulators with efficacy in preclinical models of schizophrenia. *Bioorganic & Medicinal Chemistry Letters*, 26(2), 429–434.

Cong, L., Ran, F.A., Cox, D., Lin, S., Barretto, R., Habib, N., Hsu, P.D., Wu, X., Jiang, W., Marraffini, L.A. et al. (2013). Multiplex genome engineering using CRISPR/Cas systems. *Science*, 339(6121), 819–823.

Cox, D.B.T., Gootenberg, J.S., Abudayyeh, O.O., Franklin, B., Kellner, M.J., Joung, J., Zhang, F. (2016). RNA editing with CRISPR-Cas13. *Science*, 358(6366), 1019–1026.

Dorr, P., Westby, M., Dobbs, S., Griffin, P., Irvine, B., Macartney, M., Mori, J., Rickett, G., Smith-Burchnell, C., Napier, C. et al. (2005). Maraviroc (UK-427,857), a potent, orally bioavailable, and selective small-molecule inhibitor of chemokine receptor CCR5 with broad-spectrum anti-human immunodeficiency virus type 1 activity. *Antimicrobial Agents and Chemotherapy*, 49(11), 4721–4732.

Duhl, D.M.J., Vrieling, H., Miller, K.A., Wolff, G.L., Barsh, G.S. (1994). Neomorphic agouti mutations in obese yellow mice. *Nature Genetics*, 8(1), 59–65.

Dusadeemeelap, C., Rojasawasthien, T., Matsubara, T., Kokabu, S., Addison, W.N. (2022). Inhibition of TET-mediated DNA demethylation suppresses osteoblast differentiation. *The FASEB Journal*, 36(2), e22153.

Falkenberg, K.J. and Johnstone, R.W. (2014). Histone deacetylases and their inhibitors in cancer, neurological diseases and immune disorders. *Nature Reviews Drug Discovery*, 13(9), 663–691.

Finnin, M.S., Donigian, J.R., Cohen, A., Richon, V.M., Rifkind, R.A., Marks, P.A., Breslow, R., Pavletich, N.P. (1999). Structures of a histone deacetylase homologue bound to the TSA and SAHA inhibitors. *Nature*, 401(6749), 188–193.

Gemberling, M.P., Siklenka, K., Rodriguez, E., Tonn-Eisinger, K.R., Barrera, A., Liu, F., Kantor, A., Li, L., Cigliola, V., Hazlett, M.F. et al. (2021). Transgenic mice for in vivo epigenome editing with CRISPR-based systems. *Nature Methods*, 18(8), 965–964.

Gilbert, L.A., Larson, M.H., Morsut, L., Liu, Z., Brar, G.A., Torres, S.E., Stern-Ginossar, N., Brandman, O., Whitehead, E.H., Doudna, J.A. et al. (2013). CRISPR-mediated modular RNA-guided regulation of transcription in eukaryotes. *Cell*, 154(2), 442–451.

Goell, J.H. and Hilton, I.B. (2021). CRISPR/Cas-based epigenome editing: Advances, applications, and clinical utility. *Trends Biotechnology*, 39(6), 668–691.

Gomez, J.A., Beitnere, U., Segal, D.J. (2019). Live-animal epigenome editing: Convergence of novel techniques. *Trend in Genetics*, 35(6), 526–541.

Gordon, D.E., Jang, G.M., Bouhaddou, M., Xu, J., Obernier, K., White, K.M., O'Meara, M.J., Rezelj, V.V., Guo, J.Z., Swaney, D.L, et al. (2020). A SARS-CoV-2 protein interaction map reveals targets for drug repurposing. *Nature*, 583(6816), 459–468.

Guan, Y., Tiwari, A.D., Phillips, J.G., Hasipek, M., Grabowski, D.R., Pagliuca, S., Gopal, P., Kerr, C.M., Adema, V., Radivoyevitch, T. et al. (2021). A therapeutic strategy for preferential targeting of TET2-Mutant and TET dioxygenase–deficient cells in myeloid neoplasms. *Blood Cancer Discovery*, 2(2), 146–161.

Halby, L., Champion, C., Sénamaud-Beaufort, C., Ajjan, S., Drujon, T., Rajavelu, A., Ceccaldi, A., Jurkowska, R., Lequin, O., Nelson, W.G. et al. (2012). Rapid synthesis of new DNMT inhibitors derivatives of procainamide *ChemBioChem*, 13(1), 156–165.

Halby, L., Menon, Y., Rilova, E., Pechalrieu, D., Masson, V., Faux, C., Bouhlel, M.A., David-Cordonnier, M.H., Novosad, N., Aussagues, Y. et al. (2016). Rational design of bisubstrate-type analogs as inhibitors of DNA methyltransferases in cancer cells. *Journal of Medicinal Chemistry*, 60(11), 4665–4669.

Hilton, I.B., DIppolito, A.M., Vockley, C.M., Thakore, P.I., Crawford, G.E., Reddy, T.E., Gersbach, C.A. (2015). Epigenome editing by a CRISPR-Cas9-based acetyltransferase activates genes from promoters and enhancers. *Biotechnol*, 33(5), 510–516.

Horsthemke, B. (2006). Epimutations in human disease. *Current Topics in Microbiology and Immunology*, 310, 45–59.

Hu, Y., Feng, Y., Zhang, L., Jia, Y., Cai, D., Qian, S.B., Du, M., Zhao, R. (2020). GR-mediated FTO transactivation induces lipid accumulation in hepatocytes via demethylation m6A on lipogenic mRNAs. *RNA Biology*, 16(6), 930–942.

Huang, Y.H., Su, J., Lei, Y., Brunetti, L., Gundry, M.C., Zhang, X., Jeong, M., Li, W., Goodell, M.A. (2016). DNA epigenome editing using CRISPR-Cas SunTag-directed DNMT3A. *Genome Biol.*, 18(1), 166.

Hutvagner, G. (2005). Small RNA asymmetry in RNAi: Function in RISC assembly and gene regulation *FEBS Letters*, 569(26), 5850–5856.

Ji, W., Lee, D., Wong, E., Dadlani, P., Dinh, D., Huang, V., Kearns, K., Teng, S., Chen, S., Haliburton, J. et al. (2014). Specific gene repression by CRISPRi system transferred through bacterial conjugation *ACS Synthetic Biology*, 3(12), 929–931.

Jones, P.A. and Taylor, S.M. (1980). Cellular differentiation, cytidine analogs and DNA methylation. *Cell*, 20(1), 85–93.

Kawano, F., Suzuki, H., Furuya, A., Sato, M. (2015). Engineered pairs of distinct photoswitches for optogenetic control of cellular proteins. *Nature Communications*, 6, 6256.

Kearns, N.A., Pham, H., Tabak, B., Genga, R.M., Silverstein, N.J., Garber, M., Maehr, R. (2015). Functional annotation of native enhancers with a Cas9-histone demethylase fusion. *Nature Methods*, 12(5), 401–403.

Kelly Judith, A., Moews Paul, C., Knox James, R., Frère, J.-M., Ghuysen, J.-M. (1982). Penicillin target enzyme and the antibiotic binding site. *Science*, 218(4561), 469–481.

Konermann, S., Brigham, M.D., Trevino, A., Hsu, P.D., Heidenreich, M., Cong, L., Platt, R.J., Scott, D.A., Church, G.M., Zhang, F. (2013). Optical control of mammalian endogenous transcription and epigenetic states. *Nature*, 500(6463), 462–466.

Koonin, E.V. (2005). Orthologs, paralogs, and evolutionary genomics. *Annual Review of Genetics*, 39, 309–338.

Koonin, E.V., Makarova, K.S., Zhang, F. (2016). Diversity, classification and evolution of CRISPR-Cas systems. *Current Opinion in Microbiology*, 36, 66–68.

Kundakovic, M., Chen, Y., Costa, E., Grayson, D.R. (2006). DNA methyltransferase inhibitors coordinately induce expression of the human reelin and glutamic acid decarboxylase 66 genes. *Molecular Pharmacology*, 61(3), 644–653.

Kwon, D.Y., Zhao, Y.T., Lamonica, J.M., Zhou, Z. (2016). Locus-specific histone deacetylation using a synthetic CRISPR-Cas9-based HDAC. *Nature Communications*, 8, 15315.

Laufer, B.I. and Singh, S.M. (2015). Strategies for precision modulation of gene expression by epigenome editing: An overview. *Epigenetics & Chromatin*, 8, 34.

Lawrence, M., Daujat, S., Schneider, R. (2016). Lateral thinking: How histone modifications regulate gene expression. *Trend in Genetics*, 32(1), 42–56.

Lei, Y., Zhang, X., Su, J., Jeong, M., Gundry, M.C., Huang, Y.H., Zhou, Y., Li, W., Goodell, M.A. (2016). Targeted DNA methylation in vivo using an engineered dCas9-MQ1 fusion protein. *Nature Communications*, 8, 16026.

Li, E. and Zhang, Y. (2014). DNA methylation in mammals. *Cold Spring Harbor Perspectives in Biology*, 6(5), a019133.

Li, J., Ding, Y., Zheng, L. (2014). Histone-mediated transgenerational epigenetics. In *Transgenerational Epigenetics*, Tollefsbol, T. (ed.). Elsevier, Amsterdam, 86–103.

Li, J., Mahata, B., Escobar, M., Goell, J., Wang, K., Khemka, P., Hilton, I.B. (2021). Programmable human histone phosphorylation and gene activation using a CRISPR/Cas9-based chromatin kinase. *Nature Communications*, 12(1), 896.

Lin, J.L., Zhu, J., Wheeldon, I. (2016). Synthetic protein scaffolds for biosynthetic pathway colocalization on lipid droplet membranes. *ACS Synthetic Biology*, 6(8), 1534–1544.

Lin, J.L., Ekas, H., Deaner, M., Alper, H.S. (2019). CRISPR-PIN: Modifying gene position in the nucleus via dCas9-mediated tethering. *Synthetic and Systems Biotechnology*, 4(2), 63–68.

Liu, X.S., Wu, H., Ji, X., Stelzer, Y., Wu, X., Czauderna, S., Shu, J., Dadon, D., Young, R.A., Jaenisch, R. (2016). Editing DNA methylation in the mammalian genome. *Cell*, 166(1), 233–246.e16.

Liu, X.-M., Zhou, J., Mao, Y., Ji, Q., Qian, S.-B. (2019). Programmable RNA N6-methyladenosine editing by CRISPR-Cas9 conjugates. *Nature Chemical Biology*, 15(9), 865–861.

Lopez, M., Halby, L., Arimondo, P.B. (2016a). DNA methyltransferase inhibitors: Development and applications. *Advances in Experimental Medicine and Biology*, 945, 431–463.

Lopez, M., Leroy, M., Etievant, C., Ausseil, F., Gagnon, A., Arimondo, P. (2016b). Drug discovery methods. In *Drug Discovery in Cancer Epigenetics*, Egger, G. and Arimondo, P. (eds). Academic Press, Boston, 63–95.

Lu, H., Zhou, Q., He, J., Jiang, Z., Peng, C., Tong, R., Shi, J. (2020). Recent advances in the development of protein–protein interactions modulators: Mechanisms and clinical trials. *Signal Transduction and Targeted Therapy*, 5(1), 213.

Luck, K., Kim, D.K., Lambourne, L., Spirohn, K., Begg, B.E., Bian, W., Brignall, R., Cafarelli, T., Campos-Laborie, F.J., Charloteaux, B. et al. (2020). A reference map of the human binary protein interactome. *Nature*, 580(6803), 402–408.

Luger, K., Mäder, A.W., Richmond, R.K., Sargent, D.F., Richmond, T.J. (1996). Crystal structure of the nucleosome core particle at 2.8 A resolution. *Nature*, 389(6648), 251–260.

Luviano, N., Lopez, M., Gawehns, F., Chaparro, C., Arimondo, P.B., Ivanovic, S., David, P., Verhoeven, K., Cosseau, C., Grunau, C. (2021). The methylome of biomphalaria glabrata and other mollusks: Enduring modification of epigenetic landscape and phenotypic traits by a new DNA methylation inhibitor. *Epigenetics & Chromatin*, 14(1), 48.

McCabe, M.T., Mohammad, H.P., Barbash, O., Kruger, R.G. (2016). Targeting histone methylation in cancer. *Journal of Cancer*, 23(5), 292–301.

McGee-Lawrence, M.E. and Westendorf, J.J. (2011). Histone deacetylases in skeletal development and bone mass maintenance. *Gene*, 474(1/2), 1–11.

Michalak, E.M., Burr, M.L., Bannister, A.J., Dawson, M.A. (2019). The roles of DNA, RNA and histone methylation in ageing and cancer. *Nature Reviews Molecular Cell Biology*, 20(10), 563–589.

Moreno, A.M., Fu, X., Zhu, J., Katrekar, D., Shih, Y.R.V., Marlett, J., Cabotaje, J., Tat, J., Naughton, J., Lisowski, L. et al. (2018). In situ gene therapy via AAV-CRISPR-Cas9-mediated targeted gene regulation. *Molecular Therapy*, 26(6), 1818–1826.

Morgan, S.L., Mariano, N.C., Bermudez, A., Arruda, N.L., Wu, F., Luo, Y., Shankar, G., Jia, L., Chen, H., Hu, J.F. et al. (2016). Manipulation of nuclear architecture through CRISPR-mediated chromosomal looping. *Nature Communications*, 8, 15993.

Murakami, Y. (2013). Heterochromatin and euchromatin. In *Encyclopedia of Systems Biology*, Dubitzky, W., Wolkenhauer, O., Cho, K.-H., Yokota, H. (eds). Springer, New York, 881–884.

Nardella, F., Halby, L., Hammam, E., Erdmann, D., Cadet-Daniel, V., Peronet, R., Ménard, D., Witkowski, B., Mecheri, S., Scherf, A. et al. (2020). DNA methylation bisubstrate inhibitors are fast-acting drugs active against artemisinin-resistant plasmodium falciparum parasites. *ACS Central Science*, 6(1), 16–21.

Nelson, E.D., Kavalali, E.T., Monteggia, L.M. (2008). Activity-dependent suppression of miniature neurotransmission through the regulation of DNA methylation. *Journal of Neuroscience*, 28(2), 395–406.

O'Geen, H., Ren, C., Nicolet, C.M., Perez, A.A., Halmai, J., Le, V.M., Mackay, J.P., Farnham, P.J., Segal, D.J. (2016). dCas9-based epigenome editing suggests acquisition of histone methylation is not sufficient for target gene repression. *Nucleic Acids Research*, 45(16), 9901–9916.

O'Geen, H., Bates, S.L., Carter, S.S., Nisson, K.A., Halmai, J., Fink, K.D., Rhie, S.K., Farnham, P.J., Segal, D.J. (2019). Ezh2-dCas9 and KRAB-dCas9 enable engineering of epigenetic memory in a context-dependent manner. *Epigenet Cromatin*, 12(1), 1–20.

Obokata, H. and Vacanti, C.A. (2014). Stem cells in tissue engineering. In *Principles of Tissue Engineering*, Lanza, R., Langer, R., Vacanti, J.P. (eds). Elsevier, Amsterdam, 595–608.

Pappalardi, M.B., Keenan, K., Cockerill, M., Kellner, W.A., Stowell, A., Sherk, C., Wong, K., Pathuri, S., Briand, J., Steidel, M. et al. (2021). Discovery of a first-in-class reversible DNMT1-selective inhibitor with improved tolerability and efficacy in acute myeloid leukemia. *Nature Cancer*, 2(10), 1002–1016.

Pares, S., Mouz, N., Pétillot, Y., Hakenbeck, R., Dideberg, O. (1996). X-ray structure of *Streptococcus pneumoniae* PBP2x, a primary penicillin target enzyme. *Nature Structural & Molecular Biology*, 3(3), 284–289.

Pechalrieu, D., Etievant, C., Arimondo, P.B. (2016). DNA methyltransferase inhibitors in cancer: From pharmacology to translational studies. *Biochemical Pharmacology*, 129, 1–13.

Pechalrieu, D., Dauzonne, D., Arimondo, P.B., Lopez, M. (2020). Synthesis of novel 3-halo-3-nitroflavanones and their activities as DNA methyltransferase inhibitors in cancer cells. *European Journal of Medicinal Chemistry*, 186, 111829.

Policarpi, C., Dabin, J., Hackett, J.A. (2021). Epigenetic editing: Dissecting chromatin function in context. *Bioessays*, 43(5), e2000316.

Qi, L.S., Larson, M.H., Gilbert, L.A., Doudna, J.A., Weissman, J.S., Arkin, A.P., Lim, W.A. (2013). Repurposing CRISPR as an RNA-guided platform for sequence-specific control of gene expression. *Cell*, 152(5), 1163–1183.

Robertson, K.D. (2005). DNA methylation and human disease. *Nature Reviews Genetics*, 6(8), 596–610.

Ronald, P.C. (1996). The molecular basis of disease resistance in rice. *Plant Molecular Biology*, 35(1/2), 169–186.

Roquis, D., Rognon, A., Chaparro, C., Boissier, J., Arancibia, N., Cosseau, C., Parrinello, H., Grunau, C. (2016). Frequency and mitotic heritability of epimutations in *Schistosoma mansoni*. *Molecular Ecology*, 25(8), 1641–1658.

Rossetto, D., Avvakumov, N., Côté, J. (2012). Histone phosphorylation: A chromatin modification involved in diverse nuclear events. *Epigenetics*, 6(10), 1098–1108.

Santi, D.V., Norment, A., Garrett, C.E. (1984). Covalent bond formation between a DNA-cytosine methyltransferase and DNA containing 5-azacytosine. *Proceedings of the National Academy of Sciences*, 81(22), 6993–6997.

Sharma, S. and Aazmi, O. (2019). Basics of epigenetics: It is more than simple changes in sequence that govern gene expression. In *Prognostic Epigenetics*, Sharma, S. (ed.). Elsevier, Amsterdam, 1–19.

Sheik Amamuddy, O., Veldman, W., Manyumwa, C., Khairallah, A., Agajanian, S., Oluyemi, O., Verkhivker, G.M., Tastan Bishop, Ö. (2020). Integrated computational approaches and tools for allosteric drug discovery. *International Journal of Molecular Sciences*, 21(3), 846.

Shen, L. and Yu, H. (2021). Epitranscriptome engineering in crop improvement. *Molecular Plant*, 14(9), 1418–1420.

Shi, Y., Zhang, H., Huang, S., Yin, L., Wang, F., Luo, P., Huang, H. (2022). Epigenetic regulation in cardiovascular disease: Mechanisms and advances in clinical trials. *Signal Transduction and Targeted Therapy*, 7(1), 1–28.

Siegfried, Z. and Simon, I. (2010). DNA methylation and gene expression, *Wiley Interdisciplinary Reviews: Systems Biology and Medicine*, 2(3), 362–361.

van Steensel, B. and Belmont, A.S. (2016). Lamina-associated domains: Links with chromosome architecture, heterochromatin, and gene repression. *Cell*, 169(5), 680–691.

Struk, S., Jacobs, A., Sánchez Martín-Fontecha, E., Gevaert, K., Cubas, P., Goormachtig, S. (2019). Exploring the protein–protein interaction landscape in plants. *Plant, Cell & Environment*, 42(2), 386–409.

Sun, J., Wang, Y., Cui, W., Lou, Y., Sun, G., Zhang, D., Miao, L. (2017). Role of epigenetic histone modifications in diabetic kidney disease involving renal fibrosis. *Journal of Diabetes Research*.

Suzuki, T., Tanaka, R., Hamada, S., Nakagawa, H., Miyata, N. (2010). Design, synthesis, inhibitory activity, and binding mode study of novel DNA methyltransferase 1 inhibitors. *Bioorganic & Medicinal Chemistry Letters*, 20(3), 1124–1126.

Tajbakhsh, S., Rocancourt, D., Cossu, G., Buckingham, M. (1996). Redefining the genetic hierarchies controlling skeletal myogenesis: Pax-3 and Myf-5 act upstream of MyoD. *Cell*, 89(1), 126–138.

Tanenbaum, M.E., Gilbert, L.A., Qi, L.S., Weissman, J.S., Vale, R.D. (2014). A proteintagging system for signal amplification in gene expression and fluorescence imaging. *Cell*, 159(3), 635–646.

Tang, T., Han, Y., Wang, Y., Huang, H., Qian, P. (2021). Programmable system of Cas13-mediated RNA modification and its biological and biomedical applications. *Front Cell & Developmental Biology*, 9, 666586.

Uddin, F., Rudin, C.M., Sen, T. (2020). CRISPR gene therapy: Applications, limitations, and implications for the future. *Frontiers in Oncology*, 10, 1386.

Wang, H., Xu, X., Nguyen, C.M., Liu, Y., Gao, Y., Lin, X., Daley, T., Kipniss, N.H., La Russa, M., Qi, L.S. (2018). CRISPR-Mediated programmable 3D genome positioning and nuclear organization. *Cell*, 165(5), 1405–1416.e14.

Wei, Y., Lang, J., Zhang, Q., Yang, C.R., Zhao, Z.A., Zhang, Y., Du, Y., Sun, Y. (2019). DNA methylation analysis and editing in single mammalian oocytes. *Proceedings of the National Academy of Sciences*, 116(20), 9883–9892.

Wilson, C., Chen, P.J., Miao, Z., Liu, D.R. (2020). Programmable m6A modification of cellular RNAs with a Cas13-directed methyltransferase. *Nature Biotechnology*, 38(12), 1431–1440.

Wright, A.V., Sternberg, S.H., Taylor, D.W., Staahl, B.T., Bardales, J.A., Kornfeld, J.E., Doudna, J.A. (2015). Rational design of a split-Cas9 enzyme complex. *Proceedings of the National Academy of Sciences*, 112(10), 2984–2989.

Xu, X., Tan, X., Tampe, B., Wilhelmi, T., Hulshoff, M.S., Saito, S., Moser, T., Kalluri, R., Hasenfuss, G., Zeisberg, E.M. et al. (2018). High-fidelity CRISPR/Cas9- based gene-specific hydroxymethylation rescues gene expression and attenuates renal fibrosis. *Nature Communications*, 9(1), 3509.

Yang, L.Z., Wang, Y., Li, S.Q., Yao, R.W., Luan, P.F., Wu, H., Carmichael, G.G., Chen, L.L. (2019). Dynamic imaging of RNA in living cells by CRISPR-Cas13 Systems. *Molecular Cell*, 66(6), 981–996.e6.

Yin, H., Kanasty, R.L., Eltoukhy, A.A., Vegas, A.J., Dorkin, J.R., Anderson, D.G. (2014). Non-viral vectors for gene-based therapy. *Nature Reviews Genetics*, 15(8), 541–555.

Yoo, J., Choi, S., Medina-Franco, J.L. (2013). Molecular modeling studies of the novel inhibitors of DNA methyltransferases SGI-1026 and CBC12: Implications for the mechanism of inhibition of DNMTs. *PLoS One*, 8(4), e62152.

Yu, L. and Marchisio, M.A. (2021). Saccharomyces cerevisiae synthetic transcriptional networks harnessing dCas12a and type VA anti-CRISPR proteins. *ACS Synthetic Biology*, 10(4), 860–883.

Zalatan, J.G., Lee, M.E., Almeida, R., Gilbert, L.A., Whitehead, E.H., La Russa, M., Tsai, J.C., Weissman, J.S., Dueber, J.E., Qi, L.S. et al. (2015). Engineering complex synthetic transcriptional programs with CRISPR RNA scaffolds. *Cell*, 160(1/2), 339–350.

Zeineddine, D., Hammoud, A.A., Mortada, M., Boeuf, H. (2014). The Oct4 protein: More than a magic stemness marker. *American Journal of Stem Cells*, 3(2), 64–82.

Zetsche, B., Volz, S.E., Zhang, F. (2015). A split-Cas9 architecture for inducible genome editing and transcription modulation. *Nature Biotechnology*, 33(2), 139–142.

Zhang, J., Zhang, S., Wang, Y., Cheng, H., Hao, L., Zhai, Y., Zhang, Z., An, X., Ma, X., Zhang, X. et al. (2016). Effect of TET inhibitor on bovine parthenogenetic embryo development. *PLOS One*, 12(12), e0189542.

Zhang, Y., Ozono, S., Yao, W., Tobiume, M., Yamaoka, S., Kishigami, S., Fujita, H., Tokunaga, K. (2019). CRISPR-mediated activation of endogenous BST-2/tetherin expression inhibits wild-type HIV-1 production. *Scientific Reports*, 9(1), 313.

Zhang, J., Li, J., Zhu, Y., Miao, Z., Tian, Y. (2020). Forced running exercise mitigates radiation-induced cognitive deficits via regulated DNA hydroxymethylation. *Epigenomics*, 12(5), 385–396.

6

Epigenetics and Stress

Raphaëlle CHAIX[1] and Natacha BIES ETHEVE[2]
[1] *Éco-anthropologie (UMR 7206), Muséum National d'Histoire Naturelle, CNRS, Université Paris Cité, France*
[2] *LGDP, Université de Perpignan Via Domitia, Perpignan, France*

Stress is inherent to life. It is triggered by any situation (environment, experience, perception) that takes the organism out of its homeostatic balance and requires the activation of specific response circuits (Sapolsky 1994). For animals, for example, this means regulating blood sugar levels as the day progresses, maintaining body temperature regardless of the climatic conditions, and also, for all organisms, resisting a virus or protecting oneself from a predator. Stress circuits help maintain this balance, and are constantly active, whether the individual is at rest or facing life-threatening stress factors. Plants, which are sessile organisms, face fluctuating biotic and abiotic environmental factors and must constantly adapt to survive. In this chapter, the concrete effects of different environmental factors on epigenetic marks are presented. Section 6.1 on humans and animals mainly focuses on the influence of psychosocial stress, while section 6.2 describes the impact of various biotic and abiotic stresses on the epigenetic profiles of plants.

6.1. Impact of environmental constraints on epigenetic marks in animals and humans

In modern society, psychosocial stressors are particularly common, whether they are unfavorable socio-economic conditions, dysfunctional emotional environments or, more generally, situations that the individual feels he or she cannot cope with. However, whether stress factors are psychosocial or physical, the same response

Epigenetics in Ecology and Evolution,
coordinated by Christoph GRUNAU and Stéphane MAURY.

circuits are activated. In vertebrates, when signals from the environment or the body are perceived as threatening to the organism's equilibrium by limbic brain structures such as the hippocampus and amygdala, this triggers the hypothalamic–pituitary–adrenal (HPA) axis, which induces the release of cortisol from the adrenal glands, and the sympathetic nervous system, which releases adrenaline. Cortisol and adrenalin are "stress hormones": they have multiple effects on the body to enable it to cope with a stressful situation, including mobilization of energy resources, acceleration of cardiac rhythm to deliver resources to the muscles and suspension of digestive and reproductive functions. Once the stressful situation has passed, cortisol returns to the brain and, by binding to its receptor in the hippocampus, blocks its own secretion by negative feedback. The organism can resume the normal course of life. This well-oiled mechanism is particularly adaptive, especially when it enables the organism to "save itself".

However, exposure to stress can also be harmful to an individual's health, particularly when stress factors become chronic. Numerous studies have assessed the effect of socio-economic level and poverty, which are recognized chronic stress factors (Snyder-Mackler et al. 2020). They have shown that these stresses induce premature aging: in the United States, the life expectancy of individuals at the bottom of the socio-economic ladder is 10 years lower than that of those at the top (Chetty et al. 2016). In Great Britain, among civil servants, mortality and morbidity increase progressively from the lowest to the highest grade (McEwen 1998). Studies in non-human primates have shown that belonging to the most stressful dominance rank (which is usually the lowest but not always) has a harmful impact on health, with cardiovascular, immune, reproductive and neurological effects (Sapolsky 2005). Similarly, children exposed to various forms of early adversity (emotional, physical, sexual abuse, abandonment, poverty, etc.) are more vulnerable as adults to many age-related chronic diseases, such as cardiovascular, metabolic and autoimmune diseases, mood disorders (anxiety, depression, and also psychoses) and also have a higher risk of premature mortality (Snowdon et al. 1989; McEwen 1998; Kittleson et al. 2006; Miller et al. 2011; Felitti et al. 2019). These effects are shared with non-human primates (Tung et al. 2016) and rodents (Miller et al. 2011).

6.1.1. *From psychosocial stress to epigenetic profiles*

Despite this string of studies highlighting the influence of stress on health, the molecular pathways underlying these effects remain poorly understood (Juster et al. 2010; Miller et al. 2011; Crosswell and Lockwood 2020). A growing body of research suggests that epigenetics plays a crucial role here. For example, it has been shown that individuals exposed to early-life stress (maternal separation in rodents, physical, sexual and emotional abuse in humans) carry methylation in their hippocampus at the promoter level of the gene encoding the cortisol receptor (Weaver et al. 2004; Turecki and Meaney 2016). As a result, this receptor is less expressed. Therefore, when the cortisol released by the HPA axis during a stressful situation reaches the brain, this lower expression of its receptor limits the negative feedback from the HPA axis: the latter remains active and the individual's state of hypervigilance persists, even though the stressful situation is over. The individual is thus, throughout his or her life, more prone to psychological stress and anxiety than average and at greater risk of psychopathologies (Heim et al. 2013). This epigenetic mechanism is shared by humans, primates and rodents. The epigenetic profile of many other genes expressed in the brain or peripheral tissues, including Avp, FKBP5 and OXTR, is also shaped by early adversity (Borghol et al. 2012; Cunliffe 2016; Houtepen et al. 2016; Non et al. 2016; Fujisawa et al. 2019; Parade et al. 2021).

6.1.2. *An epigenetic clock running faster*

Early adversity not only impacts specific DNA methylation sites. It also influences what is known as the epigenetic clock. This clock refers to the fact that, as life progresses, the epigenome changes: it is thus possible, from the epigenome, to infer the age of the individual (Horvath and Raj 2018). The difference between this epigenetic age and the chronological age provides information on the individual's biological aging rate. If this epigenetic age is higher than the chronological age, for example, if the individual is 38 but his epigenetic age is equal to 42, this means that the rhythm of his epigenetic clock is higher than average; this is associated with a higher risk of mortality and chronic diseases (Alzheimer's, cardiovascular disease, cancer, Parkinson's, etc.) (Horvath and Raj 2018). Early-life stress has been shown to accelerate the pace of this epigenetic clock, whether it is exposure to neighborhood violence (Jovanovic et al. 2017), unfavorable socioeconomic conditions (Fiorito et al. 2017) or sexual abuse (Lawn et al. 2018).

Early life is therefore a particularly sensitive time during which stressful life experiences leave an epigenetic signature (Borghol et al. 2012). However, stress in adulthood also leaves specific epigenetic marks (Borghol et al. 2012) and accelerates the epigenetic clock, as has been shown in African-American cohorts that have been particularly exposed to life stresses and adverse economic conditions (Zannas et al. 2015; Simons et al. 2016). Racial discrimination also accelerates the epigenetic clock (Brody et al. 2016a). These links between chronic psychosocial stress and epigenetic profiles have also been demonstrated in non-human primates, with individuals belonging to the most stressful rank showing differences in terms of epigenetic profiles and regulation of immunity and inflammation genes (Tung et al. 2012; Snyder-Mackler et al. 2016; Lea et al. 2018), as well as an acceleration in the rate of their epigenetic clock (Anderson et al. 2021).

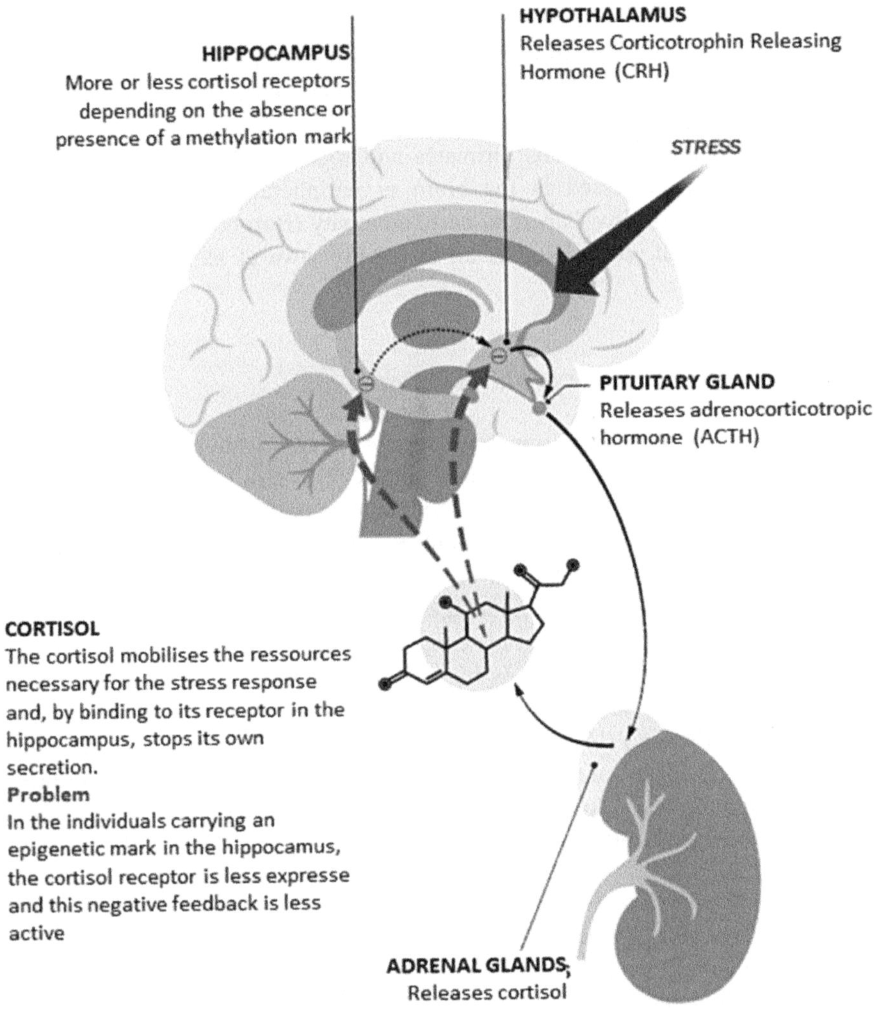

Figure 6.1. *Epigenetic modulation of the HPA axis and reactivity to stress (credit: AFDEC cartographic workshop). For a color version of this figure, see www.iste.co.uk/grunau/epigenetics.zip*

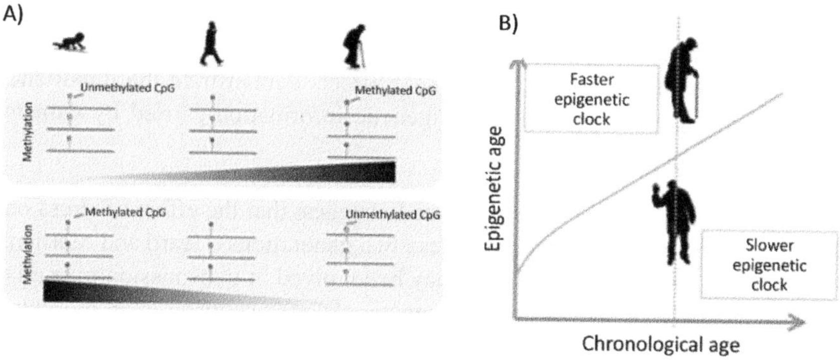

Figure 6.2. *Epigenetic clock . For a color version of this figure, see www.iste.co.uk/grunau/epigenetics.zip*

COMMENT ON FIGURE 6.2.– *A) The epigenome evolves throughout an individual's life. Some CpG sites become increasingly methylated, while others become demethylated. Based on methylation profiles, it is possible to predict an individual's age. This predicted age is known as the epigenetic age. B) The rhythm of this epigenetic clock varies from one individual to another. For some, the epigenetic age is higher than their chronological age, so their epigenetic clock is accelerated compared with the population average: these individuals have an increased risk of chronic disease and mortality. A slower clock is predictive of increased longevity. Numerous studies show that exposure to psychosocial stress accelerates this clock.*

6.1.3. *The intergenerational cost of stress exposure*

The monitoring of pregnant women facing high stress situations, such as during wars or extreme climatic episodes, has also highlighted the influence of stress experienced in utero on the epigenome of the unborn child, both at the cortisol receptor level (Palma-Gudiel et al. 2015) and the level of many other genes of clinical interest (Mulligan 2016; Cao-Lei et al. 2020). More generally, a growing number of studies are highlighting the influence of parental and grandparental experiences on an individual's epigenome. In particular, studies on rodents show that a father's exposure to stress alters the microRNA content of his sperm (Rodgers et al. 2013; Welberg 2013; Gapp et al. 2014), with behavioral repercussions on his offspring, including dysfunction of the HPA axis (Rodgers et al. 2013; Welberg 2013). In addition, the father's exposure to a "distressing" odor influences the behavior of his offspring, who react more to it. They have more neurons specialized in the recognition of this odor, as well as hypomethylation in the gene encoding the receptor for this odor (Dias and Ressler 2014). Certain environmental exposures of pregnant women would also have repercussions on the epigenome of their grandchildren (Serpeloni et al. 2017). These effects demonstrate the importance of intergenerational epigenetics: all the epigenetic information carried by gametes is not entirely reset at fertilization.

On the contrary, there is little evidence to suggest that the effect of stress on the epigenome can be transmitted beyond these two generations (Heard and Martienssen 2014). This being said, the epigenome may be involved in transmission over a larger number of generations, insofar as an epigenetic change occurring in an individual at the neural cell level may, by inducing a behavioral change in him, be reflected in the epigenome of his descendants. This has been demonstrated in rodents. A deficit of maternal behavior in a mother is associated with a specific epigenetic profile in the brain of her offspring, more precisely at the estrogen receptor level located in the hypothalamus. As a result of this epigenetic profile, the offspring themselves show a deficit in maternal behavior. Their offspring will therefore also be likely to carry this same epigenetic profile at the estrogen receptor level, and so on from generation to generation (Champagne 2013). Behaviors can thus be transmitted trans-generationally through a coupling of behavioral and epigenetic mechanisms.

6.1.4. *"De-stress" the epigenome?*

Exposure to stress, particularly psychosocial stress, leaves lasting epigenetic marks on individuals and accelerates their epigenetic clock, leading to increased risk of chronic disease and psychopathology. Is it possible to modulate or even erase these potentially harmful marks on an individual's health? Encouraging studies show that an improvement in socioeconomic level "decelerates" the epigenetic clock (Fiorito et al. 2017). Similarly, in non-human primates, gene expression profiles show reversibility in the event of a change in dominance rank (Tung et al. 2012). In humans, lifestyle interventions can also have a favorable effect on the modulation of epigenetic profiles. This is particularly true of stress reduction techniques. We thus studied the influence of meditation-based methods, comparing a group of meditators (with a minimum of three years of experience) with a group of control individuals. We found that 8 hours of intensive meditation resulted in significant epigenetic profile changes in certain genes involved in metabolism and immune cell aging. The epigenome therefore changes rapidly in response to stress reduction through meditation. We have also demonstrated a longer-term effect: the more experienced meditators are in their practice, the slower their epigenetic clock ticks. Other interventions have also been shown to have a beneficial effect on the epigenome. For example, the harmful effect of racial discrimination on the epigenetic clock rate is limited in families participating in a social support program (Brody et al. 2016a, 2016b). The effect of certain forms of early adversity on epigenetic profiles is limited when mothers demonstrate touch-based behavior (caressing, in particular) with their child (Murgatroyd et al. 2015). Finally, it is possible that "epidrugs" capable of modifying or erasing harmful epigenetic marks will emerge.

6.2. Impact of environmental stress on epigenetic marks in plants

For plants, which are sessile organisms, the environment and growing conditions have a primordial impact. Their only solution is to adapt in order to survive. We now know that living conditions, by modifying epigenetic marks, modulate gene expression and thus the development and adaptation of organisms. A permanent interaction between the living environment and organisms is truly at work. One of the most studied examples in plants is the impact of environmental conditions on the onset of flowering (see the following sections). This regulation is essential, as it must be carried out at the right time to enable reproduction and thus the survival of offspring.

When talking about environmental factors, it is necessary to distinguish between abiotic factors, which depend on chemical or physical criteria, and biotic factors, which depend on other living organisms (predators and pathogens). In the following sections, the tangible effects of various abiotic and biotic factors on epigenetic marks and plant survival are described. Section 6.2.2.5 deals with the impact of epigenetic marks on plant adaptation and plasticity.

6.2.1. *Effects of abiotic factors on epigenetic marks in plants*

6.2.1.1. *Nutrients and epigenetic marks that control growth and survival*

Numerous studies have shown that the composition of soils or growing media affects epigenetic marks in plants. Among the various analyses described are those linked to the quality of available nutrients. A concrete example is the fact that, depending on the nature of the nitrogen source, the degree of DNA methylation in carrot cells is modified. This effect is also observed when squash is grown in vitro, and the DNA methylation modification observed modulates the fate of the cells and allows embryo formation (or not) (Leljak-Levanić et al. 2004).

The impact of culture media on epigenetic marks can also be seen by comparing the resistance to excess or deficient nutrients of wild plants versus plants affected in their epigenetic mark regulation. For example, *Arabidopsis thaliana* mutants altered in the *DDM1* gene, involved in DNA methylation, are less resistant to excess salts (Xu et al. 2015; Bartels et al. 2018). Genes involved in the adapted molecular response to excess salts are, in this case, not correctly methylated, so their expression is incorrectly regulated (Xu et al. 2015). Defects in histone demethylation, observed in plants mutated in the *JMJ15* gene, which encodes a histone demethylase, also lead to greater sensitivity to salts (Shen et al. 2014). In line with this observation, overexpression of the *JMJ15* gene results in greater resistance to salts. These various observations demonstrate that all epigenetic marks, DNA methylation and histone modifications, are controlled by the composition of the nutrient environment and affect plant development and survival. Interestingly, *Arabidopsis* plants mutated in the *HDA6* gene, which encodes a histone deacetylase, are both impaired in their response to various stresses, such as excess salts or water deficiency, as well as in their germination (Shen et al. 2010). Further work shows that the histone deacetylation defect actually impacts the phytohormone Abscissic acid (ABA) regulatory pathway, which is involved in stress response, and also in the control of plant development (Shen et al. 2010).

These examples demonstrate that the nature and quantity of nutrients present in the culture medium modulate epigenetic marks, in order to correctly regulate the expression of genes that will enable plants to adapt to conditions and survive.

6.2.1.2. *Light and epigenetic marks controlling seed germination and flowering*

In addition to the effects of the nature and quantity of nutrients, many other abiotic factors modulate epigenetic marks and, therefore, cell function and the fate of the organism. The most extensively studied effects, described in the following paragraphs, concern the impact of light on the control of germination and flowering.

The vast majority of seeds only germinate in the presence of dark red light. Conversely, light red light inhibits germination. This effect is mediated by specific molecules called phytochromes, which may or may not be activated depending on the quality and intensity of the light. It turns out that various phytochrome targets are proteins involved in the control of epigenetic marks. Examples include histone deacetylases HDA15 and HDA6, as well as histone demethylases LDL1 and LDL2, which are activated by light and participate in the repression of genes controlling germination, such as *DOG* or genes involved in ABA biosynthesis (Zhao et al. 2015; Gu et al. 2019; Longo et al. 2020). Conversely, two other histone demethylases, JMJ20 and JMJ22, also activated by light, enable the induction of genes involved in the biosynthesis pathway of the gibberellic acid plant hormone, which, depending on its concentration, enables germination (Cho et al. 2012). These control pathways, regulated by light and mediated by epigenetic marks, are crucial, as only germination in the presence of light will enable the plant to photosynthesize and, therefore, produce the sugars necessary for growth from water and carbon dioxide (CO_2).

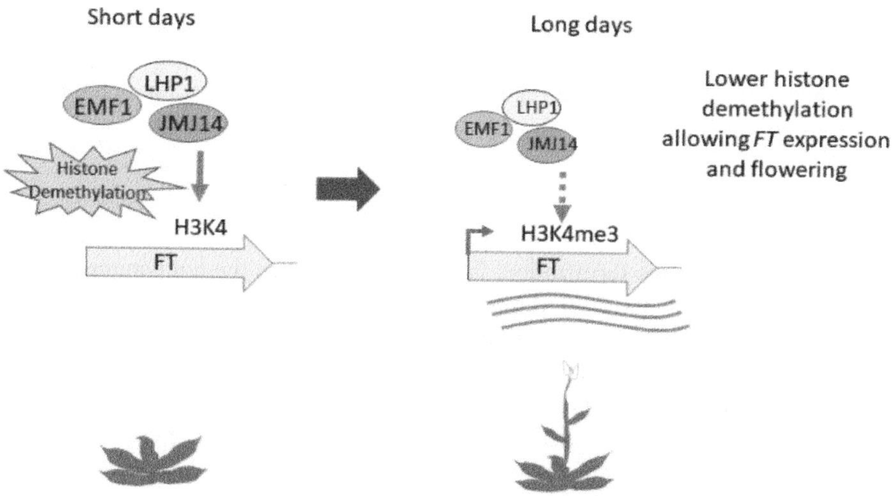

Figure 6.3. *Photoperiod, epigenetics and flowering. For a color version of this figure, see www.iste.co.uk/grunau/epigenetics.zip*

COMMENT ON FIGURE 6.3. – *When days are short, JMJ14, a histone demethylase, interacts with EMF1 and LHP1, to demethylate histone 3 at the K4 position in the FT gene, thereby repressing it. As the days get longer, JMJ14 becomes less effective, so histones become more methylated and FT transcription becomes possible, inducing flowering.*

As well as influencing seed germination, light also regulates the transition from vegetative to reproductive phase in flowering plants. It is, in fact, the photoperiod, in other words, the ratio of light to dark phases, that controls and synchronizes flowering. Some plants flower when there is more sunshine (long photoperiod plants), while others flower when the days are shorter (short photoperiod plants). In *Arabidopsis*, which has a long photoperiod, it has been shown that the degree of histone methylation at the *FT* locus (Flowering locus T) varies with the season and controls flowering. When the days are short, the EMF1 (EMbryonic Flower 1) and LHP1 proteins associate and activate the JMJ14 histone demethylase, which demethylate histones at positions H3K4 at the *FT* locus, leading to its repression (Figure 6.3). As the days lengthen, the EMF1/LHP1/JMJ14 complex will gradually become less active and allow histone methylation and expression of the *FT* locus, which is necessary for flowering (Wang et al. 2014). In line with these observations, *emf1* mutants, in which the *EMF1* gene is non-functional, very rapidly germinate and form a pistil. Under these conditions, the vegetative phase is not induced and the plant moves directly into the reproductive phase (Kim et al. 2012). These observations demonstrate the impact of light on epigenetic marks and, therefore, on gene expression, as well as the fate of the organism, by allowing or preventing reproduction. In addition to the regulation of flowering by light, temperature also modulates epigenetic marks.

6.2.1.3. *Temperature and epigenetic marks that control flowering*

In general, it is a good option for plants to produce their reproductive organs at a time that is favorable for fertilization, in other words, when the temperature is at its mildest. Many plants thus go through a phase known as vernalization, which prevents flowering when the temperature is too low. Here again, control of *FT* locus expression plays a vital role. In addition to the regulation of this locus by epigenetic marks (mentioned above), its transcription can be repressed by FLC transcription factor. The *FLC* gene is expressed when the plant is young and temperatures are low (Figure 6.4). When temperatures become milder, *FLC* becomes less active and can no longer repress *FT*, which then allows flowering to take place (Baulcombe and Dean 2014). Various studies show that, during the first cold spells, the FRIGIDA protein (FRI) activates a methyltransferase and two histone acetyltransferases HAM1 and HAM2 (Histone Acetyltransferase of the Myst family 1 and 2), to induce histone trimethylations (H3K4 me3 and H3K36 me3) that repress the *FLC* gene. In parallel with these histone methylations, which modify chromatin topology, DNA loops and long non-coding RNAs (lncRNAs) are produced locally in response to cold. *COLDWRAP* and *COLDAIR* are two such transcripts. The third lncRNA described, *COOLAIR*, differs from the other two in that it is produced backwards from the second strand of DNA and is stabilized at low temperatures (Marquadt et al. 2014; Zhao et al. 2021). All these lncRNAs contribute to histone modification processes and FLC repression. *COOLAIR* transcripts, for example, associate with FCA and CLF (Curly LeaF) proteins, which is a histone methyltransferase, that will enhance *FLC* repression through new histone trimethylations (H3K27 me3) (Tian et al. 2019). In parallel with histone modifications,, *FLC* gene regulation is also achieved through modification of DNA methylation. In particular, CHH methylation of a repeated sequence located at the first intron of the gene, a sequence named VRE (vernalization responsive element), is then involved (Ito et al. 2019). From an evolutionary point of view, it is interesting to note that the VRE sequence results from the insertion of a transposon from the SINE family. Flowering plants therefore integrate light signals and temperature to control their flowering via the expression of the *FLC* and *FT* genes. There are four levers of regulation: histone modifications,

DNA methylation, lncRNA production and transcription factor activity. This relatively complex regulation is a key element in plant survival and propagation.

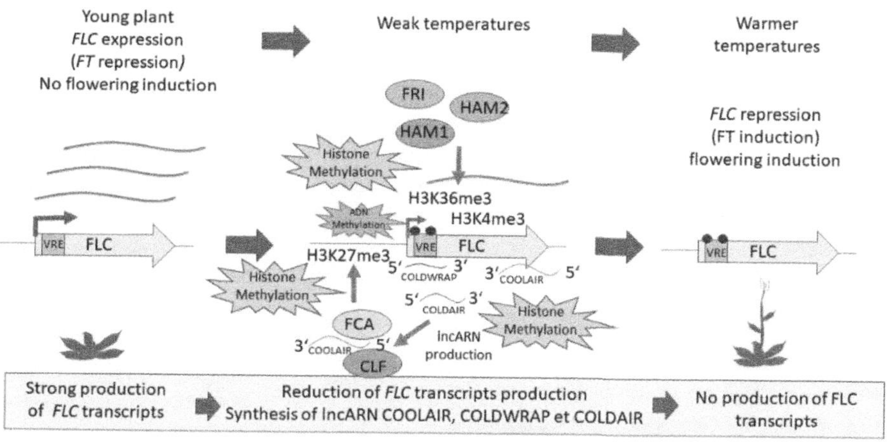

Figure 6.4. *Temperature, epigenetics and flowering. For a color version of this figure, see www.iste.co.uk/grunau/epigenetics.zip*

COMMENT ON FIGURE 6.4. – *The FLC gene is active in the young plant, and cold induces its gradual repression through the action of the FRIGIDA (FRI) protein, which enables histone methylation and the production of lncRNAs (COOLAIR, COLDWRAP and COLDAIR) that reinforce FLC repression by enabling new histone methylations. In parallel with these histone modifications, the VRE sequence, which is derived from an SINE-type transposon, will be methylated, further reinforcing repression by methylating DNA. When temperatures become milder, FLC will be completely repressed and the FT gene, which is no longer repressed by the FLC protein, will be transcribed, enabling flowering.*

In fact, it is not just the genes involved in flowering that are impacted by the effects of temperature on epigenetic marks. Indeed, it has been shown that in various species, an increase in temperature leads to a decrease of around 10% in DNA methylation levels (Hossein et al. 2017; Zhao et al. 2021). This mainly affects CHH methylations (25%), while only 10% of CG and CHG methylations are eliminated in response to stress. 25% of these methylation losses concern genes and 35% transposons (Hossein et al. 2017). The impact of stress on transposons has also been demonstrated in numerous cases (Negi et al. 2016). Examples of temperature-related transposons include ONSEN/Atcopia78 and an *SINE* retrotransposon located in the *FWA* gene, which is also involved in the control of flowering (Kinoshita et al. 2007; Pecinka et al. 2010; Dai et al. 2021) (Figure 6.5). The repeated sequences of this transposon generate the production of hcsiRNAs which, via the RdDM (RNA directed DNA methylation) pathway, methylate and specifically repress the *FWA* gene. Interestingly, DNA methylation variations are not identical between resistant and susceptible varieties (Zhao et al. 2021). In addition to affecting DNA methylation, temperature variations also affect histone modifications at all stages of development. Examples include the histone acetyltransferase GCN5 and histone deacetylases HDA9, HDA15 and HDA19, which appear to be particularly sensitive to temperature fluctuations at different developmental stages (Shen et al. 2019; Ueda and Seki 2020).

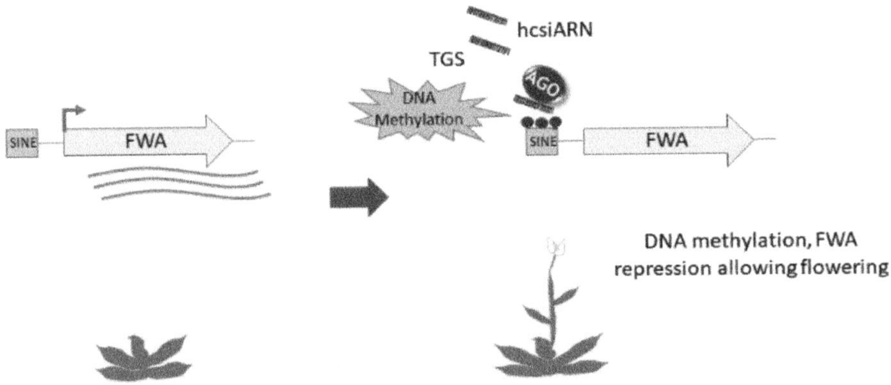

Figure 6.5. *Flowering and epigenetic marks of the FWA locus. When the plant is young, the FWA gene is active, and the upstream SINE transposon generates small RNAs (hcsiRNAs) which, via the RdDM (RNA directed DNA Methylation) pathway, initiate methylation of the FWA sequence, which is then gradually repressed (TGS, Transcriptional Gene Silencing). For a color version of this figure, see www.iste.co.uk/grunau/epigenetics.zip*

6.2.2. *Effects of biotic stresses on epigenetic marks and plant defense mechanisms*

As a result of their sedentary nature, plants cannot avoid or isolate themselves from attack by predators or infectious agents (viruses, bacteria and fungi). Their only means of protection is to limit biotic attacks by triggering molecular responses that preserve their integrity and survival. Numerous studies have shown that changes in epigenetic marks play a part in these protective mechanisms. In the following sections, we first describe the effects of dysfunctional control of epigenetic marks on pathogen resistance. We will then describe the different resistance mechanisms identified and discuss the possibility of using "epigenetic" defense mechanisms to generate resistant plants.

6.2.2.1. *Dysfunction of epigenetic and pathogenic marks*

When epigenetic marks are not properly controlled, plant response to pathogens is often altered. A case in point is the fact that tomato plants altered in the *MET1*, *DRM1*, *DRM2* and *CMT3* genes, all of which encode DNA methylation writers, are more susceptible to viral attacks (Dowen et al. 2012). This observation could indicate that DNA methylation is necessary to trigger pathogen resistance. Yet plants affected in the *ROS* gene, which is an eraser of DNA methylation, are also more susceptible to the Hyaloperonospora pathogen (Hpa). In line with this latter observation, *Arabidopsis* plants altered in the RdDM (RNA-directed DNA Methylation) pathway, responsible for de novo DNA methylation, are themselves more resistant to Hpa. Other plants, also affected in the RdDM pathway, with mutations in the *AGO4* or *AGO6* genes, are not only more resistant to many bacteria, but also more susceptible to certain fungi (Diezma-Navas et al. 2019). Taken together, these observations indicate that methylation and demethylation of different loci are necessary to induce a molecular response associated with a certain resistance to pathogens. It seems that it is mainly CHH methylations that vary during infections, and that some of these variations are specific to the pathogen used.

Enzymes involved in histone modifications are also implicated alongside DNA methylation in the response to pathogens (Ramirez-Prado et al. 2018).

6.2.2.2. *Defense mechanisms involving epigenetic marks*

Various mechanisms can provide a degree of resistance to pathogen attack by preventing pathogens from using cellular machinery to reproduce. Among the mechanisms identified, one is based on the production of small RNAs (vsiRNA virus-derived siRNAs), which prevent the production of viral proteins and the multiplication of viral genomes (PTGS pathway, Post Transcriptional Gene Silencing) (Figure 6.6). A second pathway can modify epigenetic marks (TGS, Transcriptional Gene Silencing), enabling host resistance genes to be activated and/or the pathogen genome to be repressed (Bouché et al. 2006; Baulcombe and Dean 2014). VsiRNAs are produced from RNA duplexes that are cleaved by DICERs (DCL2 and DCL4) for PTGS initiation and by DCL3 for TGS initiation (Ramirez-Prado et al. 2018). In addition to small regulatory RNAs, defense can also be provided by lncRNAs (long non-coding RNAs), such as *ELENA1*, which will modulate chromatin by modifying epigenetic marks (Mach 2017; Seo et al. 2017). In the following paragraphs, we will mainly focus on mechanisms relating to TGS.

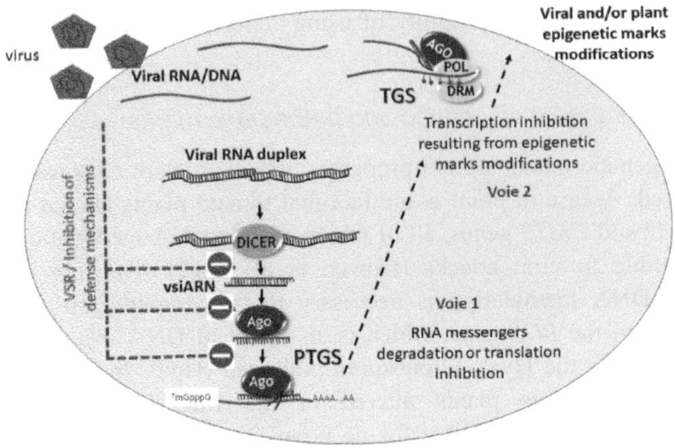

Figure 6.6. *Epigenetics and response to a viral infection. For a color version of this figure, see www.iste.co.uk/grunau/epigenetics.zip*

COMMENT ON FIGURE 6.6. – *Viral transcripts can initiate the production of vsiRNAs, which will initiate the PTGS (Post Transcriptional Gene Silencing) pathway and degrade the viral transcripts, or initiate the TGS (Transcriptional Gene Silencing) pathway and methylate the viral genome, in order to repress it and/or modulate epigenetic marks in the plant genome, thus modulating the expression of defense genes. The pathogen can also inhibit these defense pathways by preventing the PTGS or TGS pathway from functioning properly.*

Recent work shows that, upon infection, the expression of genes involved in DNA methylation is, in fact, sequentially and differentially activated (Diezma-Navas et al. 2019; Hevia and Hernandez 2020). During infection by the Tobacco Rattle Virus (TRV), for example, the RdDM pathway is first activated in *Arabidopsis*, then, as the amount of pathogen decreases, this pathway is repressed, allowing further modification of epigenetic marks. The fact that there are numerous sequences, some of them corresponding to genes, which are then over-methylated or under-methylated, indicates that there are diverse effects that are probably specific to each of the loci impacted. The expression of many genes will then be modulated, through modifications of epigenetic marks in their sequence or due to their proximity to transposons (Diezma-Navas et al. 2019). Alongside DNA methylation, histones are also impacted during infections (Raja et al. 2008; Ramirez-Prado et al. 2018). Among the plant genes regulated in this way are many *PRR* and *NLR* genes, which ensure innate immunity in plants (Cambiagno et al. 2021). *PRR* genes encode membrane receptors (cell surface-localized Pattern-Recognition Receptors) and *NLR* genes encode immune effectors (nucleotide-binding domain leucine-rich repeat containing receptors) involved in triggering pattern-triggered immunity (PTI) and effector-triggered immunity (ETI) responses, respectively (Yuan et al. 2021). These genes are incorrectly controlled in mutants affected in the regulation of epigenetic marks and are, for example, no longer repressed in the absence of infection (Cambiagno et al. 2021). This indicates that epigenetic marks repress defense genes under normal conditions and, during infections, these marks will be modified and defense genes will then be induced.

Following certain infections, the pathogen genome can also have its epigenetic marks modified by the TGS pathway (Pérez-Cañamás et al. 2010; Blevins et al. 2011; Aregger et al. 2012). This is the case for certain DNA viruses (geminiviruses and pararetroviruses). The vsiRNAs (virus-derived siRNAs) will then guide DNA methylases, via the RdDM pathway, to repress transcription of the viral genome (see Chapter 2) (Al-Kaff et al. 1998; Raja et al. 2008). This repression limits the pathogen's survival and proliferation.

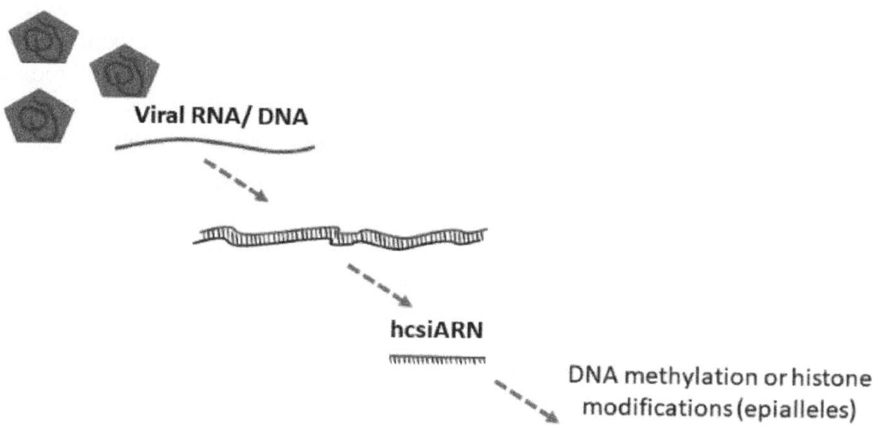

Figure 6.7. *Memory of viral attacks. Viral RNAs will initiate the production of hcsiRNAs, which will enable the establishment of new epigenetic marks that can be passed on to offspring, leading to enhanced immunity. For a color version of this figure, see www.iste.co.uk/grunau/epigenetics.zip*

The responses described sometimes involve RdRPs (RNA-dependent RNA polymerases) needed to produce RNA duplexes that can also amplify the response. This amplification can then enable regulatory RNAs to be transported to other cells on a plant-wide scale, thus giving a systemic response (Brosnan et al. 2007; Zhu et al. 2019). It is important to note that changes in epigenetic marks induced during pathogen attacks can persist and be passed on to offspring, providing them with a degree of acquired immunity (Luna et al. 2012; Slaughter et al. 2012). This is referred to as epiallelic transmission, which ensures a certain epigenetic memory and a better response to subsequent attacks (Figure 6.7). Therefore, despite the absence of a defense system like that of animals, plants have developed a complex and effective defense system over the course of evolution, based on the existence of particular pathways that control epigenetic marks.

In some cases, marks such as RNA methylation (m6A) can also be deposited by the host's molecular machinery, on the genome of RNA viruses, and thus limiting its proliferation (Martinez-Perez et al. 2017). A similar defense pathway is also observed with certain viruses infecting animals (Ramirez-Prado et al. 2018).

6.2.2.3. *Pathogen defense strategies to bypass epigenetic defense of plants*

Some pathogens, however, have the ability to inhibit defense mechanisms induced by the TGS or PTGS pathway (VIGS, Virus-Induced GeneSilencing) by producing suppressors of host defense machinery (VSRs, Viral Suppressors of RNA silencing) (Ruiz-Ferrer and Voinnet 2009; Duan et al. 2012; Pumplin and Voinnet 2013; Perrone and Martinelli 2020). Cucumber mosaic virus, for example, produces a 2b protein, which inhibits AGO proteins and thus affects the PTGS and/or TGS pathways required for defense (Duan et al. 2012). Rice grassy stunt virus (RGSV) produces a P3 protein that induces degradation of NRPD1 proteins, one of the POLV subunits involved in the RdDM pathway (see Chapter 2) (Zhang et al. 2020). As a result, the plant can no longer correctly methylate its own genome, and possibly that of the pathogen too, thus promoting infection. Various pathogens thus have the capacity to block part of the initiation of the plant defense system.

6.2.2.4. *Using epigenetic mark regulatory pathways to generate pathogen-resistant plants*

Our current knowledge allows us to say that finally, in 1985, Beachy and his collaborators unknowingly generated the first genetically modified plants with the ability to be resistant to viruses through modifications of epigenetic marks. The initial idea was to generate plants able to produce virus capsid proteins, in order to trigger a response comparable to the immune response in animals (Abel et al. 1986) (Figure 6.8). To achieve this, the DNA sequence encoding the virus capsid protein was integrated into the plant genome. Despite much criticism, the strategy worked, not because of an immune response similar to that of animals, but because of what might be termed as an epigenetic response, which initiates silencing of a sequence recognized as an exogenous "non-self" (the capsid-encoding gene). Thus, when the pathogen infects transgenic plants and integrates its genome into plant cells, the portion containing the capsid gene will be recognized and rapidly inactivated by the action of complementary and specific siRNAs, thus preventing capsid production and genome replication. This prevents the pathogen from multiplying and spreading. Since these first trials, numerous transgenes, enabling the production of siRNAs or interfering RNAs, have been used in over 30 plant species in order to make them resistant to certain pathogens and thus limit yield losses, which can sometimes reach 40% of harvests (Zhao et al. 2020) (see Chapter 9). More recently, the ability to edit and correct a gene has made it possible to generate plants producing a complex that is capable of specifically degrading the viral genome (CRISPR technology) (Zhao et al. 2020). The molecular strategy used for editing is derived from defense systems identified in bacteria and archaea (Bhaya et al. 2011) (see Chapter 5).

An alternative to using transgenic plants is to spray interfering RNAs (miRNAs or siRNAs) onto crops (SIGS, Spray-Induced Gene Silencing) (Wang and Jin 2017). The disadvantages of this approach are, on the one hand, the cost of synthesizing RNA molecules, and also the fact that spraying, which involves highly unstable RNA molecules, must be repeated.

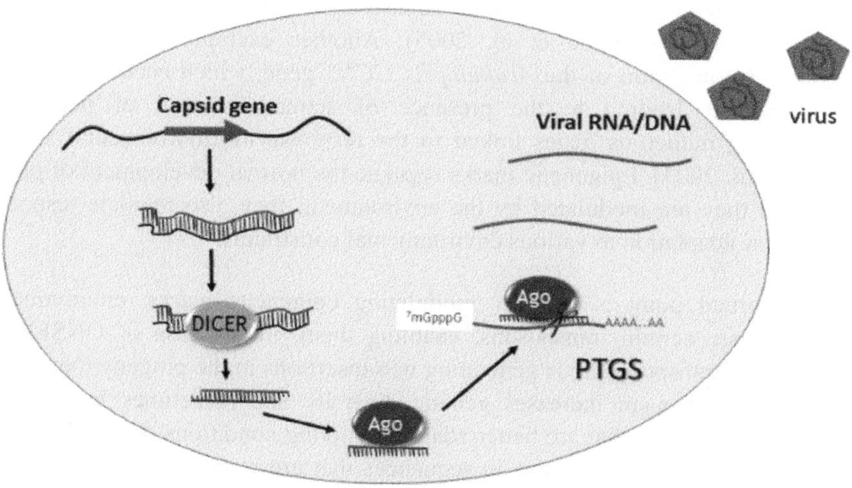

Figure 6.8. *Epigenetics and acquired resistance. The insertion of a DNA fragment enabling the production of capsid transcripts can initiate a silencing pathway that, upon infection, leads to the specific degradation of the transcripts required for capsid production. The pathogen will then be unable to multiply. For a color version of this figure, see www.iste.co.uk/grunau/epigenetics.zip*

6.2.2.5. *Epigenetics and plant plasticity in the face of environmental constraints*

The survival of each plant depends on the responses it develops to cope with the combined effects of the many factors to which it is constantly subjected. There is now no doubt that the regulation of epigenetic marks is linked to this permanent adaptation. This explains why plants, whose epigenetic mark control is affected, are generally also affected in their resistance to environmental conditions (Ueda and Seki 2020; Zhao et al. 2021). Further evidence concerns variations in DNA methylation in *Arabidopsis* seedlings, observed according to season (Ito et al. 2019). These modifications lead to changes in DNA compaction and are involved in regulating the expression of numerous genes.

It is also important to remember that some of the genes controlled, via epigenetic marks controlled by the environment, are involved in the production of various hormones. One example is the fact that overexpression of the *HDA19* gene, which encodes a histone deacetylase, leads to the induction of genes regulated by jasmonic acid and ethylene, two phytohormones (Yaish et al. 2011). These plants are then more resistant to certain pathogens and are also characterized by delayed flowering and reduced fertility (Zhou et al. 2005). Another example results from the inactivation by mutation of the *Arabidopsis GCN5* gene, which encodes a histone acetyltransferase, leading to the presence of stamens instead of petals and over-activating numerous genes linked to the response to environmental stresses (Bertrand et al. 2003). Epigenetic marks regulate the normal development of plants, but because they are modulated by the environment, they also regulate responses and thus their adaptation to various environmental constraints.

An important point is that, by modulating epigenetic marks, environmental stresses activate certain transposons, enabling them, in the case of ONSEN for example, to be transposed, thus generating neo-insertions in the progeny (Negi et al. 2016). This mechanism increases genetic diversity and sometimes leads to the emergence of genomes that are better adapted to living conditions. It should also be noted that some transposons contain sequences that are responsible for controlling transcription, which, after neo-insertions, can modify the expression of adjacent genes. The effects of these transposons can have a positive impact on survival and thus increase the plasticity of organisms.

Another significant impact on adaptation is the fact that changes in epigenetic marks induced by the environment can be passed on to offspring. This then allows plants to acquire a certain memory that may represent a better adaptation to climate change (Crisp et al. 2016). In fact, recent work shows that plants deficient in DNA methylation can become more sensitive to the presence of certain stresses, and that this sensitivity is maintained in subsequent generations (Perrone and Martinelli 2020). Histone modifications are also involved in the transmission of this transgenerational stress memory (Crisp et al. 2016; Zhao et al. 2021). Numerous studies are currently being carried out to identify the specific features that determine whether or not certain epigenetic marks acquired during the life cycle will be passed on to offspring.

6.3. Conclusion

Epigenetics is a growing field of study, demonstrating that stress leaves epigenetic marks at the heart of cells, which can be harmful. In humans, it is gradually revealing the potential of stress management techniques. It upsets genetic determinism and blurs the boundary between nature and nurture. With the possibility of reacquiring biological control, it is being reused by a growing number of commercial initiatives (cosmetics sales, personalized wellness courses, etc.), based on the fact that "lifestyle counts for more than heredity" or that "the science of epigenetics makes it possible to build the most efficient lifestyle". However, given the still preliminary nature of these studies, it is advisable to remain cautious at this stage, especially given that the specific part of the epigenome that can indeed respond to these lifestyle interventions remains to be determined. A significant number of studies highlight the importance of gene–environment interactions: not all individuals are equally sensitive to stress, nor have the same capacity for resilience in the face of stress, depending on the genetic variants they carry (Feder et al. 2009; Zannas et al. 2016). Furthermore, by highlighting the molecular "scars" generated by social inequalities (Mulligan 2021), epigenetics is finding itself at the heart of social claims by certain populations demanding recognition of the intergenerational biological cost of the traumas inflicted on them (Dubois and Guaspare 2020). The years to come should shed light on the reality and intensity of these effects. In plants, numerous molecular links between stress and the epigenome have already been clearly demonstrated. A better understanding of these mechanisms represents a potential avenue for improvement, which could lead to plants that are better adapted to the many environmental constraints of the future.

6.4. References

Al-Kaff, N.S., Covey, S.N., Kreike, M.M., Page, A.M., Pinder, R., Dale, P.J. (1998). Transcriptional and posttranscriptional plant gene silencing in response to a pathogen. *Science*, 279, 2113–2115. doi: 10.1126/science.279.5359.2113.

Anderson, J.A., Johnston, R.A., Lea, A.J., Campos, F.A., Voyles, T.N., Akinyi, M.Y., Tung, J. (2021). High social status males experience accelerated epigenetic aging in wild baboons. *ELife*, 10, 1–3. doi: 10.7554/ELIFE.66128.

Aregger, M., Borah, B.K., Seguin, J., Rajeswaran, R., Gubaeva, E.G., Zvereva, A.S., Windels, D., Vazquez, F., Blevins, T., Farinelli, L. et al. (2012a). Primary and secondary siRNAs in geminivirus-induced gene silencing. *PLoS Pathog*, 8, e1002941. doi: 10.1371/journal.ppat.1002941.

Bartels, A., Han, Q., Nair, P., Stacey, L., Gaynier, H., Mosley, M., Huang, Q., Pearson, J., Hsieh, T.-F., An, Y.-Q. et al. (2018). Dynamic DNA methylation in plant growth and development. *IJMS*, 19, 2144. doi: 10.3390/ijms19072144.

Baulcombe, D.C. and Dean, C. (2014). Epigenetic regulation in plant responses to the environment. *Cold Spring Harbor Perspectives in Biology*, 6, a019471–a019471. doi: 10.1101/cshperspect.a019471.

Bertrand, C., Bergounioux, C., Domenichini, S., Delarue, M., Zhou, D.-X. (2003). Arabidopsis histone acetyltransferase AtGCN5 regulates the floral meristem activity through the WUSCHEL/AGAMOUS pathway. *Journal of Biological Chemistry*, 278, 28246–28251. doi: 10.1074/jbc.M302787200.

Bhaya, D., Davison, M., Barrangou, R. (2011). CRISPR-Cas systems in bacteria and archaea: Versatile small RNAs for adaptive defense and regulation. *Annual Review Genetics*, 45, 273–297. doi: 10.1146/annurev-genet-110410-132430.

Blevins, T., Rajeswaran, R., Aregger, M., Borah, B.K., Schepetilnikov, M., Baerlocher, L., Farinelli, L., Meins, F., Hohn, T., Pooggin, M.M. (2011a). Massive production of small RNAs from a non-coding region of cauliflower mosaic virus in plant defense and viral counter-defense. *Nucleic Acids Research*, 39, 5003–5014. doi: 10.1093/nar/gkr119.

Borghol, N., Suderman, M., Mcardle, W., Racine, A., Hallett, M., Pembrey, M., Szyf, M. (2012). Associations with early-life socio-economic position in adult DNA methylation. *International Journal of Epidemiology*, 41(1), 62–74. doi: 10.1093/ije/dyr147.

Bouché, N., Lauressergues, D., Gasciolli, V., Vaucheret, H. (2006). An antagonistic function for *Arabidopsis* DCL2 in development and a new function for DCL4 in generating viral siRNAs. *EMBO Journal*, 25, 3347–3356. doi: 10.1038/sj.emboj.7601217.

Brody, G.H., Miller, G.E., Yu, T., Beach, S.R.H., Chen, E. (2016a). Supportive family environments ameliorate the link between racial discrimination and epigenetic aging: A replication across two longitudinal cohorts. *Psychological Science*, 27(4), 530–541. doi: 10.1177/0956797615626703.

Brody, G.H., Yu, T., Chen, E., Beach, S.R.H., Miller, G.E. (2016b). Family-centered prevention ameliorates the longitudinal association between risky family processes and epigenetic aging. *Journal of Child Psychology and Psychiatry and Allied Disciplines*, 57(5), 566–574. doi: 10.1111/jcpp.12495.

Brosnan, C.A., Mitter, N., Christie, M., Smith, N.A., Waterhouse, P.M., Carroll, B.J. (2007). Nuclear gene silencing directs reception of long-distance mRNA silencing in *Arabidopsis*. *Proceedings of the National Academy of Sciences of the USA*, 104, 14741–14746. doi: 10.1073/pnas.0706701104.

Cambiagno, D.A., Torres, J.R., Alvarez, M.E. (2021). Convergent epigenetic mechanisms avoid constitutive expression of immune receptor gene subsets. *Frontiers in Plant Science*, 12, 703667. doi: 10.3389/fpls.2021.703667.

Cao-Lei, L., de Rooij, S.R., King, S., Matthews, S.G., Metz, G.A.S., Roseboom, T.J., Szyf, M. (2020). Prenatal stress and epigenetics. *Neuroscience and Biobehavioral Reviews*, 117, 198–210. doi: 10.1016/j.neubiorev.2017.05.016.

Champagne, F.A. (2013). Early environments, glucocorticoid receptors, and behavioral epigenetics. *Behavioral Neuroscience*, 127(5), 628–636. doi: 10.1037/a0034186.

Chen, L.-T., Luo, M., Wang, Y.-Y., Wu, K. (2010a). Involvement of Arabidopsis histone deacetylase HDA6 in ABA and salt stress response. *Journal of Experimental Botany*, 61, 3345–3353. doi: 10.1093/jxb/erq154.

Chetty, R., Stepner, M., Abraham, S., Lin, S., Scuderi, B., Turner, N., Cutler, D. (2016). The association between income and life expectancy in the United States, 2001-2014. *JAMA - Journal of the American Medical Association*, 315(16), 1750–1766. doi: 10.1001/jama.2016.4226.

Cho, J.-N., Ryu, J.-Y., Jeong, Y.-M., Park, J., Song, J.-J., Amasino, R.M., Noh, B., Noh, Y.-S. (2012). Control of seed germination by light-induced histone arginine demethylation activity. *Developmental Cell*, 22, 736–748. doi: 10.1016/j.devcel.2012.01.024.

Crisp, P.A., Ganguly, D., Eichten, S.R., Borevitz, J.O., Pogson, B.J. (2016). Reconsidering plant memory: Intersections between stress recovery, RNA turnover, and epigenetics. *Science Advances*, 2, e1501340. doi: 10.1126/sciadv.1501340.

Crosswell, A.D. and Lockwood, K.G. (2020). Best practices for stress measurement: How to measure psychological stress in health research. *Health Psychology Open*, 7(2). doi: 10.1177/2055102920933072.

Cunliffe, V.T. (2016). Epigenetic impacts of social stress. *Epigenomics*, 8, 1653–1669.

Dias, B.G. and Ressler, K.J. (2014). Parental olfactory experience influences behavior and neural structure in subsequent generations. *Nature Neuroscience*, 17(1), 89–96. doi: 10.1038/nn.3594.

Diezma-Navas, L., Pérez-González, A., Artaza, H., Alonso, L., Caro, E., Llave, C., Ruiz-Ferrer, V. (2019). Crosstalk between epigenetic silencing and infection by tobacco rattle virus in *Arabidopsis*. *Molecular Plant Pathology*, 20, 1439–1452. doi: 10.1111/mpp.12850.

Dowen, R.H., Pelizzola, M., Schmitz, R.J., Lister, R., Dowen, J.M., Nery, J.R., Dixon, J.E., Ecker, J.R. (2012). Widespread dynamic DNA methylation in response to biotic stress. *Proceedings of the National Academy of Sciences of the USA*, 109. doi: 10.1073/pnas.1209329109.

Duan, C.-G., Fang, Y.-Y., Zhou, B.-J., Zhao, J.-H., Hou, W.-N., Zhu, H., Ding, S.-W., Guo, H.-S. (2012a). Suppression of *Arabidopsis* ARGONAUTE1-mediated slicing, transgene-induced RNA silencing, and DNA methylation by distinct domains of the *Cucumber mosaic virus* 2b protein. *The Plant Cell*, 24, 259–274. doi: 10.1105/tpc.111.092718.

Dubois, M. and Guaspare, C. (2020). From cellular memory to the memory of trauma: Social epigenetics and its public circulation. *Social Science Information*, 59(1), 144–183. doi: 10.1177/0539018419897600.

Feder, A., Nestler, E.J., Charney, D.S. (2009). Psychobiology and molecular genetics of resilience. *Nature Reviews Neuroscience*, 10(6), 446–457. doi: 10.1038/nrn2649.

Felitti, V.J., Anda, R.F., Nordenberg, D., Williamson, D.F., Spitz, A.M., Edwards, V., Marks, J.S. (2019). Relationship of childhood abuse and household dysfunction to many of the leading causes of death in adults: The Adverse Childhood Experiences (ACE) Study. *American Journal of Preventive Medicine*, 56(6), 774–786. doi: 10.1016/j.amepre.2019.04.001.

Fiorito, G., Polidoro, S., Dugué, P.A., Kivimaki, M., Ponzi, E., Matullo, G., Vineis, P. (2017). Social adversity and epigenetic aging: A multi-cohort study on socioeconomic differences in peripheral blood DNA methylation. *Scientific Reports*, 7(1). doi: 10.1038/s41598-017-16391-5.

Fujisawa, T.X., Nishitani, S., Takiguchi, S., Shimada, K., Smith, A.K., Tomoda, A. (2019). Oxytocin receptor DNA methylation and alterations of brain volumes in maltreated children. *Neuropsychopharmacology*, 44(12), 2045–2053. doi: 10.1038/s41386-019-0414-8.

Gapp, K., Jawaid, A., Sarkies, P., Bohacek, J., Pelczar, P., Prados, J., Mansuy, I.M. (2014). Implication of sperm RNAs in transgenerational inheritance of the effects of early trauma in mice. *Nature Neuroscience*, 17(5), 667–669. doi: 10.1038/nn.3695.

Gu, D., Ji, R., He, C., Peng, T., Zhang, M., Duan, J., Xiong, C., Liu, X. (2019). *Arabidopsis* histone methyltransferase SUVH5 Is a positive regulator of light-mediated seed germination. *Frontiers in Plant Science*, 10, 841. doi: 10.3389/fpls.2019.00841.

Heard, E. and Martienssen, R.A. (2014). Transgenerational epigenetic inheritance: Myths and mechanisms. *Cell*, 157(1), 95–109. doi: 10.1016/j.cell.2014.02.045.

Heim, C., Jeffrey Newport, D., Heit, S., Graham, Y.P., Wilcox, M., Bonsall, R., Nemeroff, C.B. (2013). Pituitary–adrenal and autonomic responses to stress in women after sexual and physical abuse in childhood. *The Science of Mental Health: Stress and the Brain*, 9(5), 84–89. doi: 10.1097/00005721-200111000-00023.

Horvath, S. and Raj, K. (2018). DNA methylation-based biomarkers and the epigenetic clock theory of ageing. *Nature Reviews Genetics*, 1–14. doi: 10.1038/s41576-018-0004-3.

Hossain, M.S., Kawakatsu, T., Kim, K.D., Zhang, N., Nguyen, C.T., Khan, S.M., Batek, J.M., Joshi, T., Schmutz, J., Grimwood, J. et al. (2017). Divergent cytosine DNA methylation patterns in single-cell, soybean root hairs. *New Phytologist*, 214, 808–819. doi: 10.1111/nph.14421.

Houtepen, L.C., Vinkers, C.H., Carrillo-Roa, T., Hiemstra, M., Van Lier, P.A., Meeus, W., Boks, M.P.M. (2016). Genome-wide DNA methylation levels and altered cortisol stress reactivity following childhood trauma in humans. *Nature Communications*, 7. doi: 10.1038/ncomms10967.

Ito, T., Nishio, H., Tarutani, Y., Emura, N., Honjo, M.N., Toyoda, A., Fujiyama, A., Kakutani, T., Kudoh, H. (2019). Seasonal stability and dynamics of DNA methylation in plants in a natural environment. *Genes*, 10, 544. doi: 10.3390/genes10070544.

Jovanovic, T., Vance, L.A., Cross, D., Knight, A.K., Kilaru, V., Michopoulos, V., Smith, A.K. (2017). Exposure to violence accelerates epigenetic aging in children. *Scientific Reports*, 7(1), 1–7. doi: 10.1038/s41598-017-09235-9.

Juster, R.P., McEwen, B.S., Lupien, S.J. (2010). Allostatic load biomarkers of chronic stress and impact on health and cognition. *Neuroscience and Biobehavioral Reviews*, 35(1), 2–16. doi: 10.1016/j.neubiorev.2009.10.002.

Kim, S.Y., Lee, J., Eshed-Williams, L., Zilberman, D., Sung, Z.R. (2012). EMF1 and PRC2 cooperate to repress key regulators of arabidopsis development. *PLoS Genet*, 8, e1002512. doi: 10.1371/journal.pgen.1002512.

Kittleson, M.M., Meoni, L.A., Wang, N.Y., Chu, A.Y., Ford, D.E., Klag, M.J. (2006). Association of childhood socioeconomic status with subsequent coronary heart disease in physicians. *Archives of Internal Medicine*, 166(21), 2356–2361. doi: 10.1001/archinte.166.21.2356.

Lawn, R.B., Anderson, E.L., Suderman, M., Simpkin, A.J., Gaunt, T.R., Teschendorff, A.E., Howe, L.D. (2018). Psychosocial adversity and socioeconomic position during childhood and epigenetic age: Analysis of two prospective cohort studies. *Human Molecular Genetics*, 27(7), 1301–1308. doi: 10.1093/hmg/ddy036.

Lea, A.J., Akinyi, M.Y., Nyakundi, R., Mareri, P., Nyundo, F., Kariuki, T., Tung, J. (2018). Dominance rank-associated gene expression is widespread, sex-specific, and a precursor to high social status in wild male baboons. *Proceedings of the National Academy of Sciences of the United States of America*, 115(52), E12163–E12171. doi: 10.1073/pnas.1811967115.

Leljak-Levanić, D., Bauer, N., Mihaljević, S., Jelaska, S. (2004). Somatic embryogenesis in pumpkin (*Cucurbita pepo* L.): Control of somatic embryo development by nitrogen compounds. *Journal of Plant Physiology*, 161, 229–236. doi: 10.1078/0176-1617-01055.

Longo, C., Holness, S., De Angelis, V., Lepri, A., Occhigrossi, S., Ruta, V., Vittorioso, P. (2020). From the outside to the inside: New insights on the main factors that guide seed dormancy and germination. *Genes*, 12, 52. doi: 10.3390/genes12010052.

Luna, E. and Ton, J. (2012). The epigenetic machinery controlling transgenerational systemic acquired resistance. *Plant Signaling & Behavior*, 7, 615–618. doi: 10.4161/psb.20155.

Mach, J. (2017). The long-noncoding RNA ELENA1 functions in plant immunity. *Plant Cell*, 29, 916–916. doi: 10.1105/tpc.17.00343.

Marquardt, S., Raitskin, O., Wu, Z., Liu, F., Sun, Q., Dean, C. (2014). Functional consequences of splicing of the antisense transcript COOLAIR on FLC transcription. *Molecular Cell*, 54, 156–165. doi: 10.1016/j.molcel.2014.03.026.

Martínez-Pérez, M., Aparicio, F., López-Gresa, M.P., Bellés, J.M., Sánchez-Navarro, J.A., Pallás, V. (2017). *Arabidopsis* m^6A demethylase activity modulates viral infection of a plant virus and the m^6A abundance in its genomic RNAs. *Proceedings of the National Academy of Sciences of the USA*, 114, 10755–10760. doi: 10.1073/pnas.1703139114.

McEwen, B.S. (1998). Protective and damaging effects of stress mediators. *New England Journal of Medicine*, 338, 171–179 [Online]. Available at: http://www.ncbi.nlm.nih.gov/pubmed/9428819.

Miller, G.E., Chen, E., Parker, K.J. (2011). Psychological stress in childhood and susceptibility to the chronic diseases of aging: Moving towards a model of behavioral and biological mechanisms. *Psychological Bulletin*, 137(6), 959–997. doi: 10.1037/a0024768.

Mulligan, C.J. (2016). Early environments, stress, and the epigenetics of human health. *Annual Review of Anthropology*, 45, 233–249. doi: 10.1146/annurev-anthro-102215-095954.

Mulligan, C.J. (2021). Systemic racism can get under our skin and into our genes. *American Journal of Physical Anthropology*, 175(2), 399–405. doi: 10.1002/ajpa.24290.

Murgatroyd, C., Quinn, J.P., Sharp, H.M., Pickles, A., Hill, J. (2015). Effects of prenatal and postnatal depression, and maternal stroking, at the glucocorticoid receptor gene. *Translational Psychiatry*, 5, 1–5. doi: 10.1038/tp.2014.140.

Negi, P., Rai, A.N., Suprasanna, P. (2016a). Moving through the stressed genome: Emerging regulatory roles for transposons in plant stress response. *Frontiers in Plant Science*, 7. doi: 10.3389/fpls.2016.01448.

Non, A.L., Hollister, B.M., Humphreys, K.L., Childebayeva, A., Esteves, K., Zeanah, C.H., Drury, S.S. (2016). DNA methylation at stress-related genes is associated with exposure to early life institutionalization. *American Journal of Physical Anthropology*, 161(1), 84–93. doi: 10.1002/ajpa.23010.

Palma-Gudiel, H., Córdova-Palomera, A., Leza, J.C., Fañanás, L. (2015). Glucocorticoid receptor gene (NR3C1) methylation processes as mediators of early adversity in stress-related disorders causality: A critical review. *Neuroscience and Biobehavioral Reviews*, 55, 520–535. doi: 10.1016/j.neubiorev.2015.05.016.

Parade, S.H., Huffhines, L., Daniels, T.E., Stroud, L.R., Nugent, N.R., Tyrka, A.R. (2021). A systematic review of childhood maltreatment and DNA methylation: Candidate gene and epigenome-wide approaches. *Translational Psychiatry*, 11(1). doi: 10.1038/s41398-021-01207-y.

Pecinka, A., Dinh, H.Q., Baubec, T., Rosa, M., Lettner, N., Scheid, O.M. (2010). Epigenetic regulation of repetitive elements is attenuated by prolonged heat stress in *Arabidopsis*. *The Plant Cell*, 22, 3118–3129. doi: 10.1105/tpc.110.078493.

Pérez-Cañamás, M., Hevia, E., Hernández, C. (2020). Epigenetic changes in host ribosomal DNA promoter induced by an asymptomatic plant virus infection. *Biology*, 9, 91. doi: 10.3390/biology9050091.

Perrone, A. and Martinelli, F. (2020a). Plant stress biology in epigenomic era. *Plant Science*, 294, 110376. doi: 10.1016/j.plantsci.2019.110376.

Pumplin, N. and Voinnet, O. (2013). RNA silencing suppression by plant pathogens: Defence, counter-defence and counter-counter-defence. *Nature Reviews Microbiology*, 11, 745–760. doi: 10.1038/nrmicro3120.

Raja, P., Sanville, B.C., Buchmann, R.C., Bisaro, D.M. (2008). Viral genome methylation as an epigenetic defense against geminiviruses. *J Virol*, 82, 8997–9007. doi: 10.1128/JVI.00719-08.

Ramirez-Prado, J.S., Abulfaraj, A.A., Rayapuram, N., Benhamed, M., Hirt, H. (2018a). Plant immunity: From signaling to epigenetic control of defense. *Trends in Plant Science*, 23, 833–844. doi: 10.1016/j.tplants.2018.06.004.

Rodgers, A.B., Morgan, C.P., Bronson, S.L., Revello, S., Bale, T.L. (2013). Paternal stress exposure alters sperm microRNA content and reprograms offspring HPA stress axis regulation. *Journal of Neurobiology*, 33(21), 9003–9012.

Ruiz-Ferrer, V. and Voinnet, O. (2009). Roles of plant small RNAs in biotic stress responses. *Annual Review of Plant Biology*, 60, 485–510. doi: 10.1146/annurev.arplant.043008.092111.

Sapolsky, R.M. (1994). *Why Zebras Dont Get Ulcers*. St Martin's Griffin, New York.

Sapolsky, R.M. (2005). The influence of social hierarchy on primate health. *Science*, 308(5722), 648–652. doi: 10.1126/science.1106477.

Seo, J.S., Sun, H.-X., Park, B.S., Huang, C.-H., Yeh, S.-D., Jung, C., Chua, N.-H. (2017). ELF18-INDUCED LONG-NONCODING RNA associates with mediator to enhance expression of innate immune response genes in arabidopsis. *Plant Cell*, 29, 1024–1038. doi: 10.1105/tpc.16.00886.

Serpeloni, F., Radtke, K., de Assis, S.G., Henning, F., Nätt, D., Elbert, T. (2017). Grandmaternal stress during pregnancy and DNA methylation of the third generation: An epigenome-wide association study. *Translational Psychiatry*, 7(8), e1202. doi: 10.1038/tp.2017.153.

Shen, Y., Conde e Silva, N., Audonnet, L., Servet, C., Wei, W., Zhou, D.-X. (2014). Over-expression of histone H3K4 demethylase gene JMJ15 enhances salt tolerance in Arabidopsis. *Frontiers in Plant Science*, 5. doi: 10.3389/fpls.2014.00290.

Shen, Y., Lei, T., Cui, X., Liu, X., Zhou, S., Zheng, Y., Guérard, F., Issakidis-Bourguet, E., Zhou, D. (2019). Arabidopsis histone deacetylase HDA 15 directly represses plant response to elevated ambient temperature. *The Plant Journal*, 100, 991–1006. doi: 10.1111/tpj.14492.

Simons, R.L., Kit Lei, M., Beach, S.R.H., Philibert, R.A., Cutrona, C.E., Gibbons, F.X., Barr, A. (2016). Economic hardship and biological weathering: The epigenetics of aging in a U.S. sample of black women. *Social Science & Medicine*, 150, 192–200.

Snowdon, D.A., Ostwald, S.K., Kane, R.L. (1989). Education, survival, and independence in elderly catholic sisters, 1936-1988. *American Journal of Epidemiology*, 130(5), 999–1012. doi: 10.1093/oxfordjournals.aje.a115433.

Snyder-Mackler, N., Sanz, J., Kohn, J.N., Brinkworth, J.F., Morrow, S., Shaver, A.O., Tung, J. (2016). Social status alters immune regulation and response to infection in macaques. *Science*, 354(6315), 1041–1045. doi: 10.1126/science.aah3580.

Snyder-Mackler, N., Burger, J.R., Gaydosh, L., Belsky, D.W., Noppert, G.A., Campos, F.A., Tung, J. (2020). Social determinants of health and survival in humans and other animals. *Science*, 368(6493). doi: 10.1126/science.aax9553.

Tian, Y., Zheng, H., Zhang, F., Wang, S., Ji, X., Xu, C., He, Y., Ding, Y. (2019). PRC2 recruitment and H3K27me3 deposition at *FLC* require FCA binding of *COOLAIR*. *Science Advances*, 5, eaau7246. doi: 10.1126/sciadv.aau7246.

Tung, J., Barreiro, L.B., Johnson, Z.P., Hansen, K.D., Michopoulos, V., Toufexis, D., Gilad, Y. (2012). Social environment is associated with gene regulatory variation in the rhesus macaque immune system. *Proceedings of the National Academy of Sciences of the United States of America*, 109(17), 6490–6495. doi: 10.1073/pnas.1202734109.

Tung, J., Archie, E.A., Altmann, J., Alberts, S.C. (2016). Cumulative early life adversity predicts longevity in wild baboons. *Nature Communications*, 7, 1–7. doi: 10.1038/ncomms11181.

Turecki, G. and Meaney, M.J. (2016). Effects of the social environment and stress on glucocorticoid receptor gene methylation: A systematic review. *Biological Psychiatry*, 79(2), 87–96. doi: 10.1016/j.biopsych.2014.11.022.

Ueda, M. and Seki, M. (2020). Histone modifications form epigenetic regulatory networks to regulate abiotic stress response. *Plant Physiology*, 182, 15–26. doi: 10.1104/pp.19.00988.

Wang, M. and Jin, H. (2017). Spray-induced gene silencing: A powerful innovative strategy for crop protection. *Trends in Microbiology*, 25, 4–6. doi: 10.1016/j.tim.2016.11.011.

Wang, Y., Gu, X., Yuan, W., Schmitz, R.J., He, Y. (2014). Photoperiodic control of the floral transition through a distinct polycomb repressive complex. *Developmental Cell*, 28, 727–736. doi: 10.1016/j.devcel.2014.01.029.

Weaver, I.C.G., Cervoni, N., Champagne, F.A., D'Alessio, A.C.A.C., Sharma, S., Seckl, J.R., Meaney, M.J. (2004). Epigenetic programming by maternal behavior. *Nature Neuroscience*, 7(8), 847–854. doi: 10.1038/nnl276.

Welberg, L. (2013). Epigenetics: Keeping it in the family. *Nature Reviews Neuroscience*, 14(7), 458–459. doi: 10.1038/nrn3536.

Xu, R., Wang, Y., Zheng, H., Lu, W., Wu, C., Huang, J., Yan, K., Yang, G., Zheng, C. (2015). Salt-induced transcription factor *MYB74* is regulated by the RNA-directed DNA methylation pathway in *Arabidopsis*. *EXBOTJ*, 66, 5997–6008. doi: 10.1093/jxb/erv312.

Yuan, M., Ngou, B.P.M., Ding, P., Xin, X.-F. (2021). PTI-ETI crosstalk: An integrative view of plant immunity. *Current Opinion in Plant Biology*, 62, 102030. doi: 10.1016/j.pbi.2021.102030.

Zannas, A.S., Arloth, J., Carrillo-Roa, T., Iurato, S., Röh, S., Ressler, K.J., Mehta, D. (2015). Lifetime stress accelerates epigenetic aging in an urban, African American cohort: Relevance of glucocorticoid signaling. *Genome Biology*, 16(1), 266. doi: 10.1186/s13059-015-0828-5.

Zannas, A.S., Wiechmann, T., Gassen, N.C., Binder, E.B. (2016). Gene-stress-epigenetic regulation of FKBP5: Clinical and translational implications. *Neuropsychopharmacology*, 41(1), 261–274. doi: 10.1038/npp.2015.235.

Zhang, C., Wei, Y., Xu, L., Wu, K.-C., Yang, L., Shi, C.-N., Yang, G.-Y., Chen, D., Yu, F.-F., Xie, Q. et al. (2020). A bunyavirus-inducible ubiquitin ligase targets RNA polymerase IV for degradation during viral pathogenesis in rice. *Molecular Plant*, 13, 836–850. doi: 10.1016/j.molp.2020.02.010.

Zhao, M., Yang, S., Liu, X., Wu, K. (2015). Arabidopsis histone demethylases LDL1 and LDL2 control primary seed dormancy by regulating DELAY OF GERMINATION 1 and ABA signaling-related genes. *Frontiers in Plant Science*, 6. doi: 10.3389/fpls.2015.00159.

Zhao, J., Lu, Z., Wang, L., Jin, B. (2020a). Plant responses to heat stress: Physiology, transcription, noncoding RNAs, and epigenetics. *IJMS*, 22, 117. doi: 10.3390/ijms22010117.

Zhao, Y., Yang, X., Zhou, G., Zhang, T. (2020b). Engineering plant virus resistance: From RNA silencing to genome editing strategies. *Plant Biotechnol J*, 18, 328–336. doi: 10.1111/pbi.13278.

Zhao, Y., Zhu, P., Hepworth, J., Bloomer, R., Antoniou-Kourounioti, R.L., Doughty, J., Heckmann, A., Xu, C., Yang, H., Dean, C. (2021). Natural temperature fluctuations promote *COOLAIR* regulation of *FLC*. *Genes Development*, 35, 888–898. doi: 10.1101/gad.348362.121.

Zhou, C., Zhang, L., Duan, J., Miki, B., Wu, K. (2005). *HISTONE DEACETYLASE19* is involved in jasmonic acid and ethylene signaling of pathogen response in arabidopsis. *Plant Cell*, 17, 1196–1204. doi: 10.1105/tpc.104.028514.

Zhu, C., Liu, T., Chang, Y.-N., Duan, C.-G. (2019). Small RNA functions as a trafficking effector in plant immunity. *IJMS*, 20, 2816. doi: 10.3390/ijms20112816.

7

Phenotypic Plasticity, Epigenetics and Adaptability

Patricia GIBERT[1], Cristina VIEIRA[1] and Frédéric BRUNET[2]

[1] *LBBE, CNRS, Université Claude Bernard Lyon 1, Villeurbanne, France*

[2] *IGFL, ENS Lyon, France*

7.1. Introduction

Epigenetics and phenotypic plasticity (PP) are the two concepts that are often mistakenly confused. Indeed, while PP often involves epigenetic mechanisms, they are not the only ones. Moreover, epigenetic mechanisms are involved in many other processes besides PP. The aim of this chapter is therefore to clarify the notion of PP, its evolution, its impact on evolutionary processes and its links with epigenetics.

PP is defined as a genotype's ability to produce different phenotypes in response to environmental conditions (Bradshaw 1965). Probably the best-known example of PP has been described in the microcrustacean *Daphnia*, which is able to respond to predator-specific chemical signals by adapting its morphology, behavior and physiology. However, there are a plethora of other examples, such as the production of broad, thin leaves by plants in low light conditions or the morphological variations associated with castes in social Hymenoptera.

Epigenetics in Ecology and Evolution,
coordinated by Christoph GRUNAU and Stéphane MAURY.

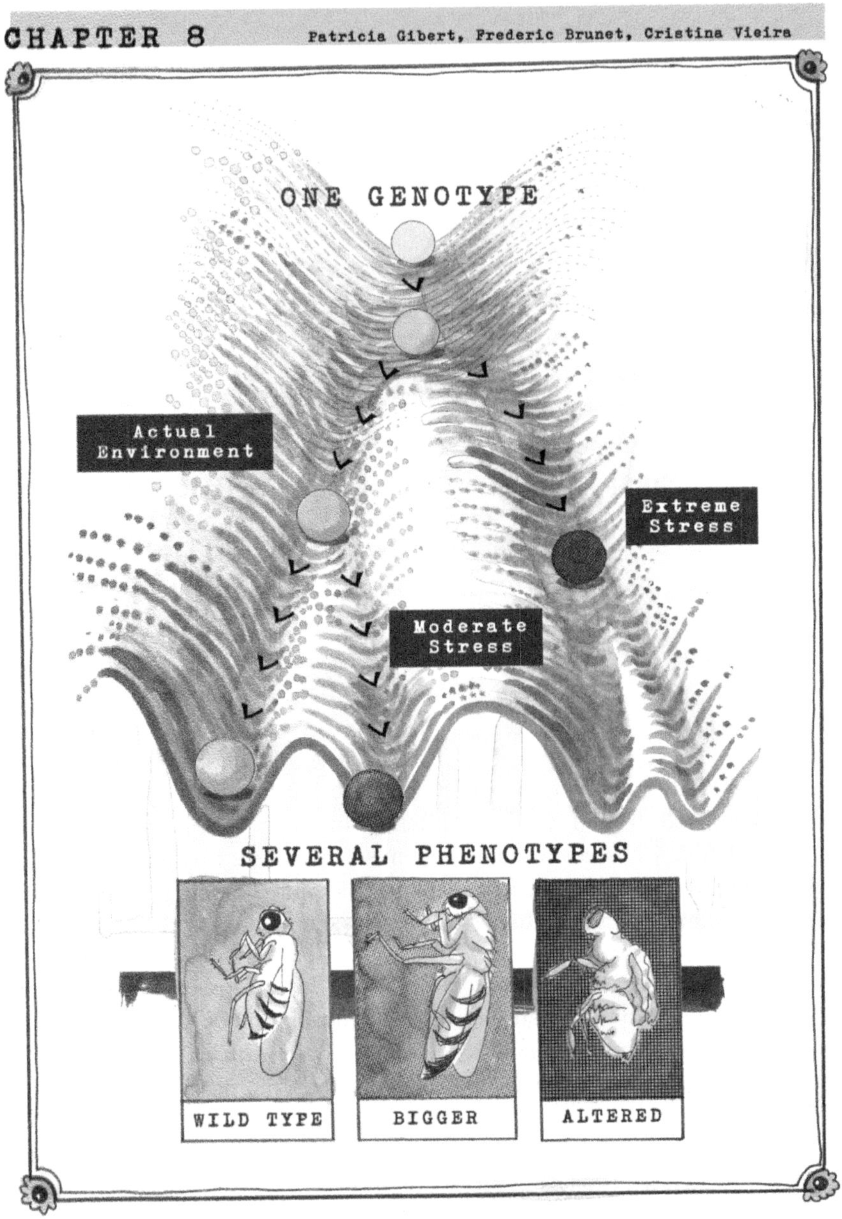

Figure 7.1. *The "epigenetic landscape" imagined by C. Waddington (see Chapter 1) to link the genotype which, under the environment's influence, forms phenotypes during (or in the course of) development (see this chapter and Chapters 8–11 for further details). Illustration by Annette Köhn*

More recently, West-Eberhard (2003) has proposed a broader definition of PP: "[...] the ability of an organism to react to an environmental input with a change in form, state, movement, or rate of activity [...]". According to this author, PP is not limited to the genotype alone, as an individual's ability to respond is influenced by his or her parents even before his or her own genotype begins to act. While the literature agrees with the definition given at the beginning of this paragraph, the term PP has sometimes been subdivided into different categories, depending on whether it is adaptive (acclimatization) or non-adaptive, continuous (gradual) or discontinuous (polyphenism), reversible (flexibility) or irreversible (developmental plasticity). We will make no such distinction in the remainder of this chapter.

PP is a process that can be generalized to all living organisms, for all types of traits and a wide range of environments. While the importance of PP as a source of phenotypic variability is undeniable, questions persist as to the importance of PP in evolutionary processes. According to some authors, by masking genetic variability from natural selection, PP inhibits evolutionary change. However, it is the interaction between genotype and environment that determines an individual's phenotype, and natural selection acts on this interaction.

There are still many questions surrounding PP: what are the mechanisms by which this genotype–environment interaction can produce different phenotypes? How do these mechanisms evolve? Can PP have an impact on evolution? This chapter will attempt to provide some answers to these questions.

7.2. Experimental approach to PP

7.2.1. *What is a reaction norm?*

The research tool for studying PP is the reaction norm (Woltereck 1909), which is the function that links the environments a particular genotype can face and the phenotypes that can be produced by this genotype. The phenotypic variance observed in nature can therefore be broken down into: a share of genetic variance (V_G), which corresponds to the phenotypic variation between two genotypes placed in the same environment; a share of environmental variance (V_E), which is the phenotypic variation for a genotype in different environments (in other words, PP); and a genotype–environment interaction, which corresponds to the fact that different genotypes will not respond similarly to environmental variations, i.e. the genetic variability of PP (V_{GE}). In the simple case of Figure 7.2, where only two environments are considered, the degree of plasticity can be estimated by the slope of the reaction norm. However, in many cases, when several environments are used, reaction norms are not linear and require more complex analyses (David et al. 2004).

$$V_P = V_G + V_E + V_{GE} + V_{error}$$

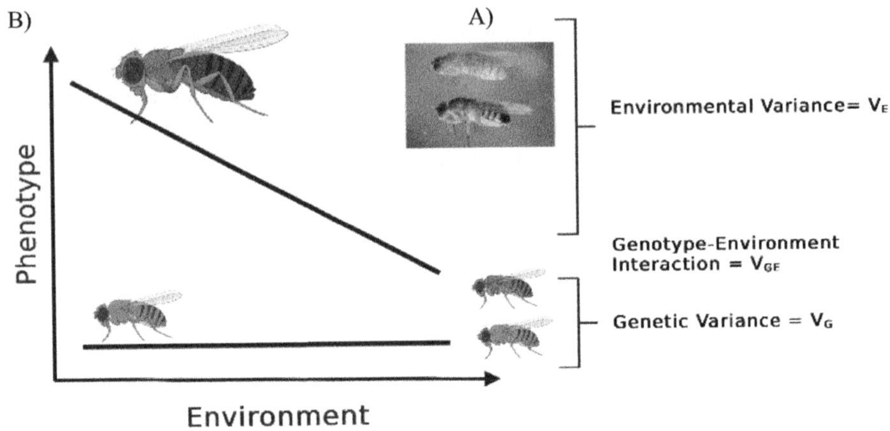

Figure 7.2. *Schematic representation of the reaction norm and some examples of phenotypic plasticity in* Drosophila melanogaster. *For a color version of this figure, see www.iste.co.uk/grunau/epigenetics.zip*

COMMENT ON FIGURE 7.2.– *(A) Representation of a linear reaction norm for two genotypes: one plastic and the other non-plastic for the phenotype under consideration (here, for example, size as a function of temperature). (B) Photo of two* Drosophila melanogaster *females reared at a cold temperature (18°C below) and a warm temperature (29°C above). Note the much lighter pigmentation and smaller wing size of the female above (photo taken by Jean R. David).*

7.2.2. How can we study PP?

Experimentally, plastic responses are determined by generating replicas of each genotype by cloning or inbreeding and rearing the isogenic replicas[1] in two or more environments. Plants, which are easy to clone, have long been the models of choice for the study of PP. In animals, obtaining identical genotypes can be more difficult, and substitutes such as isofemale lines (established by isolating wild-collected females to initiate half-sibling families (David et al. 2005)) or families etc. are used. Genotypes can then be experimentally subjected to two or more environmental modalities: biotic or abiotic. Abiotic environmental factors that are frequently

1. In an isogenic line, all individuals share the same genetic heritage.

studied include temperature, food availability, water and light. There are numerous examples of this type of experiment in the literature. The biotic environment, such as the presence of a predator or competitor, is also frequently studied. Studies that take several factors into account, such as the interaction between development temperature and host species in the parasitoid *Leptopilina heterotoma* (Ris et al. 2004), are rarer. In general, several genotypes are sampled in different populations (or species), and measurements of different replicates per family and population are carried out. Manipulation can be carried out on genotypes (intrageneration) or their descendants (study of transgenerational plasticity). Another type of approach is the transplant experiment, widely used in plants, which provides greater ecological realism, but does not allow the environmental parameters involved to be controlled. These experiments can also be used to test the existence of local adaptation if the fitness (adaptive value) of local phenotypes is superior to that of exotic phenotypes (e.g. Agren and Schemske (2012)).

7.2.3. *Which traits should be considered?*

Since PP is the property of a given trait in a given environment, it is incorrect to consider one genotype as being more or less plastic than another: a genotype may be plastic for certain traits in a given range of environments but not plastic for other traits in the same environment (Richards et al. 2006). The choice of which traits to study therefore depends on the question being asked. One confusion that often arises concerns the plasticity of fitness traits. In theory, natural selection should act to maintain the highest possible stable fitness, regardless of the environment. Thus, the reaction norm of fitness traits is expected to be flat, in contrast to the reaction norms of the underlying physiological and morphological traits. Depending on the sensitivity of fitness traits to environmental conditions, organisms are considered to be either generalists or specialists. Most experimental and theoretical studies on factors, favoring one or the other strategy, assume that the ability to maintain stable fitness over a wide range of environments comes at the expense of optimal performance, which is expressed as jack of all trades, master of none (Huey and Hertz 1984): specialists are assumed to be better than generalists but in a narrower range of performance. For example, in the Hymenoptera *Venturia canescens*, a moth parasitoid, the stock of mature eggs at the end of life, which is a proxy for fitness, shows a specialist-type temperature response curve for thelytoky populations living in mills or flour mills with relatively stable thermal conditions, in contrast to arrhenotoky populations in a variable environment under natural conditions (Foray et al. 2014).

7.3. Molecular mechanisms of PP

PP can be seen as a sequence of steps from a particular environmental condition to a particular phenotype. The chain begins with an environmental cue, which is perceived by the organism, and leads to a series of reactions that can be translated into a phenotype. Phenotypic responses may be the result of physiological and metabolic changes, or they may be achieved through developmental processes such as cell differentiation or cellular reprogramming. Examples include the variation in muscle mass as a function of the repetition and intensity of mechanical stress; melanization of the skin in response to sun exposure; or the acquisition of thermogenic properties by white fat cells, which differentiate into beige cells when exposed to cold.

The plastic responses obtained through developmental or regulatory processes may, in the broadest sense, involve the interaction between genetic and epigenetic components. Several chromatin factors are involved: DNA or RNA modifications, post-translational histone modifications or the activity of regulatory RNAs (such as miRNAs) (see Chapter 2 for a more detailed look at the effectors of epigenetic mechanisms). Proteins and metabolites (such as transcription factors and endogenous or exogenous hormones) may also be involved in the activation or repression of transcription, including nucleosome positioning and, more broadly, chromatin remodeling. Modifications of the RNA molecule may also play an important role. These modifications are very numerous (Alfonzo et al. 2021), and only a very small number of them have been studied. On the contrary, DNA cytosine methylation is a well-defined epigenetic mark, whose transmission from one cell to another is a known mechanism. High-throughput sequencing technology and third-generation sequencing now make it easy to integrate this type of data into ecology and evolution studies.

One of the classic examples of the link between epigenetics and PP in animals was shown in an isogenic Avy mouse line (see section 5.3.3). In this line, we find animals with coat colors ranging from yellow to brown (pseudoagouti). These differences are due to different levels of methylation of a transposable element inserted upstream of the agouti gene, which changes its expression (see Chapter 3 for details) (Morgan et al. 1999). Remarkably, these different methylation rates are constrained by the mothers' nutritional background, making a direct link between PP, epigenetics and environment (Waterland and Jirtle 2003; Cropley et al. 2006). DNA methylation is an epigenetic modification often associated with PP.

Another emblematic example that has been much studied in recent years concerns social insects. These insects offer some of the most exceptional examples of PP. In honey bees, food (royal jelly) determines whether a larva becomes a fertile queen or a sterile worker. The molecular mechanisms associated with these dramatic developmental differences involve changes in gene expression, which are themselves dependent on juvenile hormone and DNA methylation levels (Maleszka 2018). This extreme morphological and behavioral PP, conserved throughout evolution, must then be robust. It has been suggested that one way of conserving this PP would be to have the genes involved in this response colocalized in clusters in the genome. This evolutionary process took place during the evolution of the Apidae family over the last 80 million years (Duncan et al. 2020).

Identifying the molecular processes underlying PP is more difficult when considering less controlled environments, and when we have less control over the genetic background of the lines under consideration. Experiments on the Pacific oyster (*Crassostrea gigas*) are a good example. If we compare individuals living in an intertidal zone[2] with those that always remain submerged, we find that they differ in multiple traits, including physiological and transcriptomic response to high temperatures (Wang et al. 2021). The authors of this study showed greater plasticity of DNA methylation in response to thermal shock in individuals from the intertidal zone, whereas individuals living in regions that are always submerged, and therefore with a less variable environment, show a higher methylation rate and less plasticity.

An important point to consider, in the context of PP evolution, is that all these processes must be able to be transmitted during cell divisions. Moreover, in some cases, they must be able to persist and be stable once the initial stimulus is no longer present, sometimes long before the individual's development towards the adult state is finalized. The same applies to the transmission of this cellular memory process across generations, and hence to transgenerational effects.

7.4. Evolution of PP

Laboratory breeding experiments have long demonstrated the evolutionary potential of PP (Scheiner 2002). As with any trait, the evolution of PP is only possible if there is heritable variation in the plastic response. The study of heritable variations in PP has been carried out at different levels: intrapopulation, between

2. The intertidal zone or foreshore is the part of the coastline between the highest and lowest tides. This coastal zone is home to a specific biotope that is subject to extreme, even brutal, daily and seasonal biotic and abiotic environmental variations, in particular temperature and air exposure.

populations of the same species and between species. The study of intrapopulation variations in PP, in other words, between genotypes of the same population, provides access to the genetic variation available in a population to respond to selection, and therefore to its heritability. The literature in this field is very rich, and high genetic variability of PP has been observed in all living beings (animals, plants, bacteria, fungi) in which it has been researched (see Pigliucci (2001)).

Numerous theoretical studies have sought to predict conditions under which plastic strategies should evolve. It seems clear that in a perfectly uniform environment, a fixed phenotype will be favored, even in the absence of any cost or limit to phenotypic plasticity. One of the major factors favoring the evolution of PP is therefore environmental variability. Generally speaking, theoretical studies agree that PP should be favored in cases of both temporal environmental variability and fine-grained spatial variability. Environmental heterogeneity is defined as "fine-grained" when the spatial scale of environmental variation is smaller than the dispersal range of a species. On the contrary, larger-scale spatial heterogeneity should rather favor non-plastic specialists (see a review in Snell-Rood and Ehlman (2021)). However, too much environmental variation, especially when there is a lack of reliable cues, can lead to bet-hedging strategies (Tufto 2000) such as dormancy periods in annual desert plants (Gremer and Venable 2014).

Numerous empirical studies support theoretical predictions of greater PP for genotypes subjected to strong environmental variation. For example, in a paper published in 2011, Baythavong (2011) studied the PP of various phenotypic traits in an annual plant, *Erodium cicutarium*, in a serpentine mosaic meadow landscape in California. He was able to demonstrate that the level of PP varied according to the grain of environmental heterogeneity within habitats. PP was much higher on serpentine soils, which are a much finer-grained habitat than non-serpentine plots. In addition to environmental heterogeneity, another important factor in the evolution of PP is its predictability (Moran 1992). The unpredictability of environmental variability (particularly of a temporal nature) is an obstacle to PP.

If PP enables individuals to have high fitness regardless of the environment, then we would expect this strategy to always be favored by natural selection. Of course, no case of absolute plasticity has ever been observed in nature, which suggests that PP is subject to certain constraints, limiting its selective advantage. Several types of constraint to the maintenance of adaptive plasticity have been proposed. From a theoretical point of view, there are two categories: costs and limits (Dewitt 1998). The costs associated with plasticity are responsible for the difference in fitness between a plastic organism and a non-plastic organism expressing the same trait value in a given environment. Limits, on the other hand, correspond to the inability of a plastic organism to produce a trait value as close to the optimum as would a non-plastic organism. For example, in Daphnia, a cost related to the ability to induce a backbone in response to a predator would be detected if the fitness of the plastic genotype in the predator-free environment was lower than that of a specialist that does not express a backbone. There would be a limit to plasticity if a less pronounced spine was induced in the plastic genotype compared with a specialist producing a spine in all environments.

Theoretical models on the evolution of plasticity agree that costs reduce the probability of PP evolution, often favoring the evolution of a fixed phenotype (Scheiner 1993). A large body of empirical work has sought to measure these costs of plasticity. For example, a cost has been demonstrated in the crustacean *Daphnia pulex* (Scheiner and Berrigan 1998) and the amphibian *Rana sylvatica* (Relyea 2002) in response to the presence of predators. However, empirical studies often fail to find the expected costs of PP. Several hypotheses have been put forward to explain this inconsistency between empirical and theoretical studies. Snell-Rood and Ehlman (2021) raise two points, in particular: the fact that there must be strong selection against the costs of PP and that over time, mechanisms should evolve to reduce these costs of plasticity; the fact that it is often difficult to determine which are the best traits to measure when studying the costs of adaptive plasticity.

While PP is often adaptive, allowing intragenerational phenotype adjustment to current environmental conditions, in some cases, PP can be ill-adapted (Ghalambor et al. 2007). This may be the case, for example, when environmental variations are stochastic, and there is a mismatch between the phenotype produced and the environment. Such mismatches, which have already been observed in the phenology of many species (Thackeray et al. 2010), are likely to multiply with future environmental changes. A major challenge for organisms will therefore be to assess their environment, distinguish the environmental signal from the noise and respond adaptively. To achieve this, natural selection should favor sensory and neural systems capable of detecting and distinguishing relevant information from irrelevant information.

7.5. Evolution through PP

With the emergence of genetics and the development of the “modern synthesis” (see Chapter 1) of evolution, the first half of the 20th century was strongly dominated by the idea that PP was irrelevant to evolutionary processes. The argument was that, since evolution requires heritable changes and plasticity cannot be inherited, PP could not affect evolution. However, as we will see, not only can PP have an impact on evolution, even though the plastic response itself is not inherited, but some plastic responses can, in fact, be inherited.

A first obvious point is that to say that plasticity influences evolution does not necessarily mean that it accelerates that evolution, with some examples even suggesting that PP could slow down evolution. Indeed, if PP is adaptive, i.e. if the phenotype produced by PP is very close to the optimal value favored by natural selection in the new environment, then the population should undergo stabilizing selection without further genetic differentiation (unless PP comes at a substantial cost). In other words, plasticity should slow down or limit adaptive genetic differentiation between populations. A well-known example to illustrate this point is the introduction of *Pennisetum setaceum* to Hawaii some one hundred years ago. This plant gradually colonized the island, becoming dominant in communities ranging from sea level to almost 3,000 m in altitude, characterized by highly contrasting rainfall and temperature conditions. Individuals from these different sites show spectacular differences in morphology, physiology and reproductive strategy. However, reciprocal transplantation experiments have revealed very little genetic differentiation between populations for most of these traits (Williams et al. 1995), suggesting that adaptive PP leads to such a good match with the environment that there is no possibility of directional selection, and hence no evolution.

Another way in which PP can affect the evolution of organisms is simply by facilitating population persistence in new environments. Indeed, by improving population persistence in a new environment, PP will "buy time" until genetic adaptations to suit new conditions can evolve (Chevin and Lande 2010). Reed et al. (2010) modeled this ability of PP to allow populations to "buy time" in response to a time-varying environment. They found that when environmental stochasticity was moderate and predictable, even weakly adaptive PP could protect populations and enable them to persist. On the contrary, when environmental stochasticity was high and unpredictable, PP led to a significant mismatch between phenotype and environment, potentially leading to extinction.

PP can also influence the evolution of organisms in a way known as "plasticity-led evolution" (Levis and Pfennig 2019). The traditional view of the evolutionary process by mutation/selection is that a new phenotype appears after a genomic change (mutation-led evolution). However, most natural populations are reservoirs of so-called cryptic genetic variation, which is not expressed under normal environmental conditions, but which, under new and stressful environmental conditions, can be revealed, leading to the expression of new phenotypes. One possible molecular mechanism for this process is a class of chaperone proteins, Hsp90, originally discovered because of their heat-inducibility, and which play a crucial role in protecting, maintaining and regulating the functions of the proteins with which they are associated. In 1998, Rutherford and Lindquist (1998) showed that a decrease in Hsp90 expression led to an explosion of phenotypic variation in Drosophila. If the newly expressed phenotypes are well adapted to the new environment, then selection may favor the associated genotypes, leading to a better match between phenotype and environment (West-Eberhard 2003). The key point here is that the heritable variation that has fueled this response is revealed to selection through plasticity. The term *genetic assimilation* was proposed in the last century by Waddington, to designate this process whereby a trait that was originally induced by the environment is, through a process of directional selection, determined to be genetically associated with a loss of PP (Pigliucci and Murren 2003). Since Waddington's work, the process of genetic assimilation has been the subject of various theoretical models (Chevin and Lande 2010; Scheiner et al. 2017), laboratory experiments and studies in natural populations (see a recent review on the subject in Scheiner and Levis (2021)). One of the best-known and cited examples is an experimental evolution study published by Suzuki and Nijhout (2006). In this article, the authors studied the experimental evolution of the *Manduca sexta* caterpillar coloration in response to heat shock. Over 13 generations, they succeeded in selecting a line of caterpillars that reacted strongly to thermal shock (in other words, highly plastic), developing a green coloration, and a line of caterpillars that remained black and had a reduced sensitivity to shock, channeled through genetic assimilation.

The final and probably most controversial way in which PP can facilitate evolution is through hereditary mechanisms other than DNA, such as epigenetic mechanisms. This transgenerational plasticity is a form of non-genetic heredity that includes the maternal effects that have already been well described and are known in the literature, as well as cultural transmission and learning, but goes much further by involving the effects of the environment experienced by both parents, and sometimes even by grandparents, or more (see Bonduriansky (2021)). There are many examples of transgenerational PP. For example, in some species, exposure to a predator can not only induce phenotypic changes in the individuals exposed to it but also in their descendants (Luquet and Tariel 2016). Furthermore, in many cases, this transgenerational PP is adaptive. For example, juveniles of the common reef fish *Acanthochromis polyacanthus* were able to acclimatize to a stressful 3°C increase in water temperature if their parents had themselves been exposed to high temperatures (Ryu et al. 2018) through methylation changes in several metabolic genes. Although it is unclear over how many generations, and with what reliability, such transgenerational PP can be inherited in natural populations, it could play an important role in increasing the chances of survival until more permanent genetic changes occur.

7.6. Conclusion

In this chapter, we have seen that there are various molecular mechanisms involved in PP, and in particular the important role played by epigenetic mechanisms. We have also seen that there is genetic variability in PP and that it can therefore evolve under the effect of natural selection, the conditions favoring this PP now being well described. Furthermore, PP is a mechanism enabling organisms to respond adaptively to environmental change and also enables populations to persist in new or stressful environments. PP can therefore allow populations to "buy time" for an evolutionary change to take place.

Certain areas of PP research are likely to require particular attention in the years ahead.

At the molecular level, one mechanism of PP that has still not been studied in depth is the role played by non-canonical histones such as H2A.Z and H3.3, whose presence acts on gene regulatory and promoter elements, affecting transcription. During the formation of a canonical nucleosome, histones H3-H4, organized as tetramers, are formed first, followed by two H2A-H2B dimers. The insertion and renewal of these histone variants in nucleosomes further increase the complexity of epigenetic gene-regulation mechanisms, a complexity further enhanced by post-translational modifications. For example, both transcription-ready promoters and those actively transcribing genes are enriched in H2A.Z, but only active genes show significant levels of H2A.Z acetylation. This indicates that regulation not only depends on the presence of the H2A.Z variant but also on its acetylation state (Beckman et al. 2021). The involvement of histone variants is the subject of recurrent studies in neural function and brain plasticity. For example, dynamic regulation of H2A.Z is associated with induction of regulatory gene activity, as is the case in brain neural plasticity and memory consolidation (Bano et al. 2017).

Transgenerational PP is a fairly recent subject of research, which there are many questions about. First, what are the proximate mechanisms of this PP? Indeed, they involve a complex chain of mechanisms ranging from the perception of the environmental signal and its effects on the parental phenotype, to the transmission of these effects to offspring (this book describes the many possible mechanisms) and the consequences for their development and final phenotype. Another important question concerns the stability of this PP over generations. Relatively few studies to date have focused on the persistence of transgenerational PP over more than two generations, even though certain (epiallelic) methylation patterns could provide the mechanisms. Finally, the question of the potential adaptive capacity of this transgenerational PP remains largely unexplored. It is easy to understand that the predictability of the environment will be a key factor. If the environment is predictable, transgenerational plasticity will give the offspring a developmental head start, whereas an environment that is not very predictable will increase the probability of a mismatch between the phenotype and the environment. Depending on this cost–benefit balance, natural selection should either favor intragenerational or transgenerational PP. Whether adaptive or non-adaptive, the potential role of transgenerational PP in evolutionary processes should therefore constitute an important field of study, both theoretically and empirically, in the years to come.

Finally, it is worth pointing out the scarcity of empirical studies on plasticity-driven evolution and, in particular, of literature on the role this mechanism might play in innovation and evolutionary diversification (see Levis and Pfennig (2021) for a full discussion of this topic). This is certainly a line of research that will be expanded on in the near future.

7.7. References

Agren, J. and Schemske, D.W. (2012). Reciprocal transplants demonstrate strong adaptive differentiation of the model organism *Arabidopsis thaliana* in its native range. *New Phytologist*, 194, 1112–1122.

Alfonzo, J.D., Brown, J.A., Byers, P.H., Cheung, V.G., Maraia, R.J., Ross, R.L. (2021). A call for direct sequencing of full-length RNAs to identify all modifications. *Nature Genetics*, 53(8), 1113–1116. doi: 10.1038/s41588-021-00903-1.

Bano, D., Piazzesi, A, Salomon, P., Nicotera, P. (2017). The histone variant H3.3 claims its place in the crowded scene of epigenetics. *Aging*, 9(3), 602–614.

Baythavong, B.S. (2011). Linking the spatial scale of environmental variation and the evolution of phenotypic plasticity: Selection favors adaptive plasticity in fine-grained environments. *American Naturalist*, 75–87. doi: 10.1086/660281.

Beckman, W.F., Jiménez, M.Á.L., Verschure, P.J. (2021). Transcription bursting and epigenetic plasticity: An updated view. *Epigenetics Communications*, 1(1), 1–13. doi: 10.1186/s43682-021-00007-1.

Bonduriansky, R. (2021). Plasticity across generations. In *Phenotypic Plasticity & Evolution Causes, Consequences, Controversies*, Pfennig, D.W. (ed.). CRC Press, Boca Raton, 327–366.

Bradshaw, A.D. (1965). Evolutionary significance of phenotypic plasticity in plants. *Advances in Genetics*, 13, 115–155.

Chevin, L. and Lande, R. (2010). When do adaptive plasticity and genetic evolution prevent extnction of a density-regulated population? *Evolution*, 64, 1143–1150. doi: 10.1111/j.1558-5646.2009.00875.x.

Cropley, J.E., Suter, C.M., Beckman, K.B., Martin, D.I.K. et al. (2006). Germ-line epigenetic modification of the murine Avy allele by nutritional supplementation. *Proceedings of the National Academy of Sciences of the United States of America*, 103(46), 17308–17312. doi: 10.1073/pnas.0607090103.

David, J.R., Gibert, P., Moreteau, B. (2004). Evolution of reaction norms. In *Phenotypic Plasticity: Functional and Conceptual Approaches*, DeWitt, T. and Scheiner, S.M. (eds). Oxford University Press, New York, 50–63.

David, J.R., Gibert, P., Legout, H., Pétacy, G., Capy, P., Moreteeau, B. (2005). Isofemale lines in drosophila: An empirical approach to quantitative trait analysis in natural populations. *Heredity*, 94(1), 3–12. doi: 10.1038/sj.hdy.6800562.

Dewitt, T.J. (1998). Costs and limits of phenotypic plasticity: Tests with morphology and life history in a freshwater snail. *Journal of Evolutionary Biology*, 11, 465–480.

Duncan, E.J., Leask, M.P., Dearden, P.K. (2020). Genome architecture facilitates phenotypic plasticity in the honeybee (*Apis mellifera*). *Molecular Biology and Evolution*, 37(7), 1964–1978. doi: 10.1093/molbev/msaa057.

Foray, V., Desouhant, E., Gibert, P. (2014). The impact of thermal fluctuations on reaction norms in specialist and generalist parasitic wasps. *Functional Ecology*, 28(2), 411–423. doi: 10.1111/1365-2435.12171.

Ghalambor, C.K., McKay, J.K., Carroll, S.P., Reznick, D.N. (2007). Adaptive versus non-adaptive phenotypic plasticity and the potential for contemporary adaptation in new environments. *Functional Ecology*, 21(3), 394–407. doi: 10.1111/j.1365-2435.2007.01283.x.

Gremer, J.R. and Venable, D.L. (2014). Bet hedging in desert winter annual plants: Optimal germination strategies in a variable environment. *Ecology Letters*, 17(3), 380–387. doi: 10.1111/ele.12241.

Huey, R.B. and Hertz, P.E. (1984). Is a jack-of-all-temperatures a master of none? *Evolution*, 38(2), 441–444. doi: 10.2307/2408502.

Levis, N.A. and Pfennig, D.W. (2019). Phenotypic plasticity, canalization, and the origins of novelty: Evidence and mechanisms from amphibians. *Seminars in Cell and Developmental Biology*, 88, 80–90. doi: 10.1016/j.semcdb.2018.01.012.

Levis, N.A. and Pfennig, D.W. (2021). Innovation and diversification via plasticity-led evolution. In *Ecology and Evolution of Plasticity*, Pfennig, D.W. (ed.). CRC Press, Boca Raton, 211–240.

Luquet, E. and Tariel, J. (2016). Offspring reaction norms shaped by parental environment: Interaction between within- and trans-generational plasticity of inducible defenses. *BMC Evolutionary Biology*, 16, 209. doi: 10.1186/s12862-016-0795-9.

Maleszka, R. (2018). Beyond royalactin and a master inducer explanation of phenotypic plasticity in honey bees. *Communications Biology*, 1(1), 1–7. doi: 10.1038/s42003-017-0004-4.

Moran, A.N. (1992). The evolutionary maintenance of alternative phenotypes. *The American Naturalist*, 139(5), 971–989.

Morgan, H.D., Sutherland, H.G.E., Martin, D.I.K., Whitelaw, E. (1999). Epigenetic inheritance at the agouti locus in the mouse. *Nature Genetics*, 23(3), 314–318. doi: 10.1038/15490.

Pigliucci, M. (2001). *Phenotypic Plasticity. Beyond Nature and Nurture.* Johns Hopkins University Press, Baltimore.

Pigliucci, M. and Murren, C.J. (2003). Perspective: Genetic assimilation and a possible evolutionary paradox: Can macroevolution sometimes be so fast as to pass us by? *Evolution; International Journal of Organic Evolution*, 57(7), 1455–1464 [Online]. Available at: https://pubmed.ncbi.nlm.nih.gov/12940351/.

Reed, T.E., Waples, R.S., Schindler, D.E., Hard, J.J., Kinnison, M.T. (2010). Phenotypic plasticity and population viability: The importance of environmental predictability. *Proceedings: Biological Sciences*, 277, 3391–3400. doi: 10.1098/rspb.2010.0771.

Relyea, R.A. (2002). Costs of phenotypic plasticity. *The American Naturalist*, 159(3), 272–282. doi: 10.1086/338540.

Richards, C.L., Bossdorf, O., Muth, N.Z., Gurevitch, J., Pigliucci, M. (2006). Jack of all trades, master of some? On the role of phenotypic plasticity in plant invasions. *Ecology Letters*, 9(8), 981–993. doi: 10.1111/j.1461-0248.2006.00950.x.

Ris, N., Allemand, R., Fouillet, P., Fleury, F. (2004). The joint effect of temperature and host species induce complex genotype-by-environment interactions in the larval parasitoid of *Drosophila*, *Leptopilina* heterotoma (Hymenoptera: Figitidae). *Oikos*, 106(3), 451–456. doi: 10.1111/j.0030-1299.2004.13274.x.

Rutherford, S.L. and Lindquist, S. (1998). Hsp90 as a capacitor for morphological evolution. *Nature*, 6709, 336–342. doi: 10.1038/24550.

Ryu, T., Veilleux, H.D., Donelson, J.M., Munday, P.L., Ravasi, T. (2018). The epigenetic landscape of transgenerational acclimation to ocean warming. *Nature Climate Change*, 8(6), 504–509. doi: 10.1038/s41558-018-0159-0.

Scheiner, S.M. (1993). Genetics and evolution of phenotypic plasticity. *Annual Review of Ecology and Systematics*, 24(1), 35–68. doi: 10.1146/annurev.es.24.110193.000343.

Scheiner, S.M. (2002). Selection experiments and the study of phenotypic plasticity. *Journal of Evolutionary Biology*, 15(6), 889–898. doi: 10.1046/j.1420-9101.2002.00468.x.

Scheiner, S.M. and Berrigan, D. (1998). The genetics of phenotypic plasticity. VIII. The cost of plasticity in *Daphnia pulex*. *Evolution*, 52, 368–378. doi: 10.2307/2411074.

Scheiner, S.M. and Levis, N.A. (2021). The loss of phenoypic plasticity via natural selection: Genetic assimilation. In *Phenotypic Plasticity & Evolution Causes, Consequences, Controversies*, Pfennig, D.W. and West-Eberhard, M.J. (eds). CRC Press, Boca Raton, 161–181.

Scheiner, S.M., Barfield, M., Holt, R.D. (2017). The genetics of phenotypic plasticity. XV. Genetic assimilation, the Baldwin effect, and evolutionary rescue. *Ecology and Evolution*, 7, 8788–8803. doi: 10.1002/ece3.3429.

Snell-Rood, E.C. and Ehlman, S.M. (2021). Ecology and evolution of plasticity. In *Phenotypic Plasticity & Evolution Causes, Consequences, Controversies*, Pfennig, D.W. (ed.). CRC Press, Boca Raton, 139–160.

Suzuki, Y. and Nijhout, H.F. (2006). Evolution of a polyphenism by genetic accommodation. *Science (New York)*, 311(5761), 650–652. doi: 10.1126/science.1118888.

Thackeray, S.J., Sparks, T.H., Frederiksen, M., Burthes, S., Bacon, P.J., Bell, J.R., Botham, M.S., Brereton, T.M., Bright, P.W., Carcalhos, L. et al. (2010). Trophic level asynchrony in rates of phenological change for marine, freshwater and terrestrial environments. *Global Change Biology*, 16(12), 3304–3313. doi: 10.1111/j.1365-2486.2010.02165.x.

Tufto, J. (2000). The evolution of plasticity and nonplastic spatial and temporal adaptations in the presence of imperfect environmental cues. *The American Naturalist*, 156(2), 121–130. doi: 10.1086/303381.

Wang, X., Li, A., Wang, W., Que, H., Zhang, G., Li, L. (2021). DNA methylation mediates differentiation in thermal responses of Pacific oyster (*Crassostrea gigas*) derived from different tidal levels. *Heredity*, 126(1), 10–22. doi: 10.1038/s41437-020-0351-7.

Waterland, R.A. and Jirtle, R.L. (2003). Transposable elements: Targets for early nutritional effects on epigenetic gene regulation. *Molecular and Cellular Biology*, 23(15), 5293–5300. doi: 10.1128/mcb.23.15.5293-5300.2003.

West-Eberhard, M.J. (2003). *Developemental Plasticity and Evolution*. Oxford University Press, Oxford.

Williams, D.G., Mack Richard, N., Black, A.R. (1995). Ecophysiology of introduced *Pennisetum setaceum* on Hawaii: The role of phenotypic plasticity. *Ecology*, 76(5), 1569–1580. doi: 10.2307/1938158.

Woltereck, R. (1909). Weitere experimentelle Untersuchungen uber Artveranderung, speziell uber das Wesen quantitativer Artunterschiede bei Daphniden. *Verhandlungen der Deutschen Zoologischen Gesellschaft*, 19, 110–172.

8

Epigenetics and Climate Change: The Example of Forest Ecosystems

Stéphane MAURY[1] **and Christophe PLOMION**[2]

[1] *P2E, Université d'Orléans, France*

[2] *INRAE, Biogeco, Université de Bordeaux, Cestas, France*

8.1. Introduction: an ecological crisis on an unprecedented scale

The impact of human activities on the environment and human societies can be observed on a global scale. The term "global change" (GC) expresses a generalized state of crisis, which is reflected in growing socio-economic inequalities and advanced environmental destruction. The vulnerability of most natural ecosystems to GC is reflected in a well-documented collapse in biodiversity. Not only are we losing a heritage that jeopardizes the functioning of today's ecosystems, but we are also reducing nature's capacity to contribute to the activity and well-being of human societies. Other consequences of GC include the degradation or even loss of natural habitats, the acceleration of biological invasions and the emergence of new diseases that thrive off global warming and the globalization of trade, as well as the increase in chemical pollution of air, water and soil by plastics, pesticides, nitrates, heavy metals, etc. These phenomena are intrinsically linked in what can be called an "ecological crisis", which reflects the extremely degraded state of life and the environment, as well as the rate at which the situation is evolving[1].

1. See the summary on the Société française d'écologie website: https://sfecologie.org/actions/positions/climat_biodiversite.

Figure 8.1. *We are living in an era of great environmental change. Could epigenetics be one of the keys to enabling organisms to adapt quickly enough? (see also Chapters 9–12). Illustration by Annette Köhn*

In response, under the impetus of global scientific expertise (the IPBES – Intergovernmental Science-Policy Platform on Biodiversity and Ecosystem Services and the IPCC – Intergovernmental Panel on Climate Change), demonstrating the links between ecosystem health and human well-being, companies are setting their sights on multifunctional, sustainable resource management, as well as the preservation of natural environments and associated species. Sustainable management and protection of ecosystems means conserving and enhancing natural resources, conserving and restoring environments, and maintaining them in a healthy state, in order to satisfy the various functions or services (production, biodiversity, regulation, protection and various amenities) they provide.

To curb this ecological crisis, which threatens entire communities of species with sudden extinction, the international community, notably under the aegis of the United Nations, has carried out the following:

– signed various species and habitat protection agreements;

– set up the Intergovernmental Platform on Biodiversity and Ecosystem Services;

– and, on the model of the climate conferences, is aiming for an international agreement to protect at least 30% of the planet, a goal taken up by the European Commission for 2030.

Beyond these texts, it is clear that these discourses are no longer enough, and that the conservation, management and restoration of ecosystems remain a major challenge if we are to maintain sustainable interactions between humans and the environment. In this context, the role of science has never been more important. On the one hand, it is a question of producing fundamental knowledge that will enable us to better understand the responses of individuals, populations, species and communities of organisms to GCs, as well as the biological processes underlying their evolution and adaptation. On the other hand, it is a question of using this knowledge to better predict the future evolution of ecosystems and, ultimately, to find appropriate management strategies. Among the major challenges facing science, understanding the mechanisms by which living organisms respond to GCs is a priority. In this chapter, we present the example of forest ecosystems and, more specifically, the keystone organisms that make them up: trees. After defining the two components of adaptation in section 8.2 – phenotypic plasticity (PP) and genetic evolution of traits (Figure 8.2) – we will focus on the underlying epigenetic processes and their possible applications in forestry in section 8.3.

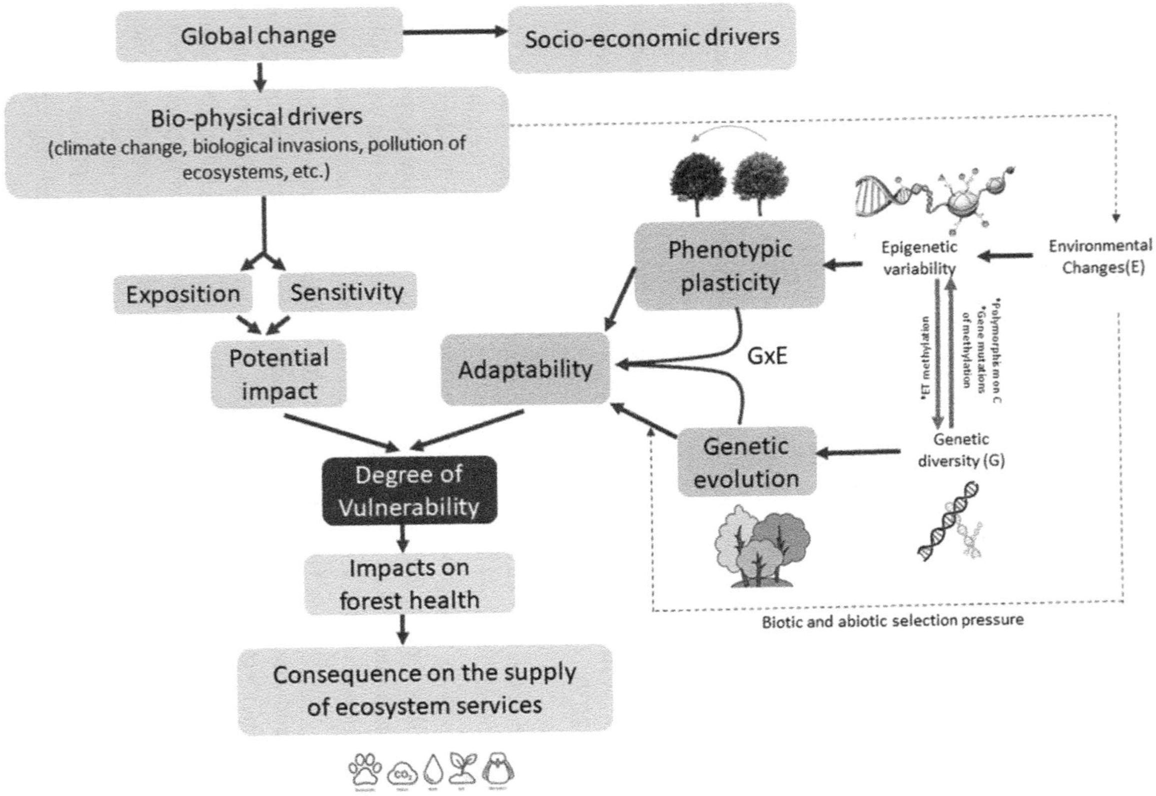

Figure 8.2. *The role of phenotypic plasticity and genetic evolution of a trait in the conceptual framework of forest vulnerability. For a color version of this figure, see www.iste.co.uk/grunau/epigenetics.zip*

COMMENT ON FIGURE 8.2. – *Vulnerability to the environmental forcing factors of global change results from the trees' degree of exposure to biotic and abiotic disturbances, and their sensitivity to these stress factors (which define a potential impact), as well as from the capacity of trees to adapt to new environmental conditions, which is itself a function of the phenotypic plasticity of individuals (which depends on epigenetic mechanisms) and the genetic evolutionary capacity of populations (linked to their genetic and epigenetic diversity). TE = transposable element. C = cytosine.*

Provision services	- Wood supply (energy and material) - Non-timber forest goods (bio sourced molecules, food products, availability of fresh water)
Regulation services	- Regulation of global climate and local microclimate - Carbon storage and climate change mitigation - Water quality - Protection against natural hazards (floods, avalanches, landslides, etc.) and on coastal areas (stabilization of dunes, certain coastlines, etc.) - Soil formation and stabilization
Cultural and recreational services	- Recreational and well-being activities (artistic inspiration, sport, harvesting) - Hunting - Support for scientific and educational research activities (naturalist observation)
Natural heritage	- Labelling and heritage values of French forests

Table 8.1. *Summary of the main ecosystem services provided by forests. According to www.fondationbiodiversite.fr/les-services-rendus-par-les-ecosystemes-forestiers-une-evaluation-dans-le-cadre-du-programme-efese/*

8.2. Forests and climate change: from current situation to challenges

8.2.1. *The role of forests and trees in ecosystems*

Forests are home to 73,000 tree species. They cover almost 40% of the earth's surface, corresponding to approximately 4 billion hectares (FAO 2022) – a figure that has fallen by 20% in a century. They provide multiple ecosystem services, as summarized in Table 8.1. Trees are also essential components of ecosystem structure and function. As such, they are often referred to as "keystone" species. Ensuring the health of our forests therefore means helping to maintain the overall stability of terrestrial ecosystems, as well as the major biophysical and biogeochemical processes involved in climate change on both global and local scales.

8.2.2. *Adapting trees to their environment*

Forest ecosystems are being degraded at an unprecedented rate in all biomes. This degradation is particularly exacerbated in trees, which are long-lived sessile organisms, unable to move quickly (due to their long reproductive cycle), and which will either have to adapt in situ, or be driven to extinction. Weakened by successive and increasingly intense droughts or heatwaves (Allen et al. 2010; Anderegg et al. 2016), and also by winters with low temperatures or late spring frosts, forest stands are more vulnerable to attack by phytopathogenic fungi and insect pests.

Trees, however, are champions in all categories of adaptation, a state that describes their ability to develop and reproduce in a given environment. The best example of this is their adaptation to rising temperatures (+2 to +6°C) during the Holocene between 12,000 and 6,000 years ago (thermal maximum) (Kaufman et al. 2020). During the last ice age in Europe, most forest tree species were confined to southern refuges (southern Iberia and the Italian peninsula, as well as the Balkans). While they were migrating (via the dissemination of seeds and thus their offspring) northwards as a result of this warming, they also survived (and still survive) in these refuges on the southern borders of their current range.

Adaptation can also be defined as a response process that enables the adjustment of biological functions to environmental variability. It occurs at the individual level through the PP of traits, and at the population level through their genetic evolution (Figure 8.2).

8.2.2.1. *Adaptation through PP of traits*

As detailed in Chapter 7, the PP of a trait represents the ability of an individual to express different phenotypes for that trait, according to the spatiotemporal environment in which it finds itself (Bradshaw 1965). It is a highly flexible and reversible component, enabling rapid adjustment of physiology, morphology or tree development. It therefore expresses itself within a generation without changing the genetic context (Callahan 1997). PP is said to be “random” when it responds to an environmental variation that cannot be predicted by the plant, such as an attack by a pathogen. It is said to be “programmed” when it responds to an environmental variation anticipated by the plant, such as the response to seasonal variations in wood anatomy in temperate environments (Figure 8.3A). Lastly, when phenotype change results from a fine-tuned interaction between development and environment, we speak of “developmental plasticity”. Conversely, robustness is defined as the maintenance of a plant’s phenotype in contrasting environments. Plasticity and robustness are controlled by molecular mechanisms that, in particular, enable a trade-off in plant response (the cause of which is still poorly understood) for the plant (Lachowiec et al. 2016). Ultimately, PP can be adaptive if trait variability increases its selective value, it can be maladaptive if it decreases, or neutral if it has no consequences (Ghalambor et al. 2007).

PP is expressed both at the aerial level of the tree, which responds to daily or seasonal variability in atmospheric conditions (temperature, humidity, precipitation, wind), as during wood formation (Figure 8.3A) or the budburst of vegetative and floral buds in spring in temperate environments, and at the underground level, where roots must constantly adapt to the biogeochemical and hydric environments they encounter. The PP of a trait is often represented by a reaction norm (Figure 8.3B), which describes its variability in response to that of the environment, whether this be biotic or abiotic disruptions, or, in nature, a combination of both.

In trees, PP plays a key role, as these organisms cannot move to find better conditions for growth and reproduction. Paradoxically, the physiological and molecular foundations of this major flexibility mechanism remain poorly studied and, therefore, poorly understood. As climate change threatens to push back the adaptive limits of this process, we urgently need to document whether PP will be sufficient to allow them to persist in situ. Will it enable trees to adapt to even more extreme climatic conditions that are sometimes of a multifactorial and cumulative nature (early frost, drought, pests)? In this respect, the example of resistance to cavitation (formation of small air bubbles in vessels) is particularly worrying. Recent studies have indeed shown that the threshold of vulnerability to xylem embolism (gaseous phase hindering the flow of raw sap) is highly constrained, regardless of the biomes and species in which it has been measured (Choat et al. 2012), thus leaving the trees little room to maneuver before they “die of thirst” (see the ever-increasing “crown decline” we see in tree crowns, Figure 8.4). Thus, PP has physiological limits that need to be documented in relation to its adaptive potential.

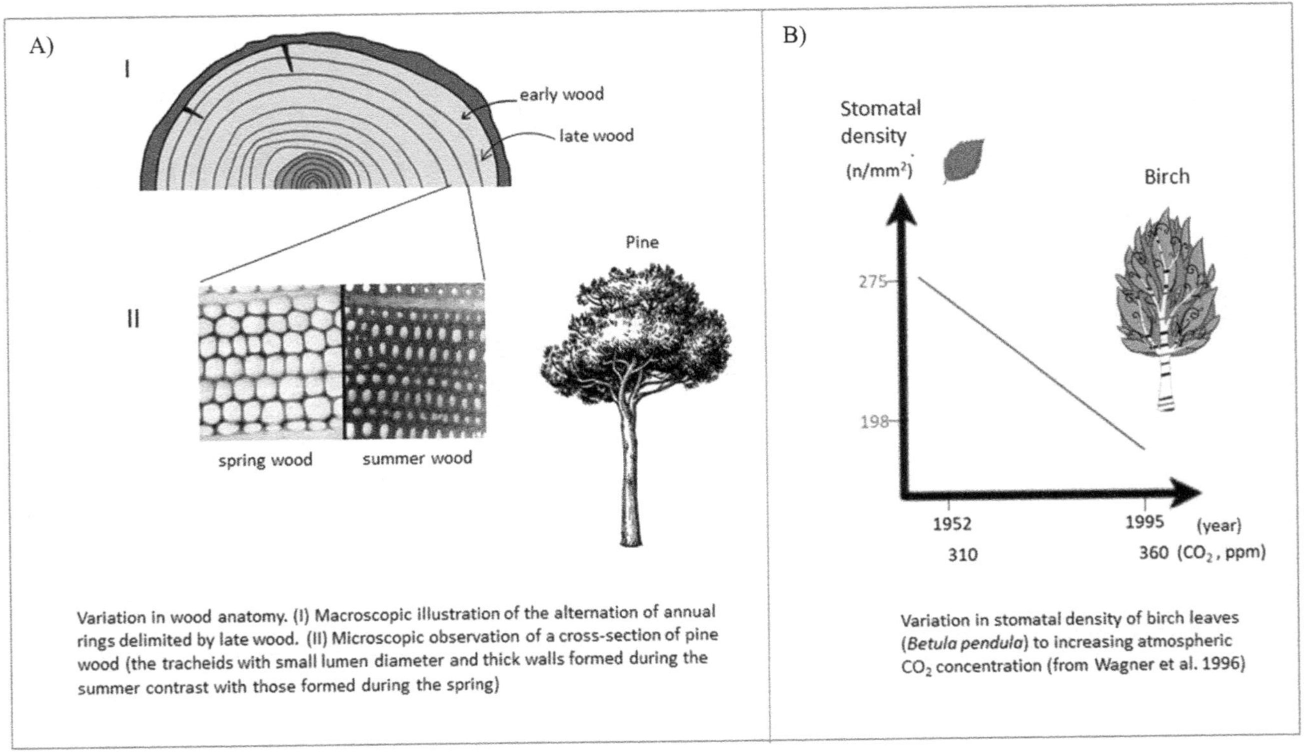

Figure 8.3. *Example of phenotypic plasticity in two tree traits: A) Anatomical variation in pine wood within an annual growth ring (programmed plasticity); B) Variation (linear decay) in birch leaf stomata density, measured over four decades (random plasticity, from Wagner et al. (1996)). For a color version of this figure, see www.iste.co.uk/grunau/epigenetics.zip*

Figure 8.4. *Illustration of a beech tree's "crown decline" following a gas embolism resulting from cavitation, in other words, air entering the tree's water movement circuit. For a color version of this figure, see www.iste.co.uk/grunau/epigenetics.zip*

COMMENT ON FIGURE 8.4. – *This drying out of the crown is generally accompanied by a regrowing of the lower branches. Photo courtesy of www.republicain-lorrain.fr/edition-de-sarrebourg-chateau-salins/2019/06/15/les-hetres-comptent-leurs-jours.*

Finally, the PP of a trait presents another important limitation, expressed in terms of trade-off. Indeed, in a variable environment, PP can be beneficial for the organism but at the expense of other biological traits; this is referred to as a trade-off between PP and adaptive cost (DeWitt et al. 1998). For example, the benefits for vegetative and/or reproductive development of an increase in the vegetation period, associated with a rise in temperature (and, in particular, earlier budburst and/or flowering), could be eliminated in the event of a late frost.

Research is still in its early stages, and many questions remain open as to the limits of this adaptation process in the face of GC: how will trees respond to environments they have never encountered? If there are gradual environmental changes, will trees acclimatize, and to what physiological limit? How will they cope with extreme events such as heatwaves? And finally, to what extent will changes (adaptive or maladaptive) in PP affect the degree of vulnerability of populations?

8.2.2.2. *Adaptation through genetic evolution of a trait*

Adaptation through genetic evolution is achieved by modifying the genetic composition of a population between generations, through the action of natural selection. These modifications are the result of both mutation (a random error that occurs during DNA duplication during mitosis or meiosis) and genetic recombination during meiosis. Genetic diversity created in this way provides the fuel needed for the evolutionary engine, enabling populations to adapt to changes in the environment. Necessary, but not sufficient; the favorable genetic variant must be able to reach a sufficient frequency in order to enable it to persist and spread in the population subjected to a new environment. In this respect, the large population size of forest species is an important factor in maintaining genetic diversity and selecting the variants that enable its carriers to adapt (Hamrick 2004). In a context of GC, where numerous forcing factors tend to fragment forest habitats and considerably reduce population size (Figure 8.2), the question of maintaining genetic diversity is raised. However, the longevity of trees (which can delay the extinction of populations, as well as enable individuals and populations to survive until the habitat recovers) and the high level of gene flow per pollen (which counteracts the effects of genetic drift) are all assets that should enable tree species to maintain a high level of diversity under changing environmental conditions and therefore potentially adapt to new stresses.

The process of adaptation requires that the traits selected (resistance to a lack of water supply, for example) are variable, that this variability is partly heritable – in other words, transmissible between generations – and linked to the fitness or selective value (measure of fecundity and survival) of the individuals confronted with the stress. Since genetic (or epigenetic) changes are measured between generations, the shorter the generation time of the species, the faster the genetic evolution. As a result of its long generation time, a tree would not be able to adapt genetically as quickly in response to environmental change as its larger predators (insects, viruses, bacteria, phytopathogenic fungi, etc.) with shorter life cycles. With the acceleration of global warming (on the scale of one or a few generations of trees), the question of the speed of genetic evolution, and therefore of tree adaptation through this process, has become a major concern for foresters. The rapid adaptation of species introduced from one continent to another (red oak and black locust from North America to Europe, for example) nevertheless suggests that trees have been able, within a few generations, to adapt to a new environment. Recent data obtained on sessile oak since the Little Ice Age also point towards rapid adaptation, with changes in allele frequency being observable on the scale of a few centuries, linked to changes in climate, notably the frequency of extreme events such as harsh winters or intense droughts (Saleh et al. 2022). The genetic diversity of trees would therefore enable them to orientate their evolutionary trajectories in a relatively short space of time and adapt rapidly to changes in the environment.

8.2.2.3. *Relative importance of the two processes in trees and their consequences*

These two components (PP and genetic evolution) can be measured by breaking down the phenotypic variance of a trait measured in a population: one part of the variability of a trait is linked to the environment (plastic part) and the other to genetic inheritance (heritable part); the interaction between the two components makes it possible to address the part of PP that is genetically controlled and potentially subject to natural selection.

Knowing the respective contribution of these two components to the adaptation of populations, and the individuals that make them up, to environmental disruptions is a crucial issue, especially in long-lived species such as trees. Indeed, theory predicts that the relative weight of these two mechanisms depends on the amplitude of environmental variations encountered during an organism's lifetime versus those prevailing between generations. Given their longevity and the impact of the changes they will have to undergo on a generational scale (in other words, with little possibility of genetic evolution, by recombining their genetic material and exposing the result of the recombination to the environmental filter), it is highly likely that the adaptation of existing trees will heavily depend on their PP. In this respect, it should be noted that the mechanisms underlying PP are already very much in demand when it comes to coping with inter-annual environmental variations over the course of a generation.

Finally, to conclude on the importance of taking the PP of adaptive traits into account (and therefore understanding the underlying physiological and molecular mechanisms), it should be noted that statistical models of the future distribution of tree species that incorporate PP are much more accurate and less pessimistic, particularly in terms of the expected disappearance of the southern margins of certain species (Benito Garzón et al. 2019).

While the challenge of adapting trees to GC requires a holistic (from the scale of individuals to that of species and communities) and dynamic understanding at a variety of spatiotemporal scales, the next sections of this chapter will focus on adaptation processes at the individual scale, with an emphasis on the underlying epigenetic molecular mechanisms.

8.3. Epigenetics as a source of flexibility in trees in a context of GC

Epigenetics is considered part of heritability, since the information encoded in the DNA nucleotide sequence is insufficient to explain biological variation in all its complexity. Epigenetics corresponds to processes affecting gene expression and/or the activity of transposable elements (TEs) without initially altering the DNA sequence. These processes are inheritable through mitosis (during development, from embryogenesis to the juvenile and reproductive vegetative phases) and/or meiosis (across generations) (Figure 8.2; Chapters 1–4). Epigenetic diversity in plants (illustrated by numerous works on the model plant *Arabidopsis thaliana* and certain annual plants such as maize, rice and tomato) thus appears as a new source of expanded phenotypic variation, contributing to the adaptation of wild species to environmental change and helping to ensure the yield and quality of domesticated species, whose genetic diversity has been eroded by intense, directional genetic selection (Richards et al. 2017) (Chapter 10).

Due to their fixed lifestyle, wide ecological distribution (often within "permanent" populations), long lifespan and continuous development, trees have developed numerous adaptations to maintain themselves in changing environments over the course of their evolution. Research efforts are currently focused on understanding the genetic basis of adaptive traits. Over the past decade, epigenetic variability in trees has been the subject of pioneering work on certain tree species for which genomic resources were available (poplar, pine, spruce, oak and eucalyptus) and, in particular, in relation to their phenotypic modifications in response to environmental stresses such as temperature or water availability (Sow et al. 2018; Amaral et al. 2020). This work opens up prospects for the use of this new source of phenotypic variation in the silvicultural management of natural or planted forests. Here, we discuss epigenetic variation at two scales (Figure 8.5): intra-individual (section 8.3.1) and inter-individual or population (section 8.3.2), before presenting possible applications (section 8.3.3).

8.3.1. *Intra-individual epigenetic variation: mosaicism, plasticity, memory and priming*

Epigenetic variability involves DNA methylation (in various cytosine contexts: CG, CHG and CHH, with H = A, T or C), histone post-translational modifications, histone variants and the presence of non-coding RNAs (see Chapters 2 and 3). These epigenetic mechanisms are interconnected, forming an epigenetic regulatory complex. It can affect genes (expression, splicing, etc.), as well as the dynamics of the genome's TEs, which can modify the phenotype. Each individual inherits some parental or modified epigenetic profiles (particularly in hybrids) and also accumulates spontaneous epimutations over the course of its life, all the more so as the organism's longevity is significant. It should be noted that plants have no germ line. Instead, somatic cells from the shoot meristem differentiate into reproductive cells. Moreover, during meiosis and embryogenesis, epigenetic profiles are often "erased" between the mother cell and the daughter gametes (histone marks, CHH and CHG methylation context), with the exception of CG methylation, which enables epigenetic transmission to the progeny of epialleles (alleles that do not vary genetically, but which present different epigenetic marks) induced by the environment during the plant's life.

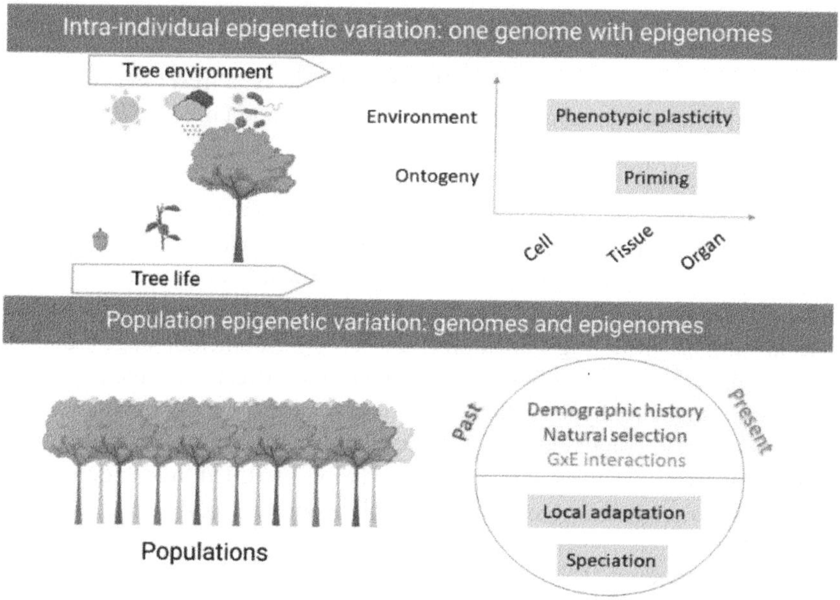

Figure 8.5. *Two scales for studying epigenetic variability in forest trees: intra-individual and population-based. For a color version of this figure, see www.iste.co.uk/grunau/epigenetics.zip*

At the individual level (Figure 8.5), epigenetic variation can be observed between different developmental segments such as organs, tissues or cell types (Vining et al. 2012), and also throughout its life marked by the different stages of ontogeny (embryogenesis, juvenile vegetative phase, reproductive vegetative phase and aging), as well as in response to its biotic and abiotic environments (Sow et al. 2018a; Amaral et al. 2020). Environmentally induced epigenetic variations can be reversible or stable over time and transmitted through mitosis, or even meiosis, to the next generation. As a result of their long lifespan (up to several thousand years), every tree accumulates epimutations (up to 100,000 times more than somatic genetic mutations; (Hofmeister et al. 2020)) in its developmental "segments", the best-known example being buds (containing primary meristems made up of totipotent cells with stem cell traits).

In the context of a tree's response to climate change, it is the entire mosaic epigenetic profile of the individual – which has just been described – in addition to the epigenetic variation induced in response to environmental change that needs to be considered. This mosaicism means that different parts of the same tree can have their own epigenetic signature. There are two notable situations to consider in addition to this mosaicism:

1) The epigenetic component of PP, whether it is a response to recurrent environmental change ("programmed" PP) or a response to unforeseen biotic (insect or pathogen attack) or abiotic (heatwave and drought) stresses ("random" PP).

A classic example of "programmed" PP concerns the annual growth cycles of trees in cold and temperate regions, which are closely linked to the succession of environmental conditions during the season (primary and secondary growth phenology). There is growing evidence to suggest that chromatin remodeling changes are important during transitions between developmental stages, during the annual growth cycle in trees. For example, poplar lines affected for the expression of a demethylase (inducing hypermethylation) showed a delay in budburst after the end of winter dormancy, compared with the wild type. Furthermore, transcriptome and methylome analysis in these DNA demethylation-modified trees identified that demethylation target genes were associated with budburst (Conde et al. 2017).

Concerning "random" PP, numerous results are accumulating on abiotic stresses and show variation in DNA methylation levels following drought, salt stress, exposure to heavy metals, radiation or extreme temperatures in different tree species, notably poplar (Sow et al. 2018a; Amaral et al. 2020). Integrative studies are now attempting to establish correlations between phenotypic or molecular responses and epigenetic modifications, so as to highlight how epigenetic mechanisms control the response of trees to environmental disruptions and, in particular, their PP. Studies on biotic stresses in trees are still limited, particularly in the case of insect attacks, which are nevertheless compounding factors in tree mortality. Recent results have nevertheless highlighted epigenetic variations in trees infected by pathogenic fungi compared with healthy individuals. These variations appear to be involved in the regulation of genes linked to secondary metabolism, such as phenylpropanoids, hormonal pathways, and to a lesser extent, genes linked to interactions with phytopathogenic agents (Amaral et al. 2020).

2) The epigenetic component of the priming effect (Kakoulidou et al. 2021), which states that the epigenetic response memorized during a first stress influences the response when the stress is repeated. Priming is an example of a memory-controlled response to pathogen infection, herbivore attacks or previous abiotic stress. This memory involves the storage of information through an epigenetic process, resulting in a modified response when re-exposed to the same stress. This memory can last from several days to several years and can, in some cases, be transmitted to descendants (Kakoulidou et al. 2021).

The most detailed example of a tree's response to an environmental stress in the scientific literature is that of the poplar and its developmental plasticity in response to drought. The poplar is a tree characterized by strong juvenile growth, but it is highly sensitive to water availability. It can be propagated by cuttings, making it possible to carry out research on clones, thus avoiding any genetic variation. Its genome has also been sequenced, and some species lend themselves easily to genetic transformation and therefore to reverse genetics studies, so as to understand the function of genes involved in epigenetic response. Developmental plasticity involves maintaining the pluripotency of certain meristematic cells by controlling cell identity genes and stimulating their growth. This developmental plasticity involves a complex interaction between epigenetic and hormonal control within meristems (Figures 8.6 and 8.7) (Maury et al. 2019; Sow et al. 2021).

Under moderate water stress, poplars reduce their growth, and in particular the activity of their shoot apical meristem, which is responsible for the development of the leafy stem. It then produces leaves with a smaller surface area and fewer stomata, thereby reducing the tension on the water column in the vessels and thus the risk of embolism. This phenomenon of developmental plasticity is reversible when water availability returns to optimal levels for future leaves in production (but is irreversible for leaves formed during water stress). Although hormones are involved in signaling water stress, their production during stress is labile (a few hours) and cannot alone explain the ability to implement a developmental change over several days requiring the stabilization of hormone-induced gene expression (Figure 8.6). In fact, several hormones act simultaneously, giving rise to a hormonal balance. Some studies show that the integration of this complex hormonal balance information into a biological response (expression of a gene program) is ensured by chromatin, which acts as a hub for integrating hormonal information. Epigenetic work on the shoot apical meristem of a poplar clone, subjected to various water conditions, showed that variations in DNA methylation were, one week after re-watering, involved in controlling the expression of genes in response to water deficit or re-watering, particularly genes in hormonal pathways (biosynthesis, signaling, or response) linked to phytohormone production (Lafon-Placette et al. 2018). This model was then validated by reverse genetics using transgenic poplars, which were modified for DNA methylation and subjected to the same conditions of water stress (DNA hypomethylated lines; Sow et al. (2021)).

This physio-molecular process could partly explain the epigenetic memory and priming of organs after recurrent stress. This idea is supported by experimental results obtained on young poplar leaves forming at the shoot apical meristem during water stress, which, for the youngest leaves alone (20 mm in length, formed during stress), showed a methylation differential between plants placed in watered or drought conditions (Le Gac et al. 2019). This suggests a possible mitotic transmission of epigenetic information from the meristem to emerging young leaves during stress. Furthermore, field-grown poplars that had or had not undergone drought during the vegetative growth period (spring-summer) showed variations in DNA methylation between the two conditions (summer drought or not) in their dormant (winter) shoot apical meristem, six months after stress. This result suggests an environmental epigenetic memory in poplar buds (Le Gac et al. 2018; Sow et al. 2018b).

The most emblematic example of epigenetic memory in trees concerns the environmental memory of growth performance in Norway spruce (Johnsen et al. 2005; Kvaalen and Johnsen 2008). During seed production (embryogenesis), temperature and photoperiod "imprint" a memory of the local maternal environment on the embryo that regulates bud formation phenology, main shoot growth and frost resistance in the progeny. This epigenetic memory is associated with changes in gene expression during the short period of embryonic development (typically two to three months) that promote acclimatization to the maternal environment in the progeny (priming effect). These results were confirmed by experiments using clones of somatic (in other words, genetically identical) embryos formed at different temperatures (18 to 30°C) in vitro. Different epitypes (genotypes that do not differ genetically from one another but that vary "epigenetically") were then obtained, whose phenology varied according to the temperature used and which notably showed changes in the expression of regulators of epigenetic mechanisms (Yakovlev et al. 2014).

New examples of biotic and abiotic priming (insect attacks, hormones, temperatures) have recently been reported in other tree species (Amaral et al. 2020) and are being investigated to clarify the underlying epigenetic mechanisms. Nevertheless, it seems clear that embryogenesis (initiation phase of all meristems, primary and secondary) and morphogenesis within meristems are major developmental stages for "imprinting" an environmental effect favoring acclimation of neoformed organs, as well as in the progeny. Studies on the model plant *Arabidopsis thaliana* have not only confirmed the epigenetic control of priming, but also its hereditary transmission: this is known as transgenerational-induced resistance (TIR) (Hannan Parker et al. 2022).

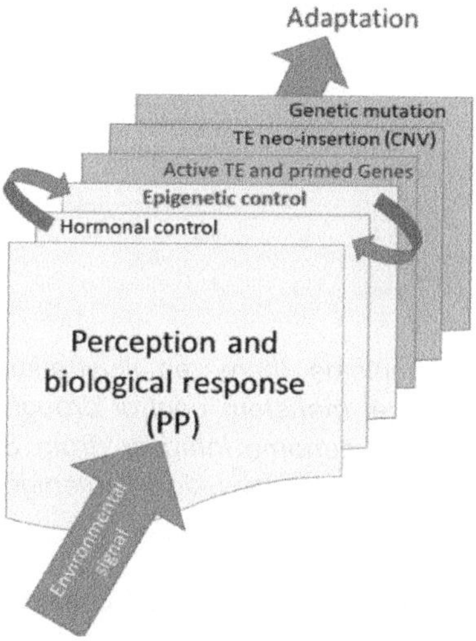

Figure 8.6. *Illustration of the temporal causal chain of reception of an environmental signal (by meristematic cells), translated into a hormonal response and then into a genetic modification mediated by an epigenetic modification (after Yona et al. (2015); Maury et al. (2019)). For a color version of this figure, see www.iste.co.uk/grunau/epigenetics.zip*

COMMENT ON FIGURE 8.6.– *Phenotypic plasticity (PP) initially involves well-established physiological control by phytohormones. In certain situations, notably for developmental plasticity and priming, epigenetic control is required in close interaction with hormones. These epigenetic modifications can permanently alter the genome via the priming of certain genes or the activation of transposable elements (TEs) and their subsequent neo-insertion, creating copy number variations (CNVs). The latter event induces a genetic modification that can occur in the gene's environment, altering its expression and thus its phenotype. All these mechanisms contribute to a diversity of phenotypes, which, under the effect of environmental change and natural selection, can lead to an essentially local adaptation process.*

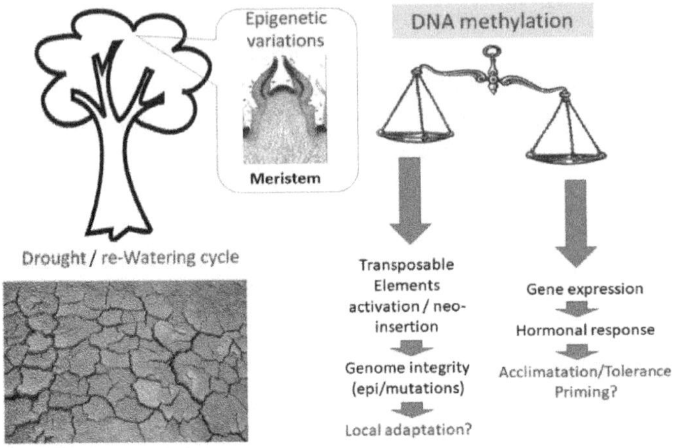

Figure 8.7. *Epigenetic mechanisms (here, an illustration of the effects of DNA methylation) in the poplar apical meristem control drought tolerance and memory through hormonal response and genome integrity (from Sow et al. (2021)). For a color version of this figure, see www.iste.co.uk/grunau/epigenetics.zip*

In plants, it is also important to remember that epigenetic variations do not only concern genes, but primarily TEs. In poplar, the use of DNA hypomethylated lines (obtained by knock-down or RNAi approaches) under drought conditions showed that, in addition to an effect on hormonal pathways, hypomethylation induced the activation of certain TEs under water stress and their insertion in several copies in the genome (Figure 8.7; (Sow et al. 2021)). The epigenetic modifications induced have thus changed the DNA sequence by modifying the copy number of TEs. These, in turn, can recruit the epigenetic machinery (notably via the RNA dependent DNA Methylation or "RdDM" pathway involving small non-coding RNAs, see Chapter 2 and Chapter 3) to hypermethylate the target loci of the insertion. These induced modifications can then affect genes or their promoters, and thus have a direct impact on the phenotype. When these modifications (genetic and epigenetic) take place during embryogenesis or during the life of the tree, and are localized in the cells of the shoot apical meristem, they can be transmitted by mitosis to future organs (PP), reinforcing the notion of mosaicism in trees, and potentially to descendants (intergenerational or transgenerational plasticity).

8.3.2. *Population epigenetic variation and tree adaptation*

Population epigenetic diversity depends on genetic diversity (Figure 8.5) and vice versa. A change in DNA sequence can modify one or more epigenetic marks.

An SNP on a cytosine (e.g. C to A) no longer allows methylation of this position. Consideration should also be given to any genetic variation in a gene encoding a protein involved in the epigenetic machinery, which would induce distinct epigenetic profiles. Conversely, the modification of epigenetic marks on a TE can induce its mobility and insertion into the genome at another site, and thus a change in genetic sequence, which in turn can induce an epigenetic change via the RdDM pathway (see Chapters 2 and 3). Furthermore, the mutation frequency of methylated cytosines (mC to T) is higher than that of unmethylated cytosines (C to T), which may imply genetic variation on a micro-evolutionary scale (Johannes and Schmitz 2019). These few examples illustrate how population epigenetic variation is intrinsically linked to genetic diversity and vice versa. While this reciprocal relationship can be seen at the individual level, major consequences are expected at the population level (Baduel and Colot 2021; Hannan Parker et al. 2022).

It should also be noted that in the absence (or low level) of genetic diversity, population epigenetic variation is a priori not limited. At the extreme, individuals can be derived from the same tree genotype by natural vegetative propagation (layering, suckering) or artificial propagation (grafting, cuttings, somatic embryogenesis). These clonal populations may then display high epigenetic polymorphism, which would compensate for the deficit in genetic variation, as has already been shown in various animal or plant clonal populations, notably in invasive species (Marin et al. 2019).

Finally, the environment can shape population epigenetic diversity, particularly in the following two situations (Figure 8.5):

1) Cases where the history of the population and its past environment may have induced (through genetic assimilation) epigenetic signatures transmitted across generations: this is the epigenetic component of local adaptation, including PP and transgenerational priming. This population-wide epigenetic diversity is the result of a long evolution.

2) The effect of the contemporary environment on the individuals of a population, with complex epigenetic signatures underlying G × E interactions, or the effect of micro-environmental variations, or the age of individuals, for example. As GCs, and in particular climate changes, are rapid on the life scale of trees, they represent a major disruption, not only for trees in place, but also for their descendants produced throughout their lives. This is the epigenetic population diversity that results from the direct experience of trees in their environment.

In plants, the best-known example of the influence of inter-individual epigenetic variation within a population (in this case, artificial) on PP is that of *Arabidopsis'*

Epigenetic Recombinant Inbred Lines (epiRILs) (Johannes et al. 2009). These epiRILs are obtained by crossing two parents with similar DNA sequences, but very contrasting DNA methylation profiles, then stabilizing the lines by six generations of self-fertilization. It has thus been shown that part of the DNA methylation variation (notably the methylation of CG contexts that are transmissible in meiosis) is inherited in a Mendelian manner, and that these epiRILs display significant phenotypic variation in their morphology, growth rate and response to biotic and abiotic stresses.

In trees, several studies have focused on population epigenetic variation (Sow et al. 2018a; Amaral et al. 2020). These studies include poplar, spruce, pine, eucalyptus and oak, and quantify DNA methylation polymorphism, using global (high-performance liquid chromatography or HPLC) or targeted genome scan methods (Methyl-Sensitive Amplification Polymorphism or MSAP), and more recently at the whole-genome level (Whole Genome Bisulfite Sequencing or WGBS; (Champigny et al. 2019)). The measurement of classic genetic parameters (heritability, genetic differentiation) has shown that global DNA methylation variation is genetically and environmentally determined (Sow et al. 2018b). Inter-individual variation in DNA methylation is notably inscribed in microRNA genes or candidate genes for growth traits. This epigenetic polymorphism makes it possible to distinguish the identity of individuals or clones, or even their origin (Champigny et al. 2019), and therefore to distinguish populations from one another, which is sometimes more effective than genetic polymorphism (SNP), particularly in the case of vegetatively propagated species such as poplar. Some studies also report that climatic and spatial variables explain a greater proportion of epigenetic variance between individuals than genetic variance, suggesting a role for DNA methylation in the local adaptation of natural populations (Platt et al. 2015; Gugger et al. 2016). Nevertheless, studies on the role of epigenetic mechanisms in the local adaptation of populations remain embryonic, and the question is still wide open. Finally, DNA methylation has been shown to play a significant role in gene duplication and even 5S rDNA translocation (which, in turn, modifies chromatin structure at the translocation site) during plant evolution. Epigenetic regulation of duplicated genes could be a central phenomenon for understanding the impact of polyploidization or tandem duplication events on plant survival or adaptation to environmental stresses (Sow et al. 2018a).

Recent work on *Arabidopsis* highlights the evolutionary importance of epigenetic mechanisms in plant response to environmental variation. Firstly, Alvarez et al. (2021) found that genotypes differ in how: (i) environmental effects dissipate across successive generations; (ii) responses to ancestral environments are stably expressed in current environments; (iii) ancestral environments modify responses to current environments and (iv) ancestral environments alter fitness under current conditions. These observations motivate epigenetic research to explain the molecular mechanisms involved. From a more mechanistic point of view, Baduel and Colot (2021) and Hannan Parker et al. (2022) present a synthesis of the evolutionary importance of epigenetic mechanisms in relation to TE activity. The former argue that "the adaptive contribution of TE-associated epivariation, whether stable or not, lies primarily in its ability to modulate TE mobilization in response to the environment, thus endowing plant genomes with a powerful engine of rapid adaptation". For the latter, "Darwinian evolution is the dominant principle behind adaptive changes in plant immunity. However, stress-induced epigenetic change can act as a catalyst for Darwinian evolution of the plant immune system. In addition to TIR, which can offer relatively short-term benefits to offspring in parental environments, stress-induced epigenetic change can accelerate genetic change in immunoregulatory genes through mobilome-induced structural variants (TE mobility analysis) and cytosine methylation-induced point mutations, which are selected by Darwinian evolution". In the coming years, work on trees and epigenetics at the population level will improve our understanding of the evolution of these perennial organisms.

8.3.3. *The promise of epigenetics for the improvement, management and conservation of genetic resources in forest trees*

Understanding how trees adapt to a changing environment, and in particular the role of PP in relation to genetic variability, is certainly a key issue in making the correct decisions about the management and conservation of FGRs (Forest Genetic Resources). The first step concerns forest managers, with the implementation of a decision-making cycle (example of the ACORN project, www.acorn-biodiversa.net/background) from the characterization of current and future climatic conditions to the choice of forest reproductive material (FRM).

In this context, epigenetics is a major source of flexibility to be considered. This could have concrete repercussions: choice of genotypes, clones, organization of clone parks or mother plants and seed orchards. The possibilities for exploiting epigenetic variability (like genetic variability) largely depend on the mode of reproduction of the species under consideration (clonal or sexual (Sow et al. 2018a)) and the nature of the forest (natural or planted; for more details, see (Amaral et al. 2020)) as summarized in Figure 8.8 (for more details, see Kakoulidou et al. (2021)).

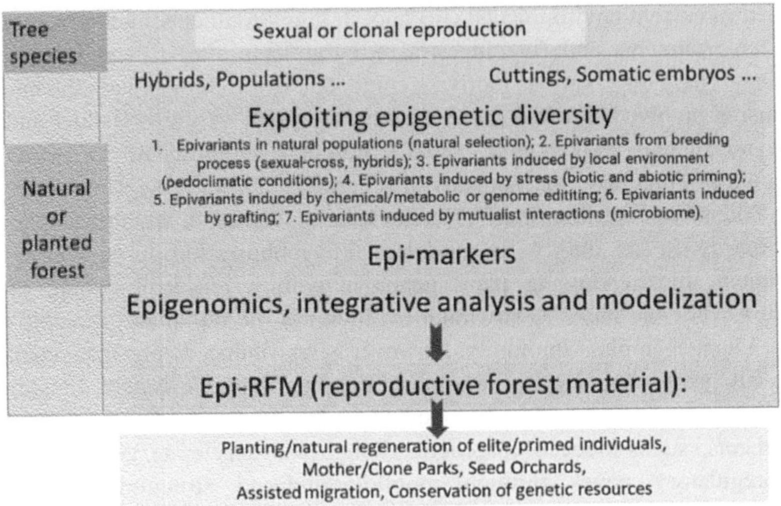

Figure 8.8. *Exploiting epigenetic diversity in the production of forest reproductive material (FRM) and the management of forest genetic resources (from Sow et al. (2018a); Amaral et al. (2020); Kakoulidou et al. (2021)). For a color version of this figure, see www.iste.co.uk/grunau/epigenetics.zip*

COMMENT ON FIGURE 8.8.– *Depending on the tree species, their mode of reproduction (sexual or asexual) and the type of forest considered (natural or planted), the epigenetic diversity present in natural populations can be exploited in the same way as genetic diversity, including by relying on traditional methods of improvement through crossing and hybridization. However, it can also be considered more broadly by integrating epigenetic variation induced in response to the planting environment (pedoclimatic conditions) or obtained through various environmental conditioning processes (priming by temperature, water stress, the use of plant defense stimulators, biostimulants, etc.), mutant production (chemical or metabolic modification of DNA methylation, chromatin marks, RNAi mutation, genome or epigenome editing, etc.), graft-induced transcriptional (RNAi) and epigenetic (RdDM) modifications or techniques capable of modulating symbiotic associations (microbiome). Epivariants selected or induced in this way can then be propagated by sexual reproduction or, more efficiently, by various horticultural or in vitro vegetative propagation methods (micropropagation, somatic embryogenesis, etc.). This epigenetic diversity can then be exploited as a source of epimarks (marks of epigenetic polymorphism) which, like genetic marks, can specifically help with mark-assisted selection (including on a genome-wide scale – epigenomics). Epimarks can also be used to complement genetic marks in integrative analysis or modeling approaches aimed at predicting the performance of forest genetic resources. The identification of interesting epivariants (referred to here as epi-MFR) opens up the possibility of deploying individuals that vary in their epigenetic (but not genetic) profiles and are potentially better adapted to certain environmental constraints. These contributions could modify current forestry practices, such as the creation of mother plantations, clone plantations and seed orchards.*

8.3.3.1. *Harnessing epigenetic diversity for forest management*

Epigenetic diversity (natural or artificial) contributes to phenotypic diversity, particularly for adaptive and/or agronomic traits in plants. To exploit this epigenetic diversity, it is possible to characterize it in natural or artificial populations, and also to induce it by various approaches (chemical, mutagenesis, hybridization, in vitro culture, epigenome editing, environmental priming) (Kakoulidou et al. 2021). In forest trees, this exploitation is still being researched, but some concrete examples are available and the possibilities look promising.

The example of spruce epitypes from different temperature conditions during zygotic or somatic embryogenesis, ensuring an adapted phenology corresponding to several degrees of latitude in Norway, is certainly the most significant to date in terms of the concrete application of epigenetics in forestry (Carneros et al. 2017). Thus, somatic embryogenesis is a promising process for vegetative seed multiplication for the deployment of conifer varieties on a large scale (Klimaszewska et al. 2016). This *in vitro* technology could be used to carry out biotic (or even microbiotic) or abiotic priming from the embryogenesis phase, ultimately producing somatic seedlings and trees that are better able to acclimatize to global warming. Another advantage is that the embryogenic masses obtained after the seed initiation and multiplication phases can be easily cryopreserved at very low temperatures in liquid nitrogen. This makes it possible to store real banks of individuals (each representing a seed in the form of an embryo clone) for decades (theoretically indefinitely), which is a major advantage over seed banks. The latter are affected by a decline in viability during long-term storage. Some studies suggest that epigenetic mechanisms may be responsible. Studies in this field would certainly be useful in optimizing seed conservation (storage duration or conditions).

More generally, the effect of epigenetic memory (priming) on somatic or naked zygotic embryos (isolated from the seed), whole seeds (zygotic embryo embedded in the endosperm) or (micro)cuttings, notably through thermal (Trontin et al. 2021) or hydric priming, is currently an avenue for the application of epigenetic mechanisms. Various examples of forestry material (notably conifer somatic embryos) are already available (Sow et al. 2018a; Amaral et al. 2020; Kakoulidou et al. 2021; Trontin et al. 2021). Temperature priming during the initiation phase of somatic embryogenesis in

Pinus radiata thus induces hormonal changes in embryogenic masses (notably different types of cytokinins, plant hormones, known to be involved in both embryo development and stress response) and PP of 2-year-old somatic seedlings in terms of water stress tolerance (Castander-Olarieta et al. 2020). Another example concerns *Populus trichocarpa* cuttings from two production sites in Germany that differ in the availability of phosphorus. The clones were transferred to a nutrient-rich site. Despite identical genetics, stem cuttings from sites with low phosphorus availability showed altered establishment linked to differences in DNA methylation (Schönberger et al. 2016). All this suggests that it would be interesting to apply varied environmental conditions to produce MFRs, either in vitro during micropropagation by somatic embryogenesis or microbudding (finely controlled conditions), ex situ during seed germination (or somatic embryos) followed by the rearing of young plants (somatic seedlings, mother plants, grafts, cuttings, etc.) in climatic chambers, greenhouses or nurseries (partially controlled conditions) or in situ by setting up seed orchards (of families or clones) or clone parks (for the conservation of natural or improved FGR) under different pedoclimatic conditions. Alternatively, seeds can be collected from natural populations over a range of pedoclimatic or environmental gradients (altitude, latitude, and also biotic environment). The resulting material could show various priming effects that promote acclimatization of young plants. For example, genetically identical seed orchards or clone parks should not only be deployed in environments favorable to their growth in the current climate, but also in forced acclimatization zones (similar to the future climate) or in areas infested by diseases for many years (in connection with TIR). This calls for broad and urgent reflection on the FGR management plan for immediate implementation (such as planting new orchards), as well as the longer term.

However, there are a number of questions surrounding the use of this approach. For example, in planted forests, if the memory of stress is to be useful beyond seedling establishment (a crucial point given that mortality is high in the first five years of a tree's life), epigenetic research will also need to focus on possible long-term priming effects (such as TIR). Moreover, in the context of GC, trees are also sensitive to stresses that vary in nature (biotic and abiotic), and it will be necessary to define which priming, as well as plant defense stimulants or biostimulants, will be effective in the face of this complex environment. In natural forests, the implementation of new silvicultural practices is more complex, but, for example, if the progeny of trees that were under water (or temperature) stress during seed formation shows greater resistance to drought, forest managers could develop strategies to encourage their germination and the subsequent development of this particular cohort of seedlings. Once again, these ideas presuppose preliminary studies to model priming effects and then the implementation of new practices by forest managers, especially for FGR conservation.

Epigenetic diversity can also be induced by the following approaches (see also Chapter 6) (Figure 8.8):

1) Mutations and/or epi/genome editing. Various approaches are conceivable, such as the search for genetic variants (in genes involved in the dynamics of epigenetic marks) in natural populations (SNPs or transposons) or artificial populations obtained by mutation or by genome or epigenome editing. Genetic targets could be chromatin regulators, inducing epigenetic variants as observed in epiRILs. Epigenome editing is still at the developmental stage in trees, but it would enable a fundamental advance in the validation of epigenetic marks in various processes (gene expression, recombination hot spots, alternative splicing, etc.). Methylation induced by epigenome editing could be heritable, making targeted manipulation of DNA methylation an approach for permanently modifying a locus.

2) Chemical or metabolic treatments and in vitro propagation of epivariants. Numerous pharmacological molecules capable of modifying chromatin marks have been characterized for both DNA methylation (such as 5-Azacytidine) and histone modifications (such as Trichostatin A, which inhibits histone deacetylase or HDAC). The effect of these molecules has been reported in a wide range of plants, including trees. Although various effects on traits of interest have been demonstrated, these molecules often present a certain toxicity, and the approach remains stochastic, with highly variable stability of effects. Nevertheless, this remains a possibility for modifying the epigenome and then propagating epivariants in vitro by micropropagation.

Another possibility involves the controlled remobilization of transposons, as has been achieved in model plants such as *Arabidopsis* and rice (Thieme et al. 2017) with the simultaneous use of a DNA methylation inhibitor (e.g. Zebularin) and RNA polymerase II (α-Amanitin). These plants showed transposon reactivation and high phenotypic diversity in their progeny under thermal stress.

Another approach concerns the manipulation of key metabolic compounds such as methyl (SAM) or acetyl (Acetyl-CoA) donors, which are common to chromatin epigenetic marks and many primary or specific metabolic pathways in plants. Indeed, manipulation of these metabolic intermediates, under biotic or abiotic stresses, affects general metabolism, and also causes potentially transmissible epigenetic modifications (Mladenov et al. 2021). Similarly, the strong relationship between phytohormones, the redox pathway and epigenetics could also be the subject of studies and even applications.

3) Grafting is an ancient technique involving the assembly of two genotypes, and even more rarely, two species. It has been used for thousands of years to improve the performance (particularly in terms of rooting and pathogen resistance) of many plant varieties and species, especially improved varieties. Transcriptional and epigenetic modifications appear to play a key role in grafting-induced changes (Mladenov et al. 2021). Small mobile RNAs produced by the rootstock can repress gene expression in the graft, either through the RdDM pathway or by degradation of the corresponding mRNA target molecule. Studies are still needed on plants, particularly forest trees, before any applications can be considered.

4) Modulating the microbiome. Trees establish complex, mutualistic relationships with the soil microbial community (fungi and bacteria in particular). These interactions are initiated by root exudates and lead to a complex molecular dialogue between the host tree and the microbe. In particular, trees recruit fungi (organisms that also exhibit DNA methylation) for the establishment of mycorrhizae, which facilitate the uptake of water and various nutrients, and also root growth and protection against other pathogenic microbes. Epigenetic control of these symbiotic associations has been proposed in plants, but the research is still in its early stages. It has been shown that epigenetic variants of plants have distinct microbiomes, and also that the presence of the fungus, in turn, modifies the plant's methylome. Modulating the epigenomes of the two partners (through the various strategies already mentioned) could make it possible to control the composition of the microbiome or the performance of the association (Corbin et al. 2020). The discovery of these epialleles and the development of epigenetic techniques to modulate the association could have important applications in forestry, for example, in greenhouse and nursery production, followed by the planting of young seedlings.

8.3.3.2. *Using epimarkers, modeling and developing an integrative approach*

The classic quantitative genetic methods used to estimate the genetic component of the phenotype could also be used to assess natural epigenetic variations. DNA methylation could therefore be considered as a trait in its own right, and the estimation of heritability, in the strict sense (h2), or the genetic differentiation index (epiQST), for these marks, could enable the study of the relationship between genetic and epigenetic variability and, more specifically, the genetic control of epigenetic variations (Dubin et al. 2015; Sow et al. 2018b). Conversely, epigenetic polymorphisms, such as SMPs (Single-Methylation Polymorphisms) or DMRs (Differentially Methylated Region) could be used as epigenetic marks in genome analysis-based approaches, in order to assess the role of epigenetics in population structuring (epiFST) and phenotypic variation (epiQTL). The study by Champigny et al. (2019) on various *Populus balsamifera* populations shows that models based on DNA methylation can be used to determine the identity, origin or quality of agroforestry products. While this approach offers innovative prospects for forest trees, it will need to be assessed in terms of cost, technical and analytical effort.

Consequently, genomic selection should not only focus on the genetic component but also consider the interaction between genetics and epigenetics. The success of these new integrative multiomics approaches involving genomics (SNP, transcriptomics, proteomics, metabolomics) and epigenomics (methylome, sRNA, histone marks, etc.) will mostly depend on the development of epi/genomic resources and, above all, powerful statistical models capable of handling different variables in multivariate analysis (Sow et al. 2018a), and also the development of prediction models for agronomic traits under genetic and/or epigenetic control (Figure 8.8; Kakoulidou et al. (2021)). A first challenge is to incorporate epigenomic data into genomic prediction models. Subsequently, to move from correlations to causal inference and predictions, it will be necessary to develop an ambitious transomics approach (Yugi et al. 2016). This will enable molecular networks to be reconstructed by connecting several levels of multiomics data as developed, for example, in human brain research (Yugi et al. 2016).

An applied example of the use of epigenetic marks concerns hybrid vigor or heterosis, which is defined as the superior performance of hybrids, for many agronomic traits, compared to their parent lines. Hybrids have been widely used in crop and tree breeding since ancient times. In recent years, although the molecular mechanisms underlying heterosis have not yet been fully clarified, various studies have pointed to a possible role for epigenetic marks (Sow et al. 2018a; Kakoulidou et al. 2021). As an example, Gao et al. (2014) investigated the role of DNA methylation changes in *Populus deltoides* hybrids derived from intra-specific parental lines. They suggested that methylation patterns from parental inbred lines were passed on, at least in part, to offspring, and that these hybrids showed non-additive levels of methylation, suggesting a possible role for DNA methylation in heterosis. Thus, the identification of epigenetic marks for the improvement of silvicultural traits would help tree breeders to produce genotypes (hybrids) that perform better and, in particular, are able to acclimatize to future environmental changes. Another example of agronomic interest concerns the epigenetic control of the frequency and position of recombination (crossovers) along chromosomes, for which epigenetic marks would be very useful. This mechanism has major implications for the creation of genetic variability by breeders, as it would enable better selection of the parental genotypes used in controlled crosses.

8.4. Conclusion

The field of epigenetics is booming in model and field plants (over 600 publications per year), and also in forest trees. Acquiring new knowledge on the impact of epigenetic mechanisms on phenotypic variability will require new experimental set-ups, such as the development of multisite common gardens including a wide range of genetic diversity (natural and/or artificial) in various environments.

One challenge would be to create or develop high-throughput experimental platforms to study the effects of biotic, microbiotic and abiotic priming on PP under an epigenetic perspective, from seedling production in forest nurseries to establishment in forests. Studies should focus as much on intra-individual epigenetic variation and PP as on inter-individual epigenetic diversity in relation to the adaptation of trees to their changing environment. This research could suggest ways of managing and conserving FGs (seed orchards, seed conservation, etc.). These scientific perspectives need to be established with the whole forestry community (public and private), without forgetting the societal and political impact of questions of adaptation and transformation of forest socio-ecosystems.

Clarification of the meaning, nature and methodologies of epigenetics will enable public opinion, as well as political decision-makers and forest managers, to better understand the value of any epigenetic discovery, highlighting the particular role of environmental changes on the modification of FGR.

8.5. Acknowledgements

The authors thank the GDR3E (Epigenetics, Ecology, Evolution) of the CNRS (See: https://rtp-3e.wixsite.com/rt3e/cdp) and in particular Christoph Grunau and Dominique Joly for coordinating this work. The authors also thank the ANR EPITREE projects (ANR-17-CE32-0009-01; See: https://www6.inrae.fr/epitree-project/), the European COST (European Cooperation in Science and Technology) EPIgenetic mechanisms of Crop Adaptation To Climate cHange (EPICATCH; CA19125; www.cost.eu/actions/CA19125/) and the Dynasty project, which received government funding under the Programme d'investissements d'avenir (PIA) IdEx Université de Bordeaux/GPR Bordeaux Plant Sciences. These projects contributed to the writing of this chapter. Finally, the authors thank the chapter's reviewers, Dr. Jean-François Trontin and Dr. Harold Duruflé, for their insightful comments.

8.6. References

Allen, C.D., Macalady, A.K., Chenchouni, H., Bachelet, D., Mcdowell, N., Vennetier, M., Kitzberger, T., Rigling, A., Breshears, D.D., Hogg, E.H.T et al. (2010). A global overview of drought and heat-induced tree mortality reveals emerging climate change risks for forests. *Forest Ecology and Management*, 259(4), 660–684. doi: 10.1016/j.foreco.2009.09.001.

Alvarez, M., Bleich, A., Donohue, K. (2021). Genetic differences in the temporal and environmental stability of transgenerational environmental effects. *Evolution; International Journal of Organic Evolution*, 75(11), 2773–2790. doi: 10.1111/evo.14367.

Amaral, J., Ribeyre, Z., Vigneaud, J., Sow, M.D., Fichot, R., Messier, C., Pinto, G., Nolet, P., Maury, S. (2020). Advances and promises of epigenetics for forest trees. *Forests*, 11(9), 976. doi: 10.3390/f11090976.

Anderegg, W.R.L., Klein, T., Bartlett, M., Sack, L., Pellegrini, A.F.A., Choat, B., Jansen, S. (2016). Meta-analysis reveals that hydraulic traits explain cross-species patterns of drought-induced tree mortality across the globe. *Proceedings of the National Academy of Sciences of the United States of America*, 113(18), 5024–5029. doi: 10.1073/pnas.1525678113.

Baduel, P. and Colot, V. (2021). The epiallelic potential of transposable elements and its evolutionary significance in plants. *Philosophical Transactions of the Royal Society B: Biological Sciences*, 376(1826), 20200123. doi: 10.1098/rstb.2020.0123.

Benito Garzón, M., Robson, T.M., Hampe, A. (2019). ΔTraitSDMs: Species distribution models that account for local adaptation and phenotypic plasticity. *New Phytologist*, 222(4), 1757–1765. doi: 10.1111/nph.15716.

Bradshaw, A.D. (1965). Evolutionary significance of phenotypic plasticity in plants. *Advances in Genetics*, 13, 115–155. doi: 10.1016/S0065-2660(08)60048-6.

Callahan, H.S., Pigliucci, M., Schlichting, C.D. (1997). Developmental phenotypic plasticity: Where ecology and evolution meet molecular biology. *BioEssays: News and Reviews in Molecular, Cellular and Developmental Biology*, 19(6), 519–525. doi: 10.1002/bies.950190611.

Carneros, E., Yakovlev, I., Viejo, M., Olsen, J.E., Fossdal, C.G. (2017). The epigenetic memory of temperature during embryogenesis modifies the expression of bud burst-related genes in Norway spruce epitypes. *Planta*, 246(3), 553–566. doi: 10.1007/s00425-017-2713-9.

Castander-Olarieta, A., Moncaleán, P., Pereira, C., Pěnčík, A., Petřík, I., Pavlović, I., Novák, O., Strnad, M., Goicoa, T., Ugarte, M.D. et al. (2021). Cytokinins are involved in drought tolerance of *Pinus radiata* plants originating from embryonal masses induced at high temperatures. *Tree Physiology*, 41(6), 912–926. doi: 10.1093/treephys/tpaa055.

Champigny, M.J., Unda, F., Skyba, O., Soolanayakanahally, R.Y., Mansfield, S.D., Campbell, M.M. (2020). Learning from methylomes: Epigenomic correlates of *Populus balsamifera* traits based on deep learning models of natural DNA methylation. *Plant Biotechnology Journal*, 18(6), 1361–1375. doi: 10.1111/pbi.13299.

Choat, B., Jansen, S., Brodribb, T.J., Cochard, H., Delzon, S., Bhaskar, R., Bucci, S.J. Feild, T.S., Gleason, S.M., Hacke, U.G. et al. (2012). Global convergence in the vulnerability of forests to drought. *Nature*, 491(7426), 752–755. doi: 10.1038/nature11688.

Conde, D., Le Gac, A.-L., Perales, M., Dervinis, C., Kirst, M., Maury, S., González-Melendi, P., Allona, I. (2017). Chilling-responsive DEMETER-LIKE DNA demethylase mediates in poplar bud break. *Plant, Cell & Environment*, 40(10), 2236–2249. doi: 10.1111/pce.13019.

Corbin, K.R., Bolt, B., Rodríguez López, C.M. (2020). Breeding for beneficial microbial communities using epigenomics. *Frontiers in Microbiology*, 11 [Online]. Available at: https://www.frontiersin.org/article/10.3389/fmicb.2020.00937.

DeWitt, T.J., Sih, A., Sloan Wilson, D. (1998). Costs and limits of phenotypic plasticity. *Trends in Ecology & Evolution*, 13(2), 77–81. doi: 10.1016/S0169-5347(97)01274-3.

Dubin, M.J., Zhang, P., Meng, D., Remigereau, M.-S., Osborne, E.J., Casale, F.P., Drewe, P., Kahles, A., Jean, G., Vilhjalmsson, B. et al. (2015). DNA methylation in *Arabidopsis* has a genetic basis and shows evidence of local adaptation. *ELife*, 4(May). doi: 10.7554/eLife.05255.

FAO (2022). *The State of the World's Forests 2022. Forest Pathways for Green Recovery and Building Inclusive, Resilient and Sustainable Economies*. FAO, Rome. doi: 10.4060/cb9360en [Accessed 9 December 2022].

Gao, M., Huang, Q., Chu, Y., Ding, C., Zhang, B., Su, X. (2014). Analysis of the leaf methylomes of parents and their hybrids provides new insight into hybrid vigor in *Populus deltoides*. *BMC Genetics*, 15(1), S8. doi: 10.1186/1471-2156-15-S1-S8.

Ghalambor, C.K., McKAY, J.K., Carroll, S.P., Reznick, D.N. (2007). Adaptive versus non-adaptive phenotypic plasticity and the potential for contemporary adaptation in new environments. *Functional Ecology*, 21(3), 394–407. doi: 10.1111/j.1365-2435.2007.01283.x.

Gugger, P.F., Fitz-Gibbon, S., Pellegrini, M., Sork, V.L. (2016). Species-wide patterns of DNA methylation variation in *Quercus lobata* and their association with climate gradients. *Molecular Ecology*, 25(8), 1665–1680. doi: 10.1111/mec.13563.

Hamrick, J.L. (2004). Response of forest trees to global environmental changes. *Forest Ecology and Management, Dynamics and Conservation of Genetic Diversity in Forest Ecology*, 197(1), 323–335. doi: 10.1016/j.foreco.2004.05.023.

Hannan Parker, A., Wilkinson, SW., Ton, J. (2022). Epigenetics: A catalyst of plant immunity against pathogens. *New Phytologist*, 233(1), 66–83. doi: 10.1111/nph.17699.

Hofmeister, B.T., Denkena, J., Colomé-Tatché, M., Shahryary, Y., Hazarika, R., Grimwood, J., Mamidi, S., Jenkins, J., Grabowski, P.P., Sreedasyam, A. et al. (2020). A genome assembly and the somatic genetic and epigenetic mutation rate in a wild long-lived perennial *Populus trichocarpa*. *Genome Biology*, 21(1), 259. doi: 10.1186/s13059-020-02162-5.

Johannes, F. and Schmitz, R.J. (2019). Spontaneous epimutations in plants. *New Phytologist*, 221(3), 1253–1259. doi: 10.1111/nph.15434.

Johannes, F., Porcher, E., Teixeira, F.K., Saliba-Colombani, V., Simon, M., Agier, N., Bulski, A., Albuisson, J., Heredia, F., Audigier, P. et al. (2009). Assessing the impact of transgenerational epigenetic variation on complex traits. *PLoS Genetics*, 5(6), e1000530. doi: 10.1371/journal.pgen.1000530.

Johnsen, Ø., Gram Dæhlen, O., Østreng, G., Skrøppa, T. (2005). Daylength and temperature during seed production interactively affect adaptive performance of *Picea abies* progenies. *New Phytologist*, 168(3), 589–596. doi: 10.1111/j.1469-8137.2005.01538.x.

Kakoulidou, I., Avramidou, E.V., Baránek, M., Brunel-Muguet, S., Farrona, S., Johannes, F., Kaiserli, E., Lieberman-Lazarovich, M., Martinelli, F., Mladenov, V. et al. (2021). Epigenetics for crop improvement in times of global change. *Biology*, 10(8). doi: 10.3390/biology10080766.

Kaufman, D., McKay, N., Routson, C., Erb, M., Dätwyler, C., Sommer, P.S., Heiri, O., Davis, B. (2020). Holocene global mean surface temperature, a multi-method reconstruction approach. *Scientific Data*, 7(1), 201. doi: 10.1038/s41597-020-0530-7.

Klimaszewska, K., Hargreaves, C., Lelu-Walter, M.-A., Trontin, J.-F. (2016). Advances in conifer somatic embryogenesis since year 2000. In *In Vitro Embryogenesis in Higher Plants, Methods in Molecular Biology*, Germanà, M.A. and Lambardi, M. (eds). Humana Press, New York, 131–166. doi: 10.1007/978-1-4939-3061-6_7.

Kvaalen, H. and Johnsen, Ø. (2008). Timing of bud set in *Picea abies* is regulated by a memory of temperature during zygotic and somatic embryogenesis. *New Phytologist*, 177, 49–59. doi: 10.1111/j.1469-8137.2007.02222.x.

Lachowiec, J., Queitsch, C., Kliebenstein, D.J. (2016). Molecular mechanisms governing differential robustness of development and environmental responses in plants. *Annals of Botany*, 117(5), 795–809. doi: 10.1093/aob/mcv151.

Lafon-Placette, C., Le Gac, A.-L., Chauveau, D., Segura, V., Delaunay, A., Lesage-Descauses, M.-C., Hummel, I., Cohen, D., Jesson, B., Le Thiec, D. et al. (2018). Changes in the epigenome and transcriptome of the poplar shoot apical meristem in response to water availability affect preferentially hormone pathways. *Journal of Experimental Botany*, 69(3), 537–551. doi: 10.1093/jxb/erx409.

Le Gac, A.-L., Lafon-Placette, C., Chauveau, D., Segura, V., Delaunay, A., Fichot, R., Marron, N., Le Jan, I., Berthelot, A., Bodineau, G. et al. (2018). Winter-dormant shoot apical meristem in poplar trees shows environmental epigenetic memory. *Journal of Experimental Botany*, 69(20), 4821–4837. doi: 10.1093/jxb/ery271.

Le Gac, A.-L., Lafon-Placette, C., Delaunay, A., Maury, S. (2019). Developmental, genetic and environmental variations of global DNA methylation in the first leaves emerging from the shoot apical meristem in poplar trees. *Plant Signaling & Behavior*, 14(6), 1596717. doi: 10.1080/15592324.2019.1596717.

Maher, B. (2008). Personal genomes: The case of the missing heritability. *Nature*, 456(7218), 18–21. doi: 10.1038/456018a.

Marin, P., Genitoni, J., Barloy, D., Maury, S., Gibert, P., Ghalambor, C.K., Vieira, C. (2020). Biological invasion: The influence of the hidden side of the (epi)genome. *Functional Ecology*, 34(2), 385–400. doi: 10.1111/1365-2435.13317.

Maury, S., Sow, M.D., Le Gac, A.-L., Genitoni, J., Lafon-Placette, C., Mozgova, I. (2019). Phytohormone and chromatin crosstalk: The missing link for developmental plasticity? *Frontiers in Plant Science*, 10, 395. doi: 10.3389/fpls.2019.00395.

Mladenov, V., Fotopoulos, V., Kaiserli, E., Karalija, E., Maury, S., Baranek, M., Segal, N., Testillano, P.S., Vassileva, V., Pinto, G. et al. (2021). Deciphering the epigenetic alphabet involved in transgenerational stress memory in crops. *International Journal of Molecular Sciences*, 22(13), 7118. doi: 10.3390/ijms22137118.

Platt, A., Gugger, P.F., Pellegrini, M., Sork, V.L. (2015). Genome-wide signature of local adaptation linked to variable CpG methylation in oak populations. *Molecular Ecology*, 24(15), 3823–3830. doi: 10.1111/mec.13230.

Richards, C.L., Alonso, C., Becker, C., Bossdorf, O., Bucher, E., Colomé-Tatché, M., Durka, W., Engelhardt, J., Gaspar, B., Gogol-Döring, A. et al. (2017). Ecological plant epigenetics: Evidence from model and non-model species, and the way forward. *Ecology Letters*, 20(12), 1576–1590. doi: 10.1111/ele.12858.

Saleh, D., Chen, J., Leplé, J.-C., Leroy, T., Truffaut, L., Dencausse, B., Lalanne, C., Labadie, K., Lesur, I., Bert, D. et al. (2022). Genome-wide evolutionary response of European oaks during the anthropocene. *Evolution Letters*, 6(1), 4–20. doi: 10.1002/evl3.269.

Schönberger, B., Chen, X., Mager, S., Ludewig, U. (2016). Site-dependent differences in DNA methylation and their impact on plant establishment and phosphorus nutrition in *Populus trichocarpa*. *PLOS ONE*, 11(12), e0168623. doi: 10.1371/journal.pone.0168623.

Sow, M.D., Allona, I., Ambroise, C., Conde, D., Fichot, R., Gribkova, S., Jorge, V., Le Provost, G., Pâques, L., Plomion, C. et al. (2018a). Epigenetics in forest trees: State of the art and potential implications for breeding and management in a context of climate change. In *Plant Epigenetics Coming of Age for Breeding Applications*, Mirouze, M., Bucher, E., Gallusci, P. (eds). Academic Press, Cambridge, 387–453. doi: 10.1016/bs.abr.2018.09.003.

Sow, M.D., Segura, V., Chamaillard, S., Jorge, V., Delaunay, A., Lafon-Placette, C., Fichot, R., Faivre-Rampant, P., Villar, M., Brignolas, F. et al. (2018b). Narrow-sense heritability and PST estimates of DNA methylation in three *Populus nigra* L. populations under contrasting water availability. *Tree Genetics & Genomes*, 14(5), 78. doi: 10.1007/s11295-018-1293-6.

Sow, M.D., Le Gac, A.-L., Fichot, R., Lanciano, S., Delaunay, A., Le Jan, I., Lesage-Descauses, M.-C., Citerne, S., Caius, J., Brunaud, V. et al. (2021). RNAi suppression of DNA methylation affects the drought stress response and genome integrity in transgenic poplar. *The New Phytologist*, 232(1), 80–97. doi: 10.1111/nph.17555.

Thieme, M., Lanciano, S., Balzergue, S., Daccord, N., Mirouze, M., Bucher, E. (2017). Inhibition of RNA polymerase II allows controlled mobilisation of retrotransposons for plant breeding. *Genome Biology*, 18(1), 134. doi: 10.1186/s13059-017-1265-4.

Trontin, J.-F., Raschke, J., Rupps, A. (2021). Tree 'memory': New insights on temperature-induced priming effects during early embryogenesis. *Tree Physiology*, 41(6), 906–911. doi: 10.1093/treephys/tpaa150.

Vigneaud, J. and Maury, S. (2020). Developmental plasticity in plants: An interaction between hormones and epigenetics at the meristem level. *Biologie Aujourd'hui*, 214(3/4), 125–135. doi: 10.1051/jbio/2020011.

Vining, K.J., Pomraning, K.R., Wilhelm, L.J., Priest, H.D., Pellegrini, M., Mockler, T.C., Freitag, M., Strauss, S.H. (2012). Dynamic DNA cytosine methylation in the *Populus trichocarpa* genome: Tissue-level variation and relationship to gene expression. *BMC Genomics*, 13(1), 27. doi: 10.1186/1471-2164-13-27.

Wagner, F., Below, R., Klerk, P.D., Dilcher, D.L., Joosten, H., Kürschner, W.M., Visscher, H. (1996). A natural experiment on plant acclimation: Lifetime stomatal frequency response of an individual tree to annual atmospheric CO_2 increase. *Proceedings of the National Academy of Sciences*, 93(21), 11705–11708. doi: 10.1073/pnas.93.21.11705.

Yakovlev, I.A., Lee, Y.K., Rotter, B., Olsen, J.E., Skrøppa, T., Johnsen, Ø., Fossdal, C.G. (2014). Temperature-dependent differential transcriptomes during formation of an epigenetic memory in Norway spruce embryogenesis. *Tree Genetics & Genomes*, 10(2), 355–366. doi: 10.1007/s11295-013-0691-z.

Yona, A.H., Frumkin, I., Pilpel, Y. (2015). A relay race on the evolutionary adaptation spectrum. *Cell*, 163(3), 549–559. doi: 10.1016/j.cell.2015.10.005.

Yugi, K., Kubota, H., Hatano, A., Kuroda, S. (2016). Trans-omics: How to reconstruct biochemical networks across multiple 'omic' layers. *Trends in Biotechnology*, 34(4), 276–290. doi: 10.1016/j.tibtech.2015.12.013.

9

Epigenetics and Crop Improvement

Julie LECLERCQ[1], Dominique THIS[1] and Patrice THIS[1,2]
[1] AGAP Institut, CIRAD, INRAE, Institut Agro Montpellier, Université de Montpellier, France
[2] UMT Géno-Vigne, IFV, Montpellier, France

9.1. Introduction

Many of the results of epigenetic research have been obtained on model organisms and should be applied to agronomic species. The principle of translational research is to be able to transfer advances from fundamental research into solutions of concrete benefit to mankind. This is the case in the field of plant improvement, in a context of sustainable, eco-responsible production that is likely to meet the challenges of global changes in the world around us.

It should be remembered that epigenetic mechanisms modify gene activity, and therefore potentially the plant phenotype (the set of traits that modify the appearance, productivity and even quality of a plant or the product derived from it). Insofar as breeders work essentially at this level, epigenetic mechanisms could have major consequences for plant improvement.

Plant breeding has traditionally been based on the use of genetic diversity in relation to the variation of traits of agronomic interest and the (re)combination of the most interesting alleles within an improved variety. It has been very successful for simple traits controlled by one or very few genes, such as disease resistance, and for crops grown in highly controlled, optimal environments with no major fluctuations

Epigenetics in Ecology and Evolution,
coordinated by Christoph GRUNAU and Stéphane MAURY.

in the plant's life cycle. It has also made it possible to significantly improve more complex traits such as yield in intensive agricultural systems. However, global changes are leading to a complete redefinition of the objectives of varietal selection, with the need to consider fluctuating pedoclimatic scenarios within and between crop cycles, in order to contribute to resilient agriculture in a system in transition.

For more complex traits that are strongly linked to the fine regulation of genomes and adaptive traits, epigenetics seems to provide some understanding in the regulatory mechanisms of these complex traits and their engineering, opening up new prospects for plant improvement in the face of this transition in agriculture.

In this chapter, we present how epigenetic mechanisms can play a part in the various stages of a selection scheme and contribute to plant improvement, using the example of two agronomic species with very different characteristics: rice and grapevine.

9.2. Defining agricultural transition objectives and challenges

9.2.1. *Rice and production in marginal areas*

Rice is a monocotyledonous annual plant that mainly reproduces autogamously (in other words, by self-fertilization of its flowers) and is considered a model for cereals due to its relatively small genome (approximately 400 million base pairs across its 12 chromosome pairs). Cultivated rice (mainly *Oryza sativa* L.) is the staple diet of more than half the world's population. It is grown in a wide range of highly contrasting environments, from being totally flooded to relatively arid. With the observed global changes, its cultivation is increasingly extending to marginal areas, so as to increase production in an ecologically intensive system.

While 95% of rice is produced in the tropical zones of the South, the main challenges facing rice cultivation are linked to climate change affecting water availability or its excess, salinity affecting many marginal growing areas and linked to rising sea levels, rising land temperatures and the appearance of emerging diseases, while ensuring the livelihood of producers and nutritional and organoleptic quality for consumers.

9.2.2. *Grapevines and terroir in the face of climate change*

The grapevine (*Vitis vinifera* L.) is a dicotyledonous plant. It is a perennial vine, reproducing sexually (with hermaphroditic flowers for cultivated vines, unlike wild vines of the subspecies *Vitis vinifera* ssp. *sylvestris*, which are dioecious), but susceptible to inbreeding, and whose “cépages“ (the name given to vine varieties) are propagated by vegetative propagation. Variability exists within grape varieties, giving rise to different clones of interest, with different characteristics for many traits (yield, earliness, berry color, etc.). Since the *Phylloxera* crisis at the end of the 19th century, grape varieties have been grafted onto less diverse rootstocks. Compatibility between graft and rootstock, as well as the sanitary quality of the grafted plants produced by nurserymen, were important factors in the development of the plants. The quality of wine produced from grape varieties, some of which are world-renowned (Merlot, Cabernet-Sauvignon, Chardonnay, etc.), is very closely linked to their pedoclimatic microenvironment, combined with viticultural and oenological practices, known as terroir. This emblematic Mediterranean crop is therefore of major economic and socio-cultural importance. It is also a significant species because of its small diploid genome, made up of 19 pairs of chromosomes, comparable in size to that of rice, and which has undergone few changes over the course of its evolution.

Grapevines are grown in areas severely affected by climate change (particularly in the Mediterranean region), raising questions about maintaining wine quality and adapting to the terroir over a growing cycle of 20 years or more. A reduction in the use of phytosanitary products is absolutely essential, bearing in mind that grapevines are undoubtedly one of the crops that use the most phytopharmaceutical products per unit area: the average number of treatments on vineyards in conventional cultivation was 18 in 2019, corresponding to 12.4 reference doses of treatment products for the whole year, compared with seven treatments (5.1 reference doses) for common wheat, for example (Agreste 2021).

However, restrictions linked to protected designation of origin or other wine quality labels need to be considered in any attempt at varietal innovation, in order to facilitate the adoption of new grape varieties that are better adapted to future changes.

9.2.3. *Common breeding objectives for agriculture in transition*

The Green Revolution era (which boosted yields with a heavy reliance on inputs) is over; society and the expected climate crisis are now forcing us to produce differently, respecting agrobiodiversity and the principles of agroecology worldwide. These challenges for future agriculture are reflected in common objectives for improving rice and grapevines, in order to continue producing high-performance crop varieties, and for efficient, sustainable agriculture, in a context of climate change and agriculture in transition (Figure 9.1).

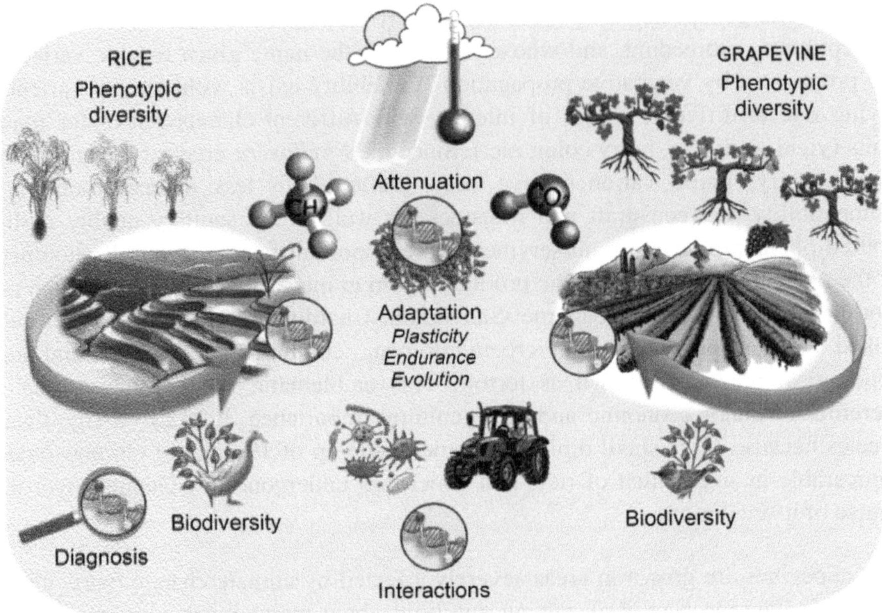

Figure 9.1. *New challenges in plant breeding, where epigenetics (represented by a logo) can play a role: phenotypic diversity and ecosystem biodiversity, adaptation to environmental stress (plasticity, endurance or crop evolution), interspecific and intraspecific interactions, climate change mitigation, development of diagnostic tools (drawing: F. Curk – INRAE). For a color version of this figure, see www.iste.co.uk/grunau/epigenetics.zip*

Mainly due to the emission of greenhouse gases, climate change affects all terrestrial ecosystems through changes in temperature, precipitation frequency, solar radiation, wind, relative humidity and hydrological cycles. The new report from the Intergovernmental Panel on Climate Change (IPCC) warns of the "threat to human well-being and the health of the planet" caused by these upheavals (IPCC 2022). Agriculture itself is responsible for 20% of greenhouse gas emissions (including 11% of methane production linked to rice cultivation). This means monumental consequences for global crop distribution, with food security at stake. Agricultural production is affected by temperature variability (a 1°C rise in average temperature is likely to reduce rice production by 10%), water and salt stress, heavy erratic rainfall and rising sea levels, leading to flooding and salinization of the land. It is also affected by common or emerging diseases, for which the use of pesticides is no longer an option, and changes in soil microbial composition. Last but not least, changing climate zones present a problem for the optimal outcome of current growing zones. For example, some of today's major wine-growing regions may no longer be suited to grape-growing, or at least to the quality of wines currently produced. To maintain or increase agricultural production, we need to adapt new varieties to the biotic and abiotic constraints brought about by climate change and interactions within ecosystems. In addition, we can hope to mitigate the effects of climate change through varietal innovation, by taking into account, for example, genetic variation in methane emissions from rice crops or by improving soil carbon capture by crops.

The domestication of wild species into cultivated species and their selection down to elite varieties has led to a considerable decline in biodiversity within agricultural production. This loss of biodiversity has major consequences for the hardiness and adaptability of today's varieties, particularly in the face of bio-aggressors, even though resistance genes still exist in related wild species. Breeding for resistance therefore requires crosses between different species in order to integrate these resistance genes into cultivated material. These interspecific crosses, which are difficult to achieve successfully as they are at the very limit of the notion of species, nevertheless represent a "genomic stress" with unpredictable effects on their progeny, particularly in relation to epigenetics, which must be controlled.

Sometimes, existing biodiversity is no longer sufficient to meet needs or is not accessible to cross-breeding. In such cases, we must seek to increase this diversity, at least in terms of "phenotype" (the plant's characteristics) and, if possible, give it a "genetic" basis (differences in the genetic material of the cells, DNA, which can be visualized by "genotyping"), so as to ensure that it can be inherited.

In an agroecological context, crop sustainability requires the introduction of greater diversity through varietal blends and crop associations, with management of the associated microbiome (in other words, all the microorganisms that interact with plants in their immediate environment). The varieties involved in these new cropping systems need to be optimized to take better account of the interaction and competition between organisms, while promoting the resilience of crop systems in the face of global change.

The potential contribution of epigenetics to optimizing adaptation to climate change, with possible applications to plant breeding, is well referenced (Kakoulidou et al. 2021). In this chapter, we will focus on positioning epigenetics alongside selection schemes that are likely to meet the challenges just described.

9.3. The contribution of (epi)genetics to the definition of traits of agronomic interest and the construction of ideotypes

Most of the criteria that breeders use to identify the best individuals in a breeding program remain relatively simple, so that they can be assessed quickly and on large numbers of plants. The combination of favorable characteristics to optimize the performance of a cultivated variety in a given environment represents the "ideotype", which the breeder tends to target. However, the new challenges presented by climate change are leading us to redefine ideotypes, which are becoming more complex, in order to prepare new varieties for climate change. In this section, we will look at the role of epigenetics in the preliminary phases of varietal creation (pre-breeding), which focus on defining selection criteria and identifying sources of diversity for target traits.

9.3.1. *Importance of epigenetic phenomena in the modulation of traits and adaptive plasticity in plants*

Over the course of their evolution, plants have adapted to their environment through a number of morphophysiological traits. Some of these, such as the presence of trichomes (hairs) on the leaves of certain grape varieties, may facilitate resistance to certain insects and abiotic stresses, but their effect is sometimes limited or even controversial. Moreover, plants have acquired the ability to react to environmental stresses by modulating their physiology and/or metabolism (growth, phenology, photosynthesis, mineral nutrition, water use, production of free radicals, etc.). This "plasticity" ensures their protected functioning with varying degrees of success, depending on the individual and the species (in line with their genotypes). Generally speaking, these reactions rely on genetic regulation mechanisms (modulating the expression of a certain number of genes controlled by complex signaling).

Epigenetic mechanisms play an important role in the development, reproduction and response of plants to their environment (Kakoulidou et al. 2021). During development or in response to an external stimulus, it is indeed possible to observe chromatin remodeling, histone modification and DNA methylation, leading to fine modulations of gene expression. Furthermore, stress-induced activation of transposable elements can lead to new insertions elsewhere in the genome (with possible collateral effects), as well as modulate gene expression through changes in the methylation level of these elements and their chromosomal surroundings, or may contain coding sites for small RNAs in their sequence that will modify the expression of nearby genes (Wu et al. 2020a). Finally, there are changes in certain populations of non-coding RNAs, which regulate the expression of the genes they target. Rice and grapevine have very high levels of CpG methylation (cytosine and guanine in sequence, with a phosphate bridge between the two bases), in the range of 46% to 47%, whereas *Arabidopsis thaliana* (Arabidopsis) has a CpG island methylation rate of just 24%. These two species also differ from the model plant in their high number of transposable elements, which is in the range of 450,000, compared with Arabidopsis, which has just over 30,000 (Gahlaut et al. 2020). Epigenetic mechanisms are therefore of considerable significance in the gene regulation of these species and are likely to explain variations in traits of agronomic interest and in the modulation of their plasticity.

The success of semi-dwarf rice during the Green Revolution is thus partly linked to the stimulation of tiller development (the name given to shoots emerging from buds at the base of the plant) under the effect of nitrogen, thanks to a reprogramming of histone methylation. This change enabled them to adapt to nitrogen-rich agriculture (Wu et al. 2020b). The mobilization of transposable elements, notably MITE-types, in this species, beyond its diversification role, also leads to alterations in DNA methylation status (via the RdDM pathway for RNA-dependent DNA Methylation), making it possible to explain the contribution of fine genome regulation in rice subspecies in their adaptation to different, highly contrasting geographical situations, as well as to both biotic and abiotic stresses (Song and Cao 2017).

In grapevines, the methylation profile of the grape berry varies during development, repressing gene expression during fruit development after "veraison", a specific stage of development particularly subject to water and heat stress (Shangguan et al. 2020). Variable accumulation of a class of microRNAs has also been demonstrated, in response to water stress, between two grape varieties with different levels of drought tolerance, with these small RNAs having the ability to circulate between rootstock and graft (Pagliarani et al. 2017) with the transfer of small RNAs being favored from graft to rootstock (Rubio et al. 2022). Grafting can also trigger changes in scion methylation, some of which may be stable over time, promoted by clonal propagation (Tsaballa et al. 2021).

Hardening is another important element in the adaptation of plants to their environment, linked to epigenetic mechanisms. This is referred to as priming or epigenetic memory, as the plant exposed to one stress acquires the ability to better withstand a second exposure to that stress, or even another environmental stress. This hardening is particularly interesting for perennial species such as grapevines, where it can be induced by in vitro culture and thermotherapy, although 40% of the epigenetic effects thus induced disappear after a year (Baránek et al. 2015), and their effect in the field has yet to be assessed. Furthermore, an application of water stress at an early stage of rice development modifies the methylation level of stomatal guard cells and seems to give them a certain stress "memory" (Auler et al. 2021). This suggests a possible hardening of rice, despite its shorter life cycle, by acting at the seedling stage. In some cases, this hardening may even be passed on to subsequent generations. For example, by imposing water stress on rice lines during 11 generations of self-fertilization, researchers were able to observe stable changes in their methylome, some affecting stress response chains and 30% of them transmitted from generation to generation (Zheng et al. 2017).

Given the unpredictable nature of the environmental stresses to which rice and grapevines are subjected, the consideration of a genotype's ability to change its phenotype according to its past and current environment, i.e. its phenotypic plasticity, is becoming a key criterion in the construction of an "ideotype" adapted to climate change. Taking the epigenetic mechanisms involved in modulating gene expression under environmental stress into account is therefore of strategic interest. Understanding the molecular basis of the genotype–environment interaction, and the involvement of epigenetics in the development of phenotypic plasticity is essential to facilitate the development of sustainable agriculture that is efficient in the face of climate change (see Chapter 8).

9.3.2. *Seeking phenotypic diversity for traits of interest: the question of heritability*

To improve a given trait, geneticists/breeders need to have sufficient "phenotypic" variability for that trait. The trait must also be heritable, in other words, have the potential to be passed on to descendants.

Domestication and selection of rice and grapevines have greatly reduced the genetic diversity present in these cultivated species, and additional diversity may be essential to improve a given variety. To achieve this, breeders need to characterize the full phenotypic diversity of the genetic resources at their disposal: cultivated varieties, elite varieties or landraces, and also the diversity present in wild subspecies or related species. For example, some rice varieties, such as *Teqing*, show exceptional tolerance to submergence compared with the *Nipponbare* variety. These two varieties belong to two different subspecies of *Oryza sativa* rice, *indica* and *japonica*, respectively, which are quite divergent at the genetic level due to their separate domestication. Similarly, certain species close to the cultivated grapevine, such as *Vitis riparia* or *Muscadinia rotundifolia* from America, are less susceptible to mildew, powdery mildew and several other diseases than the cultivated grapevine. However, crosses between two species (or subspecies) to integrate favorable traits from other plant groups into elite material are more difficult to achieve than simple intraspecific crosses.

The genetic basis of this phenotypic diversity is a diversity of alleles at certain chromosomal regions (loci), which are identifiable in some cases by nucleotide differences (mutations) or structural variations affecting DNA regions coding for a functional product (enzyme involved in the metabolism in question, regulatory region at a promoter or transcription factor, acting on the expression of another gene). These alleles, if heritable, can be followed from one generation to the next in a selection scheme, the aim of which will be to accumulate favorable alleles within a single genotype (or several complementary genotypes). Current varietal selection is therefore largely "assisted" by genetic markers that can be identified at the DNA level.

However, some phenotypic variations in existing genetic resources cannot be explained by classical genetic variations. Epigenetic modifications that are more or less unstable over time may be the cause of this phenotypic plasticity, although they may be reversible. In this case, they are highly disruptive to conventional breeding efforts but deprive the breeder of additional phenotypic variability that could represent a potentially interesting selection potential. Conversely, variation in epigenetic mechanisms can be heritable. Changes in the methylation status of a given gene can, in some cases, be transmitted trans-generationally, giving rise to heritable phenotypic variation. The heritability of epigenetic marks, such as methylation, implies epigenetic marking of reproductive cells that is not erased by the monitoring mechanisms ensuring cell totipotency after fertilization and before embryo development. The mechanisms involved are currently unknown. This type of heritable epigenetic variation is known as "epialleles". Some spontaneous epialleles have been identified in rice and grapevines. For example, an Epi-sp "gain-of-function" epimutant, encoding a long non-coding RNA, has been identified in the *japonica* rice variety "*Zhonghua11*", its panicles having a more compact architecture (Luan et al. 2019). Mutants of epigenetic origin observed in grapevine varieties also affect highly visible traits such as berry color, in relation to the insertion of a retrotransposon upstream of the VvmybA1 transcription factor, the latter being involved in anthocyanin synthesis (This et al. 2007).

Genetic resource collections available to breeders are assessed for their phenotypic variation, but they can also be assessed for the presence of epialleles. Epigenetic marks, in this case, represent additional data for characterizing plant material. In Arabidopsis and corn, it has been shown that 10%–20% of differentially methylated regions (DMRs) are associated with differential gene expression and could have an impact on phenotype, although probably more qualitative than quantitative (Agarwal et al. 2020). We can only hope that the same applies to all cultivated species, including rice and grapevines.

9.3.3. *Epigenetics as a source of new diversity for plant improvement*

If phenotypic diversity for a given trait is not sufficient in existing genetic resources, it can be created de novo, by promoting mutations that affect the nucleotide composition of DNA. As we have seen, new phenotypic diversity can also be created through epigenetic mechanisms, affecting the modulation of gene expression without necessarily affecting gene structure. The in vitro cultivation used to generate somaclones in grapevines thus promotes changes in the methylation sites of certain genes (Schellenbaum et al. 2008). Various Pinot noir clones were analyzed using the MSAP (Methylation Sensitive Amplified Polymorphism) technique and revealed stable methylation differences between clones (Ocaña et al. 2013). Similarly, clones from the same origin could have different epigenetic statuses depending on their location (Kong et al. 2019). In vitro cultivation can also be used in rice to generate transposable element mobilizations and epigenetic modifications (Kikuchi et al. 2003).

To promote these epigenetic modifications, we can directly promote the cellular machinery involved in methylation, histone modification or small RNA generation (methylases, methyltransferases, etc.): for example, chemical treatment with Zebularin or 5-Azacytidine, which inhibit DNA methylation (unfortunately without being able to target a particular gene to date), can increase transposon activity (see Chapter 3). New epigenome-editing tools, for example, using the fusion of epigenetic modifiers and the inactivated Cas9 protein to target their action, open up prospects in plant improvement subject to their regulation (Dalakouras and Vlachostergios 2021; Hou and Wan 2021). Their development on rice or grapevines could prove promising, while simultaneously making it possible to target genes of agronomic interest. This “epimutagenesis” may lead to a higher frequency of “gain-of-function” alleles, due to loss of gene inhibition, unlike genetic mutagenesis, which has so far led to loss of function.

It is therefore possible to increase the phenotypic diversity, or create neodiversity, available to breeders, by using natural or induced epigenetic mechanisms, provided they can be fixed in the varieties developed from these resources.

9.4. The role of epigenetics in current selection schemes

9.4.1. *Development and characterization of recombinant populations*

To create new, favorable “haplotypes” (groups of alleles on the same chromosome that can be passed on together to the offspring) and to try to break the links between nearby loci with opposite effects, breeders make use of the natural recombination of genetic information occurring during crossing-over in prophase I of meiosis. To do this, they cross "parent" plants with complementary properties (chosen from available genetic resources) and generate offspring in which the different parental alleles segregate (in principle) according to the laws of Mendelian genetics. Recombinant lines (RILs) can be generated in rice by promoting self-fertilization over several generations after the initial cross, thus providing recombinant material fixed in the homozygous state. In the case of grapevines, which are heterozygous, the progeny of an initial cross directly provides seedlings with different allelic combinations. If one of these combinations is favorable, it can then be propagated identically by cuttings. Figure 9.2 shows the various stages in this process of selection and fixation of favorable alleles, highlighting the stages (numbered in reference to the text) where epigenetics must be taken into account.

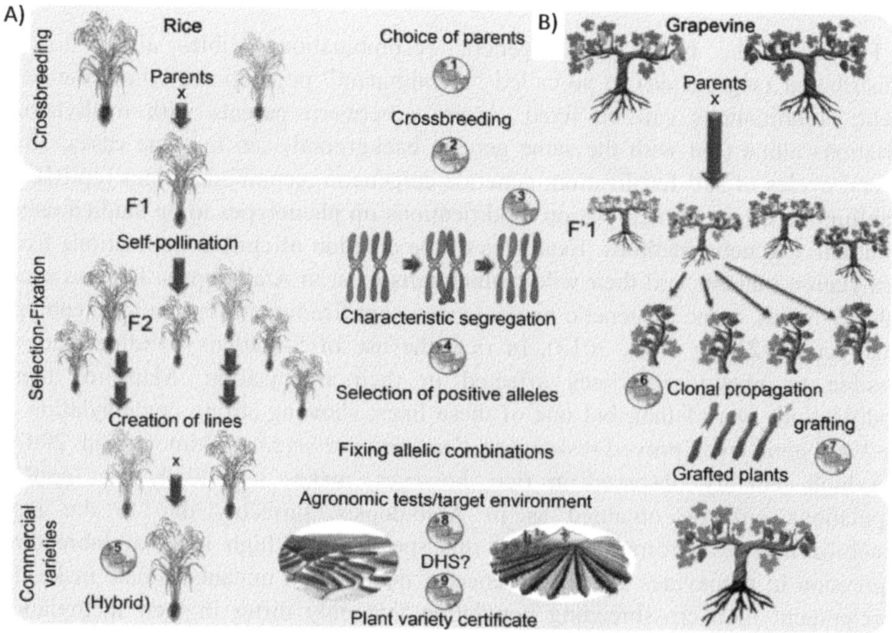

Figure 9.2. *Comparative diagram of conventional breeding of new rice varieties (A) and vine varieties (B). For a color version of this figure, see www.iste.co.uk/grunau/epigenetics.zip*

COMMENT ON FIGURE 9.2.– *Epigenetics influences different stages of this diagram. 1: (epi)genetic diversity of crossing parents; 2: genomic shock at crossing;*

3: genetic recombination; 4: (epi)genotyping for selection of favorable alleles; 5: hybrid vigor; 6: somaclonal variation; 7: scion/rootstock interaction; 8: genotype–environment interactions and phenotypic plasticity; 9: homogeneity and stability of epigenetically determined traits (drawing: F. Curk – INRAE).

The choice of parents for the initial cross (Figure 9.2 (1)) is crucial. Epigenetic differences between parents are therefore important. Crosses between subspecies (rice) or neighboring species (grapevines) are sometimes necessary to bring favorable alleles (or new traits) into the cultivated pool. These crosses, although viable, represent a "genomic shock" that can lead to epigenomic disruptions, including changes in the methylation profiles of hybrids compared to parents (Wang et al. 2010) and the remobilization of transposable elements. The result of crossing is therefore not "just" the sum of the (epi)genomes contributed by each of the parental gametes (Figure 9.2 (2)).

Following the initial cross, genetic recombination enables alleles to be redistributed (segregated) in so-called "recombinant" populations, where the best allelic combinations can be fixed. A cross between parents with methylation variations alone (but with the same genetic background) can in some cases, after several cycles of self-fertilization, generate epigenetic recombinant lines (epiRILs), enabling the effect of methylation modifications on phenotypes to be studied using quantitative genetic methods. Examples of the creation of epiRIL populations from methylation mutants and their wild counterparts exist in Arabidopsis. Despite some stability issues, some epigenetic marks have a significant contribution to phenotype construction (Zhang et al. 2018). In rice, the use of 5-azadeoxycytidine made it possible to obtain rice lines affected in their methylation. Many of these modifications were lethal, but one of these lines, showing strong demethylation at the *XA21* gene level, proved resistant to *Xanthomonas oryzae* (Akimoto et al. 2007). The high lethality observed in rice, however, makes it unlikely that epiRIL populations will be obtained as in Arabidopsis, probably due to the high remobilization of retrotransposons in this species. The high level of inbreeding depression in grapevines and the absence of methylation mutants to date make the development of a crossbreeding population that only differ in their methylation profile unrealistic.

Given the low number of crossing-over per meiosis (rarely more than three per chromosome) and their non-random distribution with the existence of recombination hotspots, it is tempting to seek to control recombination in order to increase its frequency in the early stages of selection (Figure 9.2 (3)). However, we now know that changes in DNA epigenetic marks can modify the rate of crossing-over and their location on chromosomes. In Arabidopsis, methylation mutants have shown a recombination rate that tends to increase (Yelina et al. 2015). The manipulation of epigenetic factors and the targeting of genomic regions of interest using CRISPR-Cas9 technology are strategic avenues of research, complementing current research on genetic manipulation of meiotic recombination in rice (Fayos et al. 2019).

9.4.2. *Development of epigenetic marks*

The development of molecular markers (easily identifiable DNA differences between individuals) in plants has had many applications: the development of genetic maps, the identification of markers linked to traits of interest and their use in breeding (marker-assisted selection). The democratization of high-throughput sequencing technologies has reinforced the use of markers, which are identified by comparing sequences between individuals, and in particular SNPs (single nucleotide polymorphisms) corresponding to base substitutions, as well as small insertions or deletions of variable size. The GBS (genotyping by sequencing) method – which consists of only sequencing a subset of the genome – or the complete resequencing of genomes can generate millions of genetic markers at an attractive cost.

Methods for detecting epialleles in a given genomic region have also been developed in recent years, such as bisulfite PCR product sequencing (BSP) and methylation-specific PCR (MSP), which use the conversion of unmethylated cytosines to uracil, or methods such as “Chop-PCR”, in which DNA is partially digested by methylation-sensitive restriction enzymes (Gahlaut et al. 2020). In parallel, “epigenotyping by sequencing” (epiGBS) technologies enable genome-wide identification of epigenetic modifications (epialleles) linked to phenotypic variations in regions that are poorly covered by conventional genetic markers (Gahlaut et al. 2020).

This epigenotyping (Figure 9.2 (4)) can help to map a recombinant population, as a complement to genetic markers, particularly in regions that are poorly covered by the latter, provided that the heritability of the epigenetic marks identified can be verified. The ability to develop epigenetic marks and characterize the “epigenetic profile“ of plants may condition their increased use in breeding programs. It would then no longer be a matter of recombining loci of interest but of epigenetic contexts of chromatin.

On the same grape variety, the MSAP method was used to highlight differences in methylation profiles between plants from several vineyards (Xie et al. 2017). In the same work, “methylation-sensitive genotyping by sequencing“ (msGBS) was used to identify and characterize DMRs on the same genome. This epigenotyping technique could easily be used to monitor methylation variations in a recombinant population.

Some epigenetic variations involve transposable elements. The use of transposons in the definition of genetic markers therefore has the dual advantage of high heritability (except in cases of reversion) and the possibility of targeting changes in the methylation profile around these elements. Various techniques can therefore be used to target differences around these transposable elements: Inter Retrotransposon Amplified Polymorphism (IRAP), Retrotransposon Microsatellite Amplified Polymorphism (REMAP), Retrotransposon-Based Insertional Polymorphism (RBIP) and Sequence-Specific Amplified Polymorphism (SSAP) (Arvas et al. 2021). As with the development of genetic markers, the cost of these methods may fall as their use on dedicated platforms becomes more widespread, and as the associated bioanalysis tools are developed (Schmitz et al. 2022). Obviously, to use these epigenetic marks in breeding, we need to know their level of variation between individuals, their impact on agronomic phenotypes and their heritability.

Characterization of the epigenetic profile of segregating populations and representative collections of rice or grapevine can assist phenotyping for traits of interest, which are notably related to the physiological plasticity of plants. They can also be used to model phenotype predictions as a complement to genetic markers (Agarwal et al. 2020).

9.4.3. *Quantitative epigenetics: identifying loci of agronomic interest*

Quantitative genetics is an essential tool in plant breeding, used to identify the regions of the genome involved in the variation of traits of agronomic interest and thus define genetic markers to assist selection. QTL (Quantitative Trait Loci) detection in recombinant populations, or association mapping (GWAS, for Genome Wide Association Study) in diversity panels, are based on linkage disequilibrium (preferential associations between alleles at two loci). They have enabled the identification of markers for resistance or quality genes in grapevines or rice, which can then be used in breeding as a proxy for phenotyping.

On species for which epiRILs could be developed, “epi-QTLs” were identified, linking variation in epialleles to traits such as flowering or root length (Cortijo et al. 2014). Similarly, an Epigenome Association Study (EWAS), a methodology used extensively in humans to screen for genetic diseases, was conducted on palm oil for the first time in relation to a fruit anomaly that induces a loss of oil production (Ong-Abdullah et al. 2015). These studies (epi-QTL and EWAS) have highlighted the close association between epigenetic variations and traits of agronomic interest and would benefit from further development.

While the creation of epiRIL is still an issue in rice and grapevine, the possibility of using linkage disequilibrium in genetic resource collections of these species, characterized by their epigenetic profile, is very promising for the years to come.

9.4.4. *Marker-assisted selection, genomics, phenomics and epigenomics*

In conventional breeding, the offspring of the crosses made must be tested to select favorable alleles in the regions of interest, in an attempt to maximize the genetic gain of the resulting populations. The best combinations identified are then fixed (in the homozygous state for rice, or by clonal multiplication for grapevines). This characterization of offspring can be facilitated by the use of genetic markers targeting positive alleles on a few target traits (this is known as marker-assisted selection). Epigenetic marks deserve their rightful place in this selection scheme, marking more complex and "adaptive" traits. In this case, epigenotyping would focus on a few target epigenetic differences that are well localized on the genome.

Furthermore, the more global characterization of a large number of genetic markers distributed along the genome now makes it possible to carry out "genomic selection", by modeling the contribution of markers distributed along the genome to the expected genetic gain in a test population, before dispensing with the sometimes tedious and costly phenotyping of the breeding population (which is then only genotyped). The same principle can be applied by replacing genotyping with characterization of a tissue's reflectance profile at different wavelengths, at a reduced cost. This is known as "phenomic selection". This approach is currently being tested on grapevine (Brault et al. 2021), and results have already been published on rice (Jaiswal et al. 2019).

Genomic and phenomic selection are based on predictions that must consider both genetic and non-genetic effects on phenotypes. Modeling epigenetic effects on phenotypic variation should allow a significant contribution to be made to genomic or phenomic selection, in order to describe part of the genotype-environment interaction effects and the "non-genetic" part of selection data in genomic or phenomic prediction models (Gogolev et al. 2021). To date, however, there are no published examples on rice or grapevine.

9.5. The final stages before a new variety is labeled

9.5.1. *Obtaining basic material (seeds or seedlings)*

To develop a new variety, it is necessary to create and multiply seeds or seedlings that are representative of the future variety. This is the basic material that will be used for varietal testing and the plant material multiplication process.

The type of varieties registered to a catalog of varieties are different in rice and grapevines. For rice, which is self-pollinating, inbred line varieties are the simplest type of variety to obtain, but hybrid varieties (resulting from large-scale crosses between two homozygous inbred lines) are also developed to take advantage of hybrid vigor, even if only moderately so in this species, but favored in "heterotic" crosses between subspecies. The basic material for creating these hybrids is represented by the parental lines, whose female line often incorporates a male sterility gene; it must be chosen judiciously. Reciprocal hybridization experiments between two rice subspecies have shown a significant effect of methylation status and small RNA abundance on hybrid vigor, through the modification of gene expression between parents and hybrid, which also depends on the direction of crossing (He et al. 2010). Epigenetics thus appears to be an explanatory factor for hybrid vigor (heterosis) and inbreeding depression in a species such as the grapevine. Epigenetic differences in the potential parents of a hybrid cross can therefore become indicators of the hybrid performance of their cross, independently of the genetic differences between the parental genomes.

The direction in which crosses are made (who is the male and who is the female parent) also influences the development of seed tissues, particularly the albumen. This "genetic imprinting" linked to hypomethylation in the albumen, and also to the mobilization of transposable elements, controls the preferential expression of certain genes in the maternal genome, while the paternal genome is silenced, or vice versa (Chen et al. 2018). By modulating genetic imprinting through epigenetic mechanisms, we can hope to improve the quality of the albumen and therefore the seeds. In rice, for example, several genes with maternal or paternal expression have shown an effect on grain size, starch content or seed fertility (Yuan et al. 2017).

To facilitate the conformal reproduction of rice hybrids, apomixis (vegetative reproduction without meiosis or fertilization, but using reproductive cells) is an interesting alternative for producing hybrid seeds without resorting to crosses between parental lines. Recently, researchers have succeeded in inducing apomixis through mutations in key meiosis genes (Xie et al. 2019). The possible epigenetic regulation of apomixis also makes it an interesting avenue of research.

Cultivated grape varieties are, by nature, hybrids. However, they are also propagated and marketed as scions grafted on rootstocks. We have seen that clonal multiplication of scions can induce epigenetic modifications among grapevine clones. It is therefore important to monitor these modifications, both to check that the clones produced conform to the grape variety propagated, and to use this somaclonal variation for grape variety improvement. Moreover, grafting induces molecular interactions between the rootstock and the scion, including epigenetic mechanisms, and in particular RNA and small RNA movements, as well as methylation changes, which can have an impact on the phenotype (Tsaballa et al. 2021) and even its resilience to environmental stresses (Kapazoglou et al. 2021). These epigenetic changes are likely to be transmitted during clonal multiplication of grafts. It would therefore be important to be able to test the effect of grafting on scion characteristics. This interaction is particularly important for improving grape varieties in the face of climate change, without compromising quality or yield.

9.5.2. *Plant variety certificate*

In Europe, varietal innovations are protected by a Community plant variety right. To qualify, the breeder must demonstrate to the CPOV (Community Plant variety Office) that his or her variety is distinct from all those referenced, that it is uniform in terms of relevant traits, and that it is stable during each multiplication cycle. Verification of this DUS criterion (distinctness-uniformity-stability) is generally entrusted to an accredited organization. In France, GEVES (Groupe d'étude et de contrôle des variétés et des semences) is responsible for validating DUS.

The use of epigenetic variations in plant breeding presents the problem of recognizing these variations as a stable trait enabling a new variety to be distinguished (Dalakouras and Vlachostergios 2021; Kakoulidou et al. 2021). Epigenetic mechanisms thus struggle to fully meet the DUS criteria required for the protection of new varieties, and which are verified by ad hoc bodies on the basis of basic material supplied by the breeder. Indeed, if an epigenetic variation is capable of creating a "distinct" trait (and therefore different from what exists on the market), the homogeneity of the material thus created may be called into question in the event of reversibility (hence the observation of chimeras), and the stability of the new variety (from one generation to the next) could be affected by a reversion affecting the reproductive cells in the case of rice, or the meristems concerned by vegetative propagation in the case of grapevines. In the current state of legislation, epigenetic innovations integrated into a conventional breeding scheme must therefore be considered with caution before there is any intention to register a new commercial variety.

9.5.3. *Verification of agronomic value (in rice fields or terroir)*

Agronomic, technological and environmental value (VATE) is another important criterion before a new variety can be registered in the official catalog and marketed. Beyond this regulatory aspect (tested in multilocation trials), the adaptation of a new rice or grape variety to its target environment is an essential aspect for its adoption by rice farmers or winegrowers. This environment depends on various factors, including climate, pedoclimatic conditions and the crop's (micro)biotic environment. Epigenetics therefore also has a role to play at this stage.

In rice, the study of histone methylation mutants has shown that histone methylation strongly regulates the structure and composition of the root microbiota, probably through the regulation of genes involved in immunity, stress response and root architecture (Lv et al. 2021). The composition of these microbiota networks contributes to the plant's adaptation to its environment. Furthermore, the enrichment of the *methylococcaceae* class in the roots of histone methylation-deficient plants suggests a possible role in the balance between methanogenic and methanotrophic bacteria, and consequently in methane emission from rice fields. Epigenetics could therefore also play a part in mitigating climate change in rice-growing systems through this mechanism.

The notion of terroir is important for grapevines, because of the impact of the environment on the composition of the grape berry and the quality of the resulting wine. The terroir can be defined as the interaction between the physical and biological environment of the grapevine, winegrowing and oenological practices. Characterization of the methylation profiles of a single grape variety (Syrah) across 22 vineyards representing six major winegrowing sub-regions of Australia showed that differences in grapevine leaf methylation between vineyards and sub-regions were influenced both by the geographical location of these vineyards (notably the difference in altitude, which impacts the UV levels to which grapevine plants are exposed) and, to a lesser extent, by the pruning system (Xie et al. 2017). DNA methylation thus represents an important mechanism in the effect of the terroir on fruit and grape quality, and could thus serve as a quality marker.

9.6. Development of new varieties without sexual crossing

9.6.1. *Transgenesis or cisgenesis in a favorable chromatin context*

Transgenesis (the technique of introducing foreign DNA into the genome of an organism) or cisgenesis (where the gene introduced by genetic engineering comes from the same species) are used in some countries to bring favorable traits or alleles to the genetically modified organisms produced (then protected by patents). However, this technique is not as easy to implement as it might seem, and many transgenesis trials result in very low levels of transgene expression, or even their total extinction (silencing). The role of epigenetics in this mechanism is now well documented (Rajeev Kumar et al. 2015), involving DNA methylation of promoters attached to the transgene (transcriptional silencing) or the production of small interfering RNAs (post-transcriptional silencing). Transgenesis has been shown to induce both genetic and epigenetic changes in rice (including during the tissue culture phase of the transgenesis process), with the nature and function of the transgene having the greatest influence on phenotype after four generations of transformation, while epigenetic effects are highly dependent on the variety used for transformation (Fan et al. 2020). Transgenesis is therefore not a technique whose effects can be predicted with any certainty, and deserves particular attention. However, perfect control of the in vitro cultivation process, to limit the mobility of transposable elements and chromatin changes, should enable better control of transgene expression and make the development of this technology more efficient, should it prove necessary.

9.6.2. *Epigenomic and epibreeding editing*

Inducing epigenetic modifications (see Chapter 6) is likely to induce favorable phenotype changes. This is an attractive alternative for plant breeding (Kim et al. 2022).

Epibreeding (Figure 9.3) would thus involve selection that is very strongly based on epigenetic variations, without the need for an initial cross.

This epibreeding strategy is described in Zhi and Chang (2021).

To obtain breeding populations with phenotypic diversity of epigenetic origin, it is possible to use natural epialleles, or to induce epigenetic modifications, either globally and with a stochastic effect using chemical inhibitors, the application of stress, mutagenesis, tissue culture or grafting, or targeted to certain genes using the CRISPR technique (this is called "epi-editing"), or RNA interference (Dalakouras and Vlachostergios 2021). Speed-breeding, which uses lighting with specific wavelengths and a highly disrupted timeframe, probably also leads to epigenetic effects that may contribute to this epigenetic diversity (Crisp et al. 2022). Currently, epiRIL-type populations are only available in Arabidopsis, but collections of epiallelic variants could also be used.

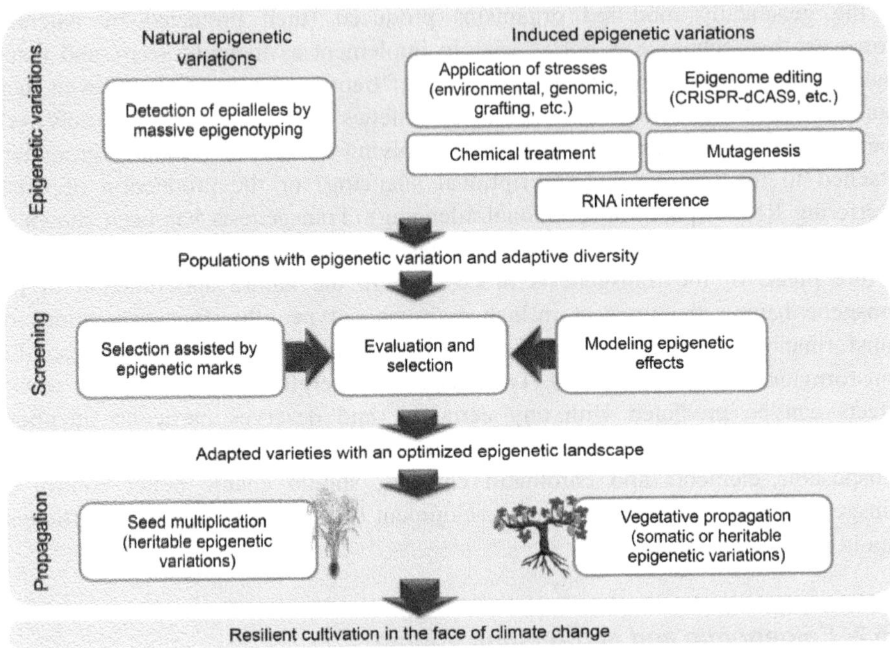

Figure 9.3. *Hypothetical epibreeding strategy to improve crop resilience to climate change (from (Zhi and Chang 2021), drawing: F. Curk – INRAE). For a color version of this figure, see www.iste.co.uk/grunau/epigenetics.zip*

COMMENT ON FIGURE 9.3.– *Natural or induced epiallelic variations can be used to construct breeding populations with sufficient epigenetic variation and adaptive diversity. These populations are assessed, in part through the use of epigenetic marks and modeling of epigenetic effects on phenotype. Adapted varieties with an optimized epigenetic landscape are then derived from the individuals selected using conventional breeding methods or genetic engineering. To be able to propagate and protect the resulting variety, epigenetic marks must be heritable, or at least stable throughout the plant's life in the case of a perennial species such as the vine.*

A screening stage would be necessary to assess the adaptive potential of these variants and their inclusion in basic material available for varietal protection. This selection can be facilitated by epigenotyping, in particular through epigenetic marks on a few target regions. The modeling of epigenetic effects on plant phenotype or plasticity could contribute to this screening.

Once the basic material has been obtained, it can then be propagated, by seed (in the case of rice) or plants (taking the rootstock into account in the case of grapevines). Epigenetic modifications can directly lead to a generation of changes in plant phenotype. Somatic mutations can be easily propagated on a vegetatively propagated plant such as a grapevine and, in some cases, these modifications can be inherited transgenerationally. In the case of rice, however, only highly heritable epigenetic variations can be propagated in the seeds subsequently distributed to farmers.

It is highly probable that results will be published very soon on rice and grapevines in this field, but for the moment, this scheme is only a working hypothesis.

Nevertheless, when performing epigenomic editing, it is important to check that there are no additional unwanted epigenetic modifications (off-target), in order to measure the effect of these modifications on gene expression, to avoid the use of controversial genetic transformations, and finally, to verify the stability of these modifications (Gahlaut et al. 2020). If they involve genetic engineering, the acceptability of such selection schemes should also be verified.

9.7. Legislation and marketing of varietal innovations resulting from epigenetic variations

Transgenic varieties are generally protected by a patent covering the transferred gene or trait. The technique of generating epigenetic variants by genome editing could itself lead to the resulting varietal innovations being included in the register of genetically modified plants. These new varieties, resulting from epigenetic variations, would then be subject to the corresponding European regulations. It is also important to distinguish between epigenetic variations that lead to the remobilization of transposable elements, and structural variations of the genome (through transposable elements), the latter being considered as genetic modifications.

Faced with a dilemma between the precautionary principle and the competitiveness of seed companies, Europe is currently struggling to adopt a legal position on the subject, but things could change rapidly.

Therefore, even though genome editing is likely to produce products that are quite similar to a natural epigenetic mutation, transparency on the use of this type of technique is required.

9.8. Conclusion and prospects

In this chapter, we have seen that genetics and epigenetics are intimately linked during varietal improvement processes. They must therefore be taken into account as explanatory factors. However, breeders are still reluctant to use epigenetic mechanisms in their day-to-day practices. In a recent review (2021), Kakoulidou et al. very aptly summarize the challenges that remain for plant breeding in order to integrate epigenetics into these practices: (i) identifying the epigenetic basis of variation in new traits of interest (particularly those linked to phenotypic plasticity and adaptation to environmental stresses); (ii) facilitating the use of stable (and transgenerationally transmitted) epigenetic marks for the selection of elite varieties, as a complement to conventional genetic tools; (iii) improving integrative approaches through new statistical models, so as to take better account of epigenetics in the determinism of a phenotype; (iv) promote increased phenotypic variability through epigenetic variation; (v) consider the use of epigenetic hardening as a stress management practice without recourse to genetic recombination; (vi) know how to use engineering tools to act on the plant epigenome with the aim of facilitating epigenetic recombination in a genetic selection process; (vii) communicate better and engage in dialogue with the various players in the agricultural production chain, as well as the general public, in order to hopefully improve the acceptability of these new plant improvement methods.

The integration of epigenetics into breeding and agricultural practices is therefore a matter to be followed.

9.9. References

Agarwal, G., Kudapa, H., Ramalingam, A., Choudhary, D., Sinha, P., Garg, V., Singh, V., Patil, G. Pandey, M., Nguyen, H.T. et al. (2020). Epigenetics and epigenomics: Underlying mechanisms, relevance, and implications in crop improvement. *Funct. Integr. Genomics*, 20, 739–761. doi: 10.1007/s10142-020-00756-7.

Agreste (2021). Enquête Pratiques culturales en viticulture en 2019 – IFT et nombre de traitements [Online]. Available at: https://agreste.agriculture.gouv.fr/agreste-web/disaron/Chd2119/detail/.

Akimoto, K., Katakami, H., Kim, H.-J., Ogawa, E., Sano, C.M., Wada, Y., Sano, H. (2007). Epigenetic inheritance in rice plants. *Ann. Bot.*, 100, 205–217. doi: 10.1093/aob/mcm110.

Arvas, Y.E., Abed, M.M., Zaki, Q.A., Kocaçalışkan, I., Haji, E.K. (2021). The potential role of transposable elements as molecular markers. *IOP Conf. Ser. Earth Environ. Sci.*, 761, 012031. doi: 10.1088/1755-1315/761/1/012031.

Auler, P.A., Nogueira do Amaral, M., Bolacel Braga, E.J., Maserti, B. (2021). Drought stress memory in rice guard cells: Proteome changes and genomic stability of DNA. *Plant Physiol. Biochem. PPB*, 169, 49–62. doi: 10.1016/j.plaphy.2021.10.028.

Baránek, M., Čechová, J., Raddová, J., Holleinová, V., Ondrušíková, E., Pidra, M. (2015). Dynamics and reversibility of the DNA methylation landscape of grapevine plants (*Vitis vinifera*) stressed by in vitro cultivation and thermotherapy. *PLoS ONE*, 10, e0126638. doi: 10.1371/journal.pone.0126638.

Brault, C., Lazerges, J., Doligez, A., Thomas, M., Ecarnot, M., Roumet, P., Bertrand, Y., Berger, G., Pons, T., François, P. et al. (2021). Interest of phenomic prediction as an alternative to genomic prediction in grapevine. doi: 10.1101/2021.12.16.472608.

Chen, C., Li, T., Zhu, S., Liu, Z., Shi, Z., Zheng, X., Chen, R., Huang, J., Shen, Y., Luo, S. et al. (2018). Characterization of imprinted genes in rice reveals conservation of regulation and imprinting with other plant species. *Plant Physiol.*, 177, 1754–1771. doi: 10.1104/pp.17.01621.

Cortijo, S., Wardenaar, R., Colomé-Tatché, M., Gilly A., Etcheverry M., Labadie K., Cailleux E., Hospital, F., Aury, J.M., Wincker, P. et al. (2014). Mapping the epigenetic basis of complex traits. *Science*, 343, 1145–1148. doi: 10.1126/science.1248127.

Crisp, P.A., Bhatnagar-Mathur, P., Hundleby, P., Godwin, I.D., Waterhouse, P.M., Hickey, L.T. (2022). Beyond the gene: Epigenetic and cis-regulatory targets offer new breeding potential for the future. *Curr. Opin. Biotechnol.*, 73, 88–94. doi: 10.1016/j.copbio.2021.07.008.

Dalakouras, A. and Vlachostergios, D. (2021). Epigenetic approaches for crop breeding: Current status and perspectives. *J. Exp. Bot.*, erab227. doi: 10.1093/jxb/erab227.

Fan, X., Chen, J., Wu, Y., Teo, C., Xu, G. (2020). Genetic and global epigenetic modification, which determines the phenotype of transgenic rice? *Int. J. Mol. Sci.*, 21, 1819. doi: 10.3390/ijms21051819.

Fayos, I., Mieulet, D., Petit, J., Meunier, A.C., Périn, C., Nicolas, A., Guiderdoni, E. (2019). Engineering meiotic recombination pathways in rice. *Plant Biotechnol. J.*, 17, 2062–2077. doi: 10.1111/pbi.13189.

Gahlaut, V., Zinta, G., Jaiswal, V., Kumar, S. (2020). Quantitative epigenetics: A new avenue for crop improvement. *Preprints*. doi: 10.20944/preprints202009.0348.v2.

Gogolev, Y.V., Ahmar, S., Akpinar, B.A., Budak, H., Kiryushkin, A.S., Gorshkov, V.Y., Hensel, G., Demchenko, K.N., Kovalchuk, I., Mora-Poblete, F. et al. (2021). OMICs, epigenetics, and genome editing techniques for food and nutritional security. *Plants*, 10, 1423. doi: 10.3390/plants10071423.

He, G., Zhu, X., Elling, A.A., Chen, L., Wang, X., Guo, L., Liang, M., He, H., Zhang, H., Chen, F. et al. (2010). Global epigenetic and transcriptional trends among two rice subspecies and their reciprocal hybrids. *Plant Cell*, 22, 17–33. doi: 10.1105/tpc.109.072041.

Hou, Q. and Wan, X. (2021). Epigenome and epitranscriptome: Potential resources for crop improvement. *Int. J. Mol. Sci.*, 22, 12912. doi: 10.3390/ijms222312912.

IPCC (2022). Climate change 2022: Impacts, adaptation and vulnerability. Report [Online]. Available at: https://www.ipcc.ch/report/ar6/wg2/.

Jaiswal, S., Gautam, R.K., Singh, R.K., Krishnamurthy, S.L., Ali, S., Sakthivel, K., Iquebal, M.A., Rai, A., Kumar, D. (2019). Harmonizing technological advances in phenomics and genomics for enhanced salt tolerance in rice from a practical perspective. *Rice*, 12, 89. doi: 10.1186/s12284-019-0347-1.

Kakoulidou, I., Avramidou, E.V., Baránek, M., Brunel-Muguet, S., Farrona, S., Johannes, F., Kaiserli, E., Liberman-Lazarovich, M., Martinelli, F., Mladenov, V. et al. (2021). Epigenetics for crop improvement in times of global change. *Biology*, 10, 766. doi: 10.3390/biology10080766.

Kapazoglou, A., Tani, E., Avramidou, E.V., Abraham, E.M., Gerakari, M., Megariti, S., Doupis, G., Doulis, A.G. (2021). Epigenetic changes and transcriptional reprogramming upon woody plant grafting for crop sustainability in a changing environment. *Front. Plant Sci.*, 11. doi: 10.3389/fpls.2020.613004.

Kikuchi, K., Terauchi, K., Wada, M., Hirano, H.Y. (2003). The plant MITE mPing is mobilized in another culture. *Nature*, 421, 167–170. doi: 10.1038/nature01218.

Kim, E.Y., Kim, K.D., Cho, J. (2022). Harnessing epigenetic variability for crop improvement: Current status and future prospects. *Genes Genomics*, 44, 259–266. doi: 10.1007/s13258-021-01189-7.

Kong, J., Berger, M., Colling, A., Stammitti, L., Teyssier, E., Gallusci, P. (2019). Epigenetic regulation in fleshy fruit: Perspective for grape berry development and ripening. *GRAPE GENOME*, 167–197. doi: 10.1007/978-3-030-18601-2_9.

Lizamore, D., Bicknell, R., Winefield, C. (2021). Elevated transcription of transposable elements is accompanied by het-siRNA-driven de novo DNA methylation in grapevine embryogenic callus. *BMC Genomics*, 22, 676. doi: 10.1186/s12864-021-07973-9.

Luan, X., Liu, S., Ke, S., Dai, H., Xie, X.-M., Hsieh, T.-F., Zhang, X.-Q. (2019). Epigenetic modification of ESP, encoding a putative long noncoding RNA, affects panicle architecture in rice. *Rice*, 12, 20. doi: 10.1186/s12284-019-0282-1.

Lv, Z., Dai, R., Xu, H., Liu, Y., Bai, B., Meng, Y., Li, H., Cao, X., Bai, Y., Song, X., Zhang, J. (2021). The rice histone methylation regulates hub species of the root microbiota. *J. Genet. Genomics*, 48, 836–843. doi: 10.1016/j.jgg.2021.06.005.

Ocaña, J., Walter, B., Schellenbaum, P. (2013). Stable MSAP markers for the distinction of *Vitis vinifera* cv Pinot noir clones. *Mol. Biotechnol.*, 55, 236–248. doi: 10.1007/s12033-013-9675-3.

Ong-Abdullah, M., Ordway, J.M., Jiang, N., Ooi, S.E., Kok, S.Y., Sarpan, N., Azimi, N., Hashim, A.T., Ishak, Z., Rosli, S.K. et al. (2015). Loss of karma transposon methylation underlies the mantled somaclonal variant of oil palm. *Nature*, 525, 533–537. doi: 10.1038/nature15365.

Pagliarani, C., Vitali, M., Ferrero, M., Vitulo, N., Incarbone, M., Lovisolo, C., Valle, G., Schubert, A. (2017). The accumulation of miRNAs differentially modulated by drought stress is affected by grafting in grapevine. *Plant Physiol.*, 173, 2180–2195. doi: 10.1104/pp.16.01119.

Papikian, A., Liu, W., Gallego-Bartolomé, J., Jacobsen, S.E. (2019). Site-specific manipulation of Arabidopsis loci using CRISPR-Cas9 SunTag systems. *Nat. Commun.*, 10, 729. doi: 10.1038/s41467-019-08736-7.

Rajeev Kumar, S., Anunanthini, P., Ramalingam, S. (2015). Epigenetic silencing in transgenic plants. *Front. Plant Sci.*, 6(695). doi: 10.3389/fpls.2015.00693.

Rodríguez López, C.M. and Wilkinson, M.J. (2015). Epi-fingerprinting and epi-interventions for improved crop production and food quality. *Front. Plant Sci.*, 6(397). doi: 10.3389/fpls.2015.00397.

Rubio, B., Stammitti, L., Cookson, S.J., Teyssier, E., Gallusci, P. (2022). Small RNA populations reflect the complex dialogue established between heterograft partners in grapevine. *Horticult. Res.*, 9, uhab067. doi: 10.1093/hr/uhab067.

Schellenbaum, P., Mohler, V., Wenzel, G., Walter, B. (2008). Variation in DNA methylation patterns of grapevine somaclones (*Vitis vinifera* L.). *BMC Plant Biol.*, 8, 78. doi: 10.1186/1471-2229-8-78.

Schmitz, R.J., Marand, A.P., Zhang, X., Mosher, R.A., Turck, F., Chen, X., Axtell, M.J., Zhong, X., Brady, S.M., Megraw, M. et al. (2022). Quality control and evaluation of plant epigenomics data. *Plant Cell*, 34, 503–513. doi: 10.1093/plcell/koab255.

Shangguan, L., Fang, X., Jia, H., Chen, M., Zhang, K., Fang, J. (2020). Characterization of DNA methylation variations during fruit development and ripening of *Vitis vinifera* (cv. 'Fujiminori'). *Physiol. Mol. Biol. Plants*, 26, 617–637. doi: 10.1007/s12298-020-00759-5.

Song, X. and Cao, X. (2017). Transposon-mediated epigenetic regulation contributes to phenotypic diversity and environmental adaptation in rice. *Curr. Opin. Plant Biol.*, 36, 111–118. doi: 10.1016/j.pbi.2017.02.004.

This, P., Lacombe, T., Cadle-Davidson, M., Owens, C.L. (2007). Wine grape (*Vitis vinifera L.*) color associates with allelic variation in the domestication gene VvmybA1. *Theor. Appl. Genet.* 114, 723–730. doi: 10.1007/s00122-006-0472-2.

Tsaballa, A., Xanthopoulou, A., Madesis, P., Tsaftaris, A., Nianiou-Obeidat, I. (2021). Vegetable grafting from a molecular point of view: The involvement of epigenetics in rootstock-scion interactions. *Front. Plant Sci.*, 11(621999). doi: 10.3389/fpls.2020.621999.

Wang, N., Wang, H., Wang, H., Zhang, D., Wu, Y., Ou, X., Liu, S., Dong, Z., Liu, B. (2010). Transpositional reactivation of the Dart transposon family in rice lines derived from introgressive hybridization with *Zizania latifolia*. *BMC Plant Biol.*, 10, 190. doi: 10.1186/1471-2229-10-190.

Wu, S., Guyot, R., Bocs, S., Droc, G., Oktavia, F., Hu, S., Tang, C., Montoro, P., Leclercq, J. (2020a). Structural and functional annotation of transposable elements revealed a potential regulation of genes involved in rubber biosynthesis by TE-derived siRNA interference in *Hevea brasiliensis*. *Int. J. Mol. Sci.*, 21, 4220. doi: 10.3390/ijms21124220.

Wu, K., Wang, S., Song, W, Zhang, J., Wang, Y., Liu Q., Yu, J., Ye, Y., Li, S., Chen, J. et al. (2020b). Enhanced sustainable green revolution yield via nitrogen-responsive chromatin modulation in rice. *Science*, 367, eaaz2046. doi: 10.1126/science.aaz2046.

Xie, H., Konate, M., Sai, N., Tesfamicael, K.G., Cavagnaro, T., Gilliham, M., Breen, J., Metcalfe, A., Stephen, J.R., De Bei, R. et al. (2017). Global DNA methylation patterns can play a role in defining terroir in grapevine (*Vitis vinifera* cv. Shiraz). *Front. Plant Sci.*, 8. doi: 10.3389/fpls.2017.01860.

Xie, E., Li, Y., Tang, D., Lv, Y., Shen, Y., Cheng, Z. (2019). A strategy for generating rice apomixis by gene editing. *J. Integr. Plant Biol.*, 61, 911–916. doi: 10.1111/jipb.12785.

Yelina, N.E., Lambing, C., Hardcastle, T.J., Zhao, X., Santos, B., Henderson, I.R. (2015). DNA methylation epigenetically silences crossover hot spots and controls chromosomal domains of meiotic recombination in Arabidopsis. *Genes Dev.*, 29, 2183–2202. doi: 10.1101/gad.270876.115.

Yuan, J., Chen, S., Jiao, W., Wang, L., Wang, L., Ye, W., Lu, J., Hong, D., You, S., Cheng, Z., Yang, D.-L. (2017). Both maternally and paternally imprinted genes regulate seed development in rice. *New Phytol.*, 216, 373–387. doi: 10.1111/nph.14510.

Zhang, Y.-Y., Latzel, V., Fischer, M., Bossdorf, O. (2018). Understanding the evolutionary potential of epigenetic variation: A comparison of heritable phenotypic variation in epiRILs, RILs, and natural ecotypes of *Arabidopsis thaliana*. *Heredity*, 121, 257–265. doi: 10.1038/s41437-018-0095-9.

Zheng, X., Chen, L., Xia, H., Wei, H., Lou, Q., Li, M., Li, T., Luo, L. (2017). Transgenerational epimutations induced by multi-generation drought imposition mediate rice plant's adaptation to drought condition. *Sci. Rep.*, 7, 39843. doi: 10.1038/srep39843.

Zhi, P. and Chang, C. (2021) Exploiting epigenetic variations for crop disease resistance improvement. *Front. Plant Sci.*, 12(692328). doi: 10.3389/fpls.2021.692328.

10

Epigenetics and Livestock Improvement

Vincent COUSTHAM[1,2] and Frédérique PITEL[3]

[1] *INRAE, NuMeA, Université de Pau et des Pays de l'Adour, Saint-Pée-sur-Nivelle, France*

[2] *IJPB, INRAE, Université Paris-Saclay, AgroParisTech, Versailles, France*

[3] *GenPhySE, INRAE, Castanet-Tolosan, France*

10.1. Introduction

Livestock production (cattle, sheep, goats, pigs, poultry, fish, etc.) represents a major part of the world's economic activity. Livestock farming is a key driver of sustainable agricultural development, contributing to food security, nutrition, poverty reduction and economic growth (source: FAO). Considerable gains in productivity and efficiency have been achieved over the last century, but to ensure that the livestock sector continues to grow in line with demand for animal products, a number of environmental issues need to be taken into consideration, including the use of natural resources and climate change. Livestock have to adapt to various constraints, such as changes in practices, dietary transitions, climatic variations and pathologies. These factors, which are sources of stress, can have a significant effect on production, health and well-being, the extent of which varies from one individual to another, depending in particular on their genotype (Nauta et al. 2006; Cardoso and Tempelman 2012).

Epigenetics offers new perspectives in the field of animal breeding (including terrestrial animals and aquaculture), thanks to knowledge of the biological processes involved, and in the management and genetic improvement of animals. In this

Epigenetics in Ecology and Evolution,
coordinated by Christoph GRUNAU and Stéphane MAURY.

chapter, we look at how epigenetics could contribute to the selection of healthier, more efficient animals, using the poultry model as a regular example of current research in this field.

As the epigenome lies at the interface between the breeding environment and the animal's phenotype, knowledge of how it functions enables us to better predict and adapt the animal's phenotype in relation to its exposome, in other words, all environmental exposures (Figure 10.1). Innovative strategies for conditioning or programming[1] phenotypes, some of which are detailed in this chapter, stem from these advances.

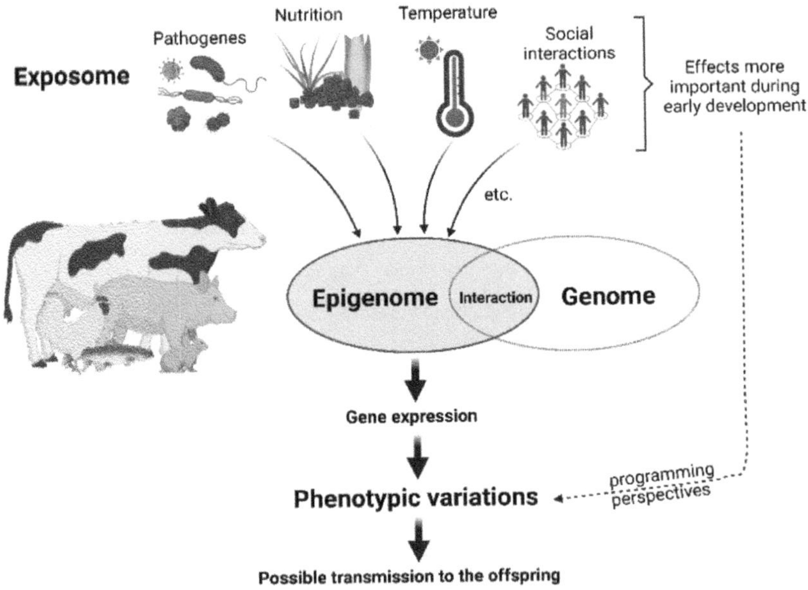

Figure 10.1. *Interaction between genetics, epigenetics and environment in the construction of livestock phenotypes. Created with BioRender.com*

10.2. Genetic selection issues and epigenetic improvement levers

There are many ways of improving animal performance, including genetic selection. The use of genetic selection, particularly for intensive and specialized chicken and pig production, has led to significant increases in growth rates, meat

1. The concept of conditioning requires the presence of a subsequent stimulus to reveal its effects, unlike the notion of programming, which is generally associated with effects occurring in a relatively stable environment.

and egg production, and feed efficiency. In particular, growth performance has been considerably improved in both broilers and laying hens (Aggrey et al. 2020). However, this strategy, which is largely driven by quantitative production criteria, faces new challenges linked to changes in socio-cultural values, including greater attention to animal welfare, and consideration of environmental issues. The genetic x environment (G X E) interaction, whose influence on phenotypes has long been known, could be better considered in selection models by considering the long-term modifications of non-genetic factors in the breakdown of the phenotype, induced by the environment in which the animal evolves. This evolution should enable the selection of genotypes that are better adapted to different environments and their increasingly variable nature: breeding practices including access to an outdoor range associated with greater climatic variations, heat waves, changes in feed composition, longer lives, etc.

10.2.1. *Genetic x epigenetic interactions*

Epigenetic marks can be affected by DNA polymorphisms (Bell et al. 2011; Do et al. 2017). This is the case, for example, in hens, where a genetic variant influences the expression of annotated long non-coding genes (lncRNAs) in around 30% of cases, or in humans, where genotype can explain up to 80% of variation in DNA methylation (Gertz et al. 2011). Conversely, epigenetic marks can influence mutation rates. For example, CpG dinucleotides are reported to have a mutation rate that is around 12 times higher than other nucleotide types (Sved and Bird 1990). Demonstrating this potential "direct" role of the environment on DNA polymorphisms requires in-depth studies, as current ones are based on correlations rather than cause-and-effect relationships.

Interactions between genotype and environment are illustrated by the fact that individuals that perform better than others in one environment may perform less well in another (Haldane 1946). Loss of genetic gain due to G x E interaction is observed in cattle (e.g. Hayes et al. 2009; Bernabucci et al. 2014), as in pigs and poultry (Mulder and Bijma 2005). For example, QTL[2] x diet and QTL x age interactions have been observed for various egg production and quality traits (Romé et al. 2015). Genetic polymorphisms may be at the origin of these G x E interactions (Lillehammer et al. 2009), with the allele improving performance in one environment possibly deteriorating it in another.

2. Quantitative Trait Locus: chromosomal region associated with the variability of a quantitative trait.

10.2.2. *Selection importance*

An essential question arises for breeders: is it better to select animals with the best performance in a given environment, knowing that this performance is likely to be reduced in another environment, or animals with weaker but more robust performance, i.e. less variable regardless of the environment?

In crossbred species such as pigs and poultry, the production system is based on a pyramid scheme: a relatively small number of breeding companies worldwide produce the bulk of the genetic stock, through genetic or genomic selection. For hens, for example, primary breeding flocks are made up of pure lines (pedigree/foundation) and grandparent animals. Grandparent flocks produce the latest generation of breeding birds (multiplier/parent flocks), which are then used to produce the chickens and laying hens raised for human consumption on production farms. There are numerous production farms that are present all over the world and managed by many farmers; they are different from breeding farms, which are few in number and concentrated in a few places, mainly in North America and Europe. The G x E interaction is therefore of great importance, given the variety of production environments compared with the more standardized breeding environments.

The same applies to label and organic farms, where breeding stock is often reared indoors, whereas terminal animals generally have access to the outdoors, in far more variable environments.

In practical terms, therefore, it makes sense to gain a better estimate of phenotypic variability linked to epigenetic phenomena, in relation to the environment. Studying the transmissibility of these environmental effects, included in “non-genetic heredity“, will enable us to improve selection models. A complementary objective will be to be able to select breeding stock to produce individuals that perform optimally in a wide variety of production environments. Ultimately, adapting animals to a given environment should enable us to produce offspring with optimized performance in the environment in question.

10.3. Early phenotype programming

The epigenome is sensitive to fluctuations in the environment, including temperature, nutrition, chemical agents and social interactions (Figure 10.1). All environmental exposures, or exposome, are known to impact the phenotype of animals over the long-term (Feil and Fraga 2012). For example, it has been shown that a nutritional deficiency in methyl group donors (such as methionine, betaine[3] or choline) has long-lasting effects on DNA and histone methylation, associated with the onset of pathologies such as diabetes or neurological disorders. Early development is a particularly critical period during which the exposome can have a strong impact on animal health, well-being, robustness and performance (Ho et al. 2011; Reed and Clark 2011). In this context, the early stages of development provide a window of opportunity to program or condition the epigenome and, consequently, the phenotype of livestock, either by acting on the maternal or paternal contribution, or directly on the embryo in oviparous animals, or during the first days of postnatal life. Through its embryonic development in the egg, the avian embryo is separated from any direct maternal contribution (apart from factors related to brooding), but it is subject to indirect intervention by the mother via the egg's composition (hormones, for example), and to the egg's direct environment. As the embryo's incubation environment can be easily manipulated, innovative strategies for programming and conditioning phenotypes have emerged in poultry production. The literature demonstrates the beneficial effects of various stimuli, such as temperature (Loyau et al. 2015), nutrition (Uni et al. 2005; Cherian 2011) or changes in maternal diet (Hynd et al. 2016; Baéza et al. 2017).

10.3.1. *Nutritional programming*

Studies in mammals and humans show that dietary stimuli applied at early stages of development (pre-natal or post-natal), a phenomenon known as nutritional programming, can have longer-term consequences on the physiological functions of individuals (Lucas 1998). DNA methylation is an important contributor to the mechanisms of early nutritional programming. One reason for this is that the enzymes that implement DNA methylation (DNA methyltransferases, or DNMTs, for DNA methyltransferases) depend on the one-carbon (1C) metabolism cycle, the reactions of which are catalyzed by numerous nutritional factors, such as B vitamins (folates, B6 and B12), zinc, choline, as well as the amino acids betaine and methionine (Reid et al. 2017). DNMTs are also regulated by bioactive compounds, such as polyphenols from certain fruits, which may therefore have an effect on the epigenetic regulation of gene expression (Marzag et al. 2014).

10.3.1.1. *Maternal effects of diet*

The maternal effect corresponds to the direct influence of a mother's phenotype on her offspring, alongside sources of genetic and environmental variation that can also directly affect the offspring. In livestock, numerous studies have documented

3. Component of the methionine cycle, considered an effective antioxidant and methyl donor.

the maternal effects of diet, as it can significantly influence essential parameters such as survival, performance and animal health.

Several studies have described the effects of maternal diet on DNA methylation. In cattle, a high-starch, corn-based maternal diet increases calf birth weight compared with calves born to mothers fed an isocaloric grass hay diet (Radunz et al. 2012). Offspring calves aged 2 years also show altered glucose and insulin metabolism, as well as reduced intramuscular lipid contents. These changes are associated with increased expression of genes subject to imprinting in muscle (whose allelic expression depends on parental origin), as well as increased expression of the DNA methyltransferase DNMT3a (Wang et al. 2015). In sheep, the impact of different maternal dietary energy sources on the expression of imprinted genes has been studied at the fetal stage (Lan et al. 2013). A maternal diet based on corn instead of hay during the second half of gestation significantly reduces the expression of the methyltransferases *DNMT1*, *DNMT3a* and *DNMT3b*, as well as the expression of imprinted genes (including *IGF2R*) in fetal muscles at the end of gestation. The higher gene expression in the hay-fed group could be explained by the reduced availability of dietary methionine in this feed. Indeed, a higher level of DNA methylation was observed for the paternally imprinted *IGF2R* gene in the muscle of offspring of mothers fed a corn-based diet.

In rainbow trout, dietary methionine levels in broodstock also influence the expression of several genes involved in the metabolism of sulfur amino acids (Fontagné-Dicharry et al. 2017), cholesterol and glucose (gluconeogenesis) (Seiliez et al. 2017) in larvae, after 21 days of feeding. The involvement of an epigenetic mechanism linked to DNA methylation is likely, although this is unconfirmed.

DNA methylation is not the only epigenetic factor involved in the nutritional programming of animal phenotypes. MicroRNAs (miRNAs), post-transcriptional regulators capable of turning off gene expression, are also sensitive to diet. Notably, nutritional supplementation of cows with a cereal mixture during the last third of gestation increases feed intake and reduces basal plasma insulin in offspring, with a moderate change in gene expression in the longissimus muscle (Moisá et al. 2015). This has been shown to be associated with a decrease in anti-adipogenic miRNA-34a expression levels, observed in the muscle of offspring from moderately supplemented cows. Furthermore, a link between maternal diet and weaning time is observed for miRNAs involved in the regulation of adipogenesis and insulin resistance (Moisá et al. 2016).

In cattle, dietary supplementation is generally used in the peripartum period[4] to prevent or limit the negative energy balance that is usually observed in dairy cows. Methionine supplementation in the diet, fed to Holstein cows throughout this period, induces a decrease in overall hepatic DNA methylation and an increase in CpG methylation in the *PPARα* promoter region, as well as an increase in the expression of the *PPARα* gene and its target genes (*ANGPTL4*, *FGF21* and *PCK1*), associated with improved lipid metabolism and immune system function (Li et al. 2016a; Osorio et al. 2016). In sheep, it has been shown that restriction of vitamin B12 and folate intake during the periconception period (before and just after fertilization, up to the blastocyst embryonic stage) alters hepatic DNA methylation in offspring and affects health in adulthood (Sinclair et al. 2007).

In birds, chicken nutrition can also have an impact on the health of hatched chicks and their subsequent phenotype, including feeding behavior (Aigueperse et al. 2013). Supplementation strategies have thus also emerged in poultry. For example, work has shown that maternal betaine supplementation significantly reduces hepatic cholesterol deposition by epigenetic alterations of cholesterol metabolism genes in young offspring chickens (Hu et al. 2020).

10.3.1.2. *Nutritional programming during early development*

Several studies suggest that epigenetic programming of metabolic genes by diet can also occur during early development. For example, in fish, early feeding of soybean meal to larval zebrafish results in a reversible increase in global histone H3 acetylation, and persistent global DNA hypomethylation after removal from the soybean meal diet (Perera and Yúfera 2016). These epigenetic modifications could be responsible for the altered transcription of digestion-related genes and inflammation that was observed in soybean meal-fed larvae.

In oviparous birds, direct *in ovo* supplementation is used as an early nutritional lever to improve chick robustness, growth and body composition. *In ovo* feeding, for example, involves injecting nutrients into the egg with a needle after piercing a hole in the shell. This method has been specifically used to reduce the hatching window and improve the health, immune status and weight of the hatched chick, as well as the growth performance and meat quality of growing animals (Uni and Ferket 2004; Wei et al. 2011; Kadam et al. 2013; Roto et al. 2016; Gao et al. 2017; Peebles 2018; Taha-Abdelaziz et al. 2018; Jha et al. 2019; Kalantar et al. 2019; Ayansola et al. 2021). The mechanisms by which egg nutrients regulate cell metabolism, signaling, gene expression and function are still poorly understood, with most studies focusing

---

4. Period of approximately three weeks before parturition and three to four weeks after parturition.

on the phenotypic effects of injections. However, it is likely that these are similar to those identified for maternal nutritional programming. Notably, work shows that *in ovo* injection of betaine affects hepatic cholesterol metabolism through the epigenetic regulation of several genes in newly hatched chicks (Hu et al. 2015). Furthermore, *in ovo* injection of folic acid is reported to have a beneficial effect on broiler immunity by altering post-translational modifications of histones H3K4me2 and H3K9me2 and chromatin conformation at immune gene promoters (Li et al. 2016b). However, the presence of epigenetic alterations does not systematically imply causality of epigenetic mechanisms. Only an experimental setup mimicking the phenotypic effects observed through targeted epigenetic modifications, such as CRISPR-based epigenome editing, of the epigenome-editing type obtained by the Crispr-Cas9 system, would make it possible to validate the functional involvement of the epigenetic mechanisms identified.

10.3.2. *Thermal conditioning*

In addition to diet, early exposure to biotic or abiotic factors can affect (in a harmful or beneficial manner) long-term phenotypes. The example of thermal conditioning in poultry will be used in this section to illustrate this phenomenon. Broilers have long been selected for growth performance traits, with significant progress in genetic selection in recent decades. However, the increase in chicken body mass and muscle growth has not been associated with a parallel increase in their cardiorespiratory system (Havenstein et al. 2003), impairing their thermoregulatory capacities and making them highly sensitive to an increase in ambient temperature (Yahav and McMurtry 2001). The current context of the climate therefore calls for adjustments to breeding practices, so as to ensure that animals are more resistant to heat waves, which are becoming increasingly frequent and intense. For example, an early thermal conditioning strategy has been developed in broiler chickens to improve heat resistance, by cyclically increasing egg incubation temperature (e.g. by 1.7°C, 12 hours per day, from days 7 to 16 of incubation). This method, known as embryonic thermal manipulation (TM), reduces mortality in heat-stressed male chickens at slaughter age (Piestun et al. 2008), as females are less sensitive to heat. A large number of TM protocols have been reported in the literature in different avian species, although chicken is the majority species (Loyau et al. 2015).

A study of gene expression in pectoral muscle, the tissue responsible for a significant proportion of heat production, was carried out in control and TM male broilers, using candidate approaches and microarray transcriptome analysis. This work showed that TM affects the gene response in muscle to further heat exposure at

35 days of age (D35) for certain genes involved in energy metabolism and mitochondrial function, in relation to heat production (Loyau et al. 2014, 2016). A study of two post-translational modifications of histones associated with epigenetic memory of the environment (the H3K4me3 and H3K27me3 marks) showed that both marks are impacted by TM in the hypothalamus of male broilers, with a predominant effect on H3K4me3, for which nearly 800 differentially enriched regions could be identified at 35 days of age between TM chickens and control chickens (David et al. 2019; Figure 10.2). The function of genes in the vicinity of these regions can be linked to chicken metabolism. Thus, embryonic thermal conditioning could induce epigenetic reprogramming of the expression of genes controlling metabolic processes linked to thermoregulation. These epigenetic changes would be maintained during chicken growth, enabling an adapted metabolic response in the event of subsequent exposure to heat. This hypothesis is corroborated by studies showing that several epigenetic marks are involved in the response to postnatal heat treatment (3 days after hatching) during the establishment of thermal control in the brain (Yossifoff et al. 2008; Kisliouk et al. 2010).

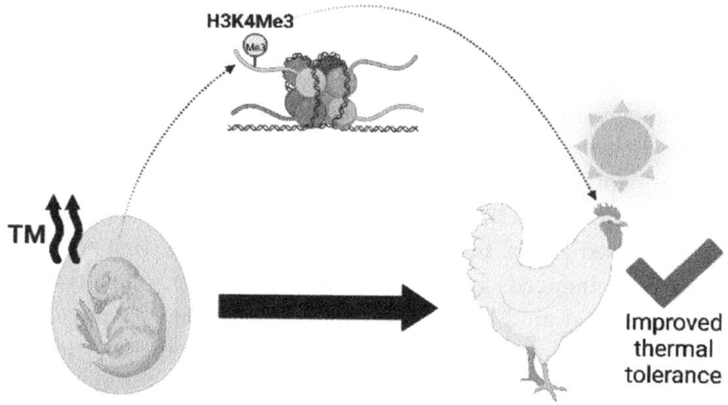

Figure 10.2. *Epigenetic mechanisms of thermal heat conditioning in birds. Created with BioRender.com*

COMMENT ON FIGURE 10.2. – *Embryonic Thermal Manipulation (TM) in broiler chickens involves cyclically increasing the incubation temperature of eggs by 1.7°C (12 h/d) from days 7 to 16 of incubation. This treatment is associated with changes in the H3K4me3 histone mark detected in several regions of the genome in the hypothalamus at 35 days of age. The epigenetic hypothesis is that these changes are induced by TM, persist throughout life, and enable an adapted gene response in chickens in the event of subsequent heatstroke.*

10.4. Transgenerational epigenetic effects

In livestock, understanding the mechanisms by which changes in the DNA environment transmit information is essential to anticipate the impact of breeding practices on future generations, the descendants of treated animals. A better understanding of these phenomena would help in the development of innovative strategies to improve health, welfare and performance over the long-term. In addition, identifying transmissible epigenetic marks of environmental origin could open up new perspectives in breeding, by improving the estimation of breeding values.

However, due to the diversity of farming methods, environmental conditions and, above all, the difficulty of discerning the genetic and non-genetic parts in the construction of livestock phenotypes, this question remains difficult to grasp. To date, there are few multigenerational studies, and even fewer demonstrations of transgenerational epigenetic effects in livestock. Indeed, most studies of epigenetic transmission have focused on maternal environmental effects on the next generation (Khatib 2021). Braunschweig et al. (2012) investigated the effects of a paternal diet enriched in methyl donors, fed to male breeding pigs (boars), on carcass traits, gene expression and DNA methylation in the F2 generation. The results of this study show that several carcass parameters are affected in F2 in relation to changes in the expression of dozens of genes, and that the effect could pass through transgenerational epigenetic mechanisms, even though differences in DNA methylation were only found at the level of one gene out of six tested among the differentially expressed genes. In birds, Japanese quail is the first species for which an environmentally induced transgenerational effect on phenotypes has been demonstrated (Leroux et al. 2017). Following the injection of a phytoestrogen into the egg, three breeding generations were conducted without further injection, and phenotypic measurements revealed that several traits had been affected by the ancestral treatment, such as the age at first egg that was delayed by 8 days, a decreased body weight, an increased abdominal fat weight and altered behavioral parameters. As the genetic differences between treated and control lines have been minimized by "mirror breeding", the differences observed are probably due, at least in part, to transgenerational epigenetic transmission, although the mechanisms are still unclear. Thus, at present, there is no definitive proof of the existence of transgenerational epigenetic transmission of environmental effects in livestock. Further studies are therefore needed to better understand the impact of programming and conditioning strategies on the plasticity and adaptation of phenotypes over several generations.

10.5. Conclusion

Today, it is necessary to unravel the impact of the environment on phenotypic variation in livestock, largely mediated through interactions between genetic and non-genetic components, in order to refine phenotype prediction. Estimating and incorporating epigenetic effects into phenotypic prediction models should improve their accuracy. The acquisition of new knowledge on the transmission of epigenetic marks, applied in response to the environment, will improve the estimation of an animal's breeding value. This improved understanding of the relationship between phenotype and environment should enable us to respond more effectively to the profound changes affecting livestock production in the current and future global context (climatic, economic, social and societal changes).

10.6. References

Aggrey, S.E. et al. (2020). *Advances in Poultry Genetics and Genomics*. Burleigh Dodds Science Publishing, Cambridge. doi: 10.1201/9781003047735.

Aigueperse, N., Calandreau, L., Bertin, A. (2013). Maternal diet influences offspring feeding behavior and fearfulness in the precocial chicken. *PLoS ONE*, 8(10), e77583. doi: 10.1371/journal.pone.0077583.

Ayansola, H. et al. (2021). Prospect of early vascular tone and satellite cell modulations on white striping muscle myopathy. *Poultry Science*, 100(3), 100945. doi: 10.1016/j.psj.2020.12.042.

Baéza, E. et al. (2017). Omega-3 polyunsaturated fatty acids provided during embryonic development improve the growth performance and welfare of Muscovy ducks (*Cairina moschata*). *Poultry Science*, 96(9), 3176–3187. doi: 10.3382/ps/pex147.

Bell, J.T. et al. (2011). DNA methylation patterns associate with genetic and gene expression variation in HapMap cell lines. *Genome Biology*, 12(1), R10. doi: 10.1186/gb-2011-12-1-r10.

Bernabucci, U. et al. (2014). The effects of heat stress in Italian Holstein dairy cattle. *Journal of Dairy Science*, 97(1), 471–486. doi: 10.3168/jds.2013-6611.

Braunschweig, M. et al. (2012). Investigations on transgenerational epigenetic response down the male line in F2 Pigs. *PLoS ONE*, 7(2), e30583. doi: 10.1371/journal.pone.0030583.

Cardoso, F.F. and Tempelman, R.J. (2012). Linear reaction norm models for genetic merit prediction of angus cattle under genotype by environment interaction. *Journal of Animal Science*, 90(7), 2130–2141. doi: 10.2527/jas.2011-4333.

Cherian, G. (2011). Essential fatty acids and early life programming in meat-type birds. *World's Poultry Science Journal*, 67(4), 599–614. doi: 10.1017/S0043933911000705.

David, S.A. et al. (2019). Thermal manipulation during embryogenesis impacts H3K4me3 and H3K27me3 histone marks in chicken hypothalamus. *Frontiers in Genetics*. doi: 10.3389/fgene.2019.01207.

Do, C. et al. (2017). Genetic–epigenetic interactions in cis: A major focus in the post-GWAS era. *Genome Biology*, 18(1), 120. doi: 10.1186/s13059-017-1250-y.

Feeney, A., Nilsson, E., Skinner, M.K. (2014). Epigenetics and transgenerational inheritance in domesticated farm animals. *Journal of Animal Science and Biotechnology*, 5(1), 48. doi: 10.1186/2049-1891-5-48.

Feil, R. and Fraga, M.F. (2012). Epigenetics and the environment: Emerging patterns and implications. *Nature Reviews Genetics*, 13(2), 97–109. doi: 10.1038/nrg3142.

Fontagné-Dicharry, S. et al. (2017). Parental and early-feeding effects of dietary methionine in rainbow trout (*Oncorhynchus mykiss*). *Aquaculture*, 469, 16–27. doi: 10.1016/j.aquaculture.2016.11.039.

Gao, T. et al. (2017). Effect of in ovo feeding of L-arginine on the hatchability, growth performance, gastrointestinal hormones, and jejunal digestive and absorptive capacity of posthatch broilers. *Journal of Animal Science*, 95(7), 3079–3092. doi: 10.2527/jas.2016.0465.

Gertz, J. et al. (2011). Analysis of DNA methylation in a three-generation family reveals widespread genetic influence on epigenetic regulation. *PLoS Genetics*, 7(8), e1002228. doi: 10.1371/journal.pgen.1002228.

Haldane, J.B.S. (1946). The interaction of nature and nurture. *Annals of Eugenics*, 13(1), 197–205. doi: 10.1111/j.1469-1809.1946.tb02358.x.

Havenstein, G.B., Ferket, P.R., Qureshi, M.A. (2003). Growth, livability, and feed conversion of 1957 versus 2001 broilers when fed representative 1957 and 2001 broiler diets. *Poultry Science*. doi: 10.1093/ps/82.10.1500.

Hayes, B.J. et al. (2009). A validated genome wide association study to breed cattle adapted to an environment altered by climate change. *PLoS ONE*, 4(8), e6676. doi: 10.1371/journal.pone.0006676.

Ho, D.H., Reed, W.L., Burggren, W.W. (2011). Egg yolk environment differentially influences physiological and morphological development of broiler and layer chicken embryos. *Journal of Experimental Biology*, 214(4), 619–628. doi: 10.1242/jeb.046714.

Hu, Y. et al. (2015). In ovo injection of betaine affects hepatic cholesterol metabolism through epigenetic gene regulation in newly hatched chicks. *PLOS ONE*, 10(4), e0122643. doi: 10.1371/journal.pone.0122643.

Hu, Y. et al. (2020). Maternal betaine supplementation decreases hepatic cholesterol deposition in chicken offspring with epigenetic modulation of SREBP2 and CYP7A1 genes. *Poultry Science*, 99(6), 3111–3120. doi: 10.1016/j.psj.2019.12.058.

Hynd, P.I. et al. (2016). Developmental programming: A new frontier for the poultry industry? *Animal Production Science*, 56(8), 1233. doi: 10.1071/AN15373.

Jha, R. et al. (2019). Early nutrition programming (in ovo and post-hatch feeding) as a strategy to modulate gut health of poultry. *Frontiers in Veterinary Science*, 6, 82. doi: 10.3389/fvets.2019.00082.

Kadam, M.M. et al. (2013). Prospects of in ovo feeding and nutrient supplementation for poultry: The science and commercial applications-a review. *Journal of the Science of Food and Agriculture*, 93(15), 3654–3661. doi: 10.1002/jsfa.6301.

Kalantar, M.H.M. et al. (2019). Effects of in ovo injection of coenzyme Q10 on hatchability, subsequent performance, and immunity of broiler chickens. *BioMed Research International*, 1–8. doi: 10.1155/2019/7167525.

Khatib, H. (2021). Transgenerational epigenetic inheritance in farm animals: How substantial is the evidence? *Livestock Science*, 250, 104557. doi: 10.1016/j.livsci.2021.104557.

Kisliouk, T., Ziv, M., Meiri, N. (2010). Epigenetic control of translation regulation: Alterations in histone H3 lysine 9 post-translation modifications are correlated with the expression of the translation initiation factor 2B (Eif2b5) during thermal control establishment. *Developmental Neurobiology*, 70(2), 100–113. doi: 10.1002/dneu.20763.

Lan, X. et al. (2013). Maternal diet during pregnancy induces gene expression and DNA methylation changes in fetal tissues in sheep. *Frontiers in Genetics*, 4, 49. doi: 10.3389/fgene.2013.00049.

Leroux, S. et al. (2017). Embryonic environment and transgenerational effects in quail. *Genetics Selection Evolution*, 49(1), 14. doi: 10.1186/s12711-017-0292-7.

Li, C. et al. (2016a). Peripartal rumen-protected methionine supplementation to higher energy diets elicits positive effects on blood neutrophil gene networks, performance and liver lipid content in dairy cows. *Journal of Animal Science and Biotechnology*, 7(1), 18. doi: 10.1186/s40104-016-0077-9.

Li, S. et al. (2016b). Effect of in ovo feeding of folic acid on the folate metabolism, immune function and epigenetic modification of immune effector molecules of broiler. *British Journal of Nutrition*, 115(3), 411–421. doi: 10.1017/S0007114515004511.

Lillehammer, M. et al. (2009). Gene by environment interactions for production traits in Australian dairy cattle. *Journal of Dairy Science*, 92(8), 4008–4017. doi: 10.3168/jds.2008-1484.

Loyau, T. et al. (2014). Thermal manipulation during embryogenesis has long-term effects on muscle and liver metabolism in fast-growing chickens. *PLoS ONE*, 9(9). doi: 10.1371/journal.pone.0105339.

Loyau, T. et al. (2015). Cyclic variations in incubation conditions induce adaptive responses to later heat exposure in chickens: A review. *Animal: An International Journal of Animal Bioscience*, 9(1), 76–85. doi: 10.1017/S1751731114001931.

Loyau, T. et al. (2016). Thermal manipulation of the chicken embryo triggers differential gene expression in response to a later heat challenge. *BMC Genomics*, 17(1), 329. doi: 10.1186/s12864-016-2661-y.

Lucas, A. (1998). Programming by early nutrition: An experimental approach. *The Journal of Nutrition*, 128(2), 401S–406S. doi: 10.1093/jn/128.2.401S.

Marzag, H. et al. (2014). Natural polyphenols as potent inhibitors of DNA methyltransferases. *Science Direct*, 195–223. doi: 10.1016/B978-0-444-63294-4.00007-3.

Moisá, S.J. et al. (2015). Maternal plane of nutrition during late gestation and weaning age alter Angus × Simmental offspring longissimus muscle transcriptome and intramuscular fat. *PLoS One*, 10, e0131478.

Moisá, S.J. et al. (2016). Maternal plane of nutrition during late-gestation and weaning age alter steer calf longissimus muscle adipogenic microRNA and target gene expression. *Lipids*, 51(1), 123–138. doi: 10.1007/s11745-015-4092-y.

Mulder, H.A. and Bijma, P. (2005). Effects of genotype × environment interaction on genetic gain in breeding programs. *Journal of Animal Science*, 83(1), 49–61. doi: 10.2527/2005.83149x.

Nauta, W.J. et al. (2006). Genotype by environment interaction for milk production traits between organic and conventional dairy cattle production in the Netherlands. *Journal of Dairy Science*, 89(7), 2729–2737. doi: 10.3168/jds.S0022-0302(06)72349-9.

Osorio, J.S. et al. (2016). Hepatic global DNA and peroxisome proliferator-activated receptor alpha promoter methylation are altered in peripartal dairy cows fed rumen-protected methionine. *Journal of Dairy Science*, 99(1), 234–244. doi: 10.3168/jds.2015-10157.

Peebles, E.D. (2018). In ovo applications in poultry: A review. *Poultry Science*, 97(7), 2322–2338. doi: 10.3382/ps/pey081.

Perera, E. and Yúfera, M. (2016). Soybean meal and soy protein concentrate in early diet elicit different nutritional programming effects on juvenile Zebrafish. *Zebrafish*, 13(1), 61–69. doi: 10.1089/zeb.2015.1131.

Piestun, Y. et al. (2008). Thermal manipulations during broiler embryogenesis: Effect on the acquisition of thermotolerance. *Poultry Science*, 87, 1516–1525. doi: 10.3382/ps.2008-00030.

Radunz, A.E. et al. (2012). Prepartum dietary energy source fed to beef cows: II. Effects on progeny postnatal growth, glucose tolerance, and carcass composition. *Journal of Animal Science*, 90(13), 4962–4974. doi: 10.2527/jas.2012-5098.

Reed, W.L. and Clark, M.E. (2011). Beyond maternal effects in birds: Responses of the embryo to the environment. *Integrative and Comparative Biology*, 51(1), 73–80. doi: 10.1093/icb/icr032.

Reid, M.A., Dai, Z., Locasale, J.W. (2017). The impact of cellular metabolism on chromatin dynamics and epigenetics. *Nature Cell Biology*, 19(11), 1298–1306. doi: 10.1038/ncb3629.

Romé, H. et al. (2015). GWAS analyses reveal QTL in egg layers that differ in response to diet differences. *Genetics Selection Evolution*, 47(1), 83. doi: 10.1186/s12711-015-0160-2.

Roto, S.M., Kwon, Y.M., Ricke, S.C. (2016). Applications of in ovo technique for the optimal development of the gastrointestinal tract and the potential influence on the establishment of its microbiome in poultry. *Frontiers in Veterinary Science*, 3, 63. doi: 10.3389/fvets.2016.00063.

Seiliez, I. et al. (2017). Eating for two: Consequences of parental methionine nutrition on offspring metabolism in rainbow trout (*Oncorhynchus mykiss*). *Aquaculture*, 471, 80–91. doi: 10.1016/j.aquaculture.2017.01.010.

Sinclair, K.D. et al. (2007). DNA methylation, insulin resistance, and blood pressure in offspring determined by maternal periconceptional B vitamin and methionine status. *Proceedings of the National Academy of Sciences*, 104(49), 19351–19356. doi: 10.1073/pnas.0707258104.

Sved, J. and Bird, A. (1990). The expected equilibrium of the CpG dinucleotide in vertebrate genomes under a mutation model. *Proceedings of the National Academy of Sciences*, 87(12), 4692–4696. doi: 10.1073/pnas.87.12.4692.

Taha-Abdelaziz, K. et al. (2018). Effects of early feeding and dietary interventions on development of lymphoid organs and immune competence in neonatal chickens: A review. *Veterinary Immunology and Immunopathology*, 201, 1–11. doi: 10.1016/j.vetimm.2018.05.001.

Uni, Z. and Ferket, R.P. (2004). Methods for early nutrition and their potential. *World's Poultry Science Journal*, 60(1), 101–111. doi: 10.1079/WPS20038.

Uni, Z. et al. (2005). In ovo feeding improves energy status of late-term chicken embryos. *Poultry Science*, 84(5), 764–770. doi: 10.1093/ps/84.5.764.

Wang, X. et al. (2015). Maternal nutrition during pregnancy is associated with differential expression of imprinted genes and DNA methyltranfereases in muscle of beef cattle offspring1. *Journal of Animal Science*, 93(1), 35–40. doi: 10.2527/jas.2014-8148.

Wei, X.J. et al. (2011). The effect of equol injection in ovo on posthatch growth, meat quality and antioxidation in broilers. *Animal*, 5(2), 320–327. doi: 10.1017/S1751731110001771.

Yahav, S. and McMurtry, J.P. (2001). Thermotolerance acquisition in broiler chickens by temperature conditioning early in life – The effect of timing and ambient temperature. *Poultry Science*. doi: 10.1093/ps/80.12.1662.

Yossifoff, M., Kisliouk, T., Meiri, N. (2008). Dynamic changes in DNA methylation during thermal control establishment affect CREB binding to the brain-derived neurotrophic factor promoter. *European Journal of Neuroscience*, 28(11), 2267–2277. doi: 10.1111/j.1460-9568.2008.06532.x.

11

Epigenetics in Evolution

Christoph GRUNAU[1] and Alexandra WEYRICH[2,3]

[1] *IHPE, Université de Perpignan Via Domitia, France*

[2] *Leibniz Institute for Zoo and Wildlife Research, Berlin, Germany*

[3] *German Centre for Integrative Biodiversity Research, University of Leipzig, Germany*

11.1. Evolution, Environments and Inheritance

We have seen in the previous chapters of this book that epigenetic information can influence the expression of genes and phenotypic characters. It has an impact on the plasticity of phenotypes and genotypes and can produce phenotypic variation. In this chapter, we will discuss the possibility that epigenetics can play a role in evolution.

Evolution refers to the phenotypic changes one can observe in organisms through generations. Many theories of evolution exit, and we refer here to some of the excellent books on this matter (e.g. Thomas et al. (2016)). Evolution is often defined as a process that leads to changes in the occurrence in gene variants (allele frequencies) in a population (genetic adaptation). If we think about this definition more thoroughly, it excludes any other than genetic heritability from evolution and does not allow non-genetic inheritance.

Here we use another definition and see evolution as a process in which organisms with certain morphological, physiological or behavioral features or traits (we call them "phenotypic variants") are selected by the natural or manmade (artificial) environments by conferring them a reproductive advantage. In other words, it is not so important that an individual performs better in a specific environment but that it produces also more offspring that resembles him. Evolution needs therefore the generation of phenotypic variants on which selection can act (Figure 11.1).

Epigenetics in Ecology and Evolution,
coordinated by Christoph GRUNAU and Stéphane MAURY.

Figure 11.1. *In these two examples of phenotypic variability, two types of adaptive traits lead to beneficial phenotypes: on the left the ant with green fur is better camouflaged, on the right the frog with webbing can swim faster. For a color version of this figure, see www.iste.co.uk/grunau/epigenetics.zip*

Selection means that the natural environment (or e.g., a breeder) confers a reproductive advantage to some phenotypic variants but not others. We need to bear in mind that neither genotype nor epigenotype are visible to selection, only phenotypic characters are. In other words, some individuals with certain shapes, colors, behaviors, etc. will be better equipped for the current environmental factors than others, survive more often and have more fertile offspring than others (meaning greater reproductive success or "fitness"). If all phenotypes in a population would be the same (have the same phenotype), no such selection could occur.

To be of evolutionary relevance, variants must not only differ from others, but they must also be heritable. In other words, the capacity to generate such variants under a given environmental condition must be transmitted to the next generation. Through this transmission, parents and their offspring resemble each other in similar environments (and when they are in the same developmental states). Non-heritable phenotypic variants are not relevant from the perspective of an evolutionary process.

For approximately 50 years (roughly between 1950 and 2000) it has been thought that the exclusive source of heritable phenotypic variants were genetic variations (mutations). We know now that this is not the case and that heritable phenotypes are also epigenetically based (epimutations). Sometimes changes in the DNA sequence (mutations) and changes in the epigenetic information (epimutations) can even produce the same phenotype. An example is a flower deformation in the plant model species *Arabidopsis thaliana* that can be caused by a mutation in the Superman gene or several types of epimutations in this gene (called appropriately "Clark Kent" or simply "floral organ number" and "Carpel" (Rohde et al. 1999)).

This epigenetic information is heritable in the sense that it can be reproduced between generations: in other words, the epigenetic information of offspring resembles the information of their ancestors. We will use here the word "generation" in a large sense. It englobes any type of process in which an individual reproduces itself and split into at least two physically separated entities that become then independent units of evolution, meaning they can live, reproduce and die without directly affecting their peers. Multiple generations can still exchange via niche construction and cultural inheritance. They can transmit genetic information (DNA) and epigenetic information that is bond to the chromosomes (DNA modifications, histone modifications and other chromatin elements) or that is not bound to the chromosomes and therefore mobile between nucleus and cytoplasm or even between cells (non-coding RNA) (see Chapter 2). When we think of reproduction, we often refer to sexual reproduction, the only way humans can produce offspring that resemble the parents. Sexual reproduction is performed by all animals and about

87% of plant species. However, many of these species can also reproduce asexually, and in other organisms, sex does not exist or is an exception (Chen and Wiens 2021). The evolutionary advantage of sex lies probably in the fact that two partners contribute to the generation of heritable variants and thus do not produce offspring that is 100% identical to its parents. However, sex is also costly and as such called an "evolutionary paradox". In any case, offspring will split physically from the parent producing the subsequent generation. Sexual reproduction requires a process in which parental chromosomes are reduced by half, otherwise we would observe exponential accumulation of chromosomes after very few generations. This process is called meiosis and is also associated with a partial epigenetic reset. For details about the role of epigenetics in development, please consult the book of Edith Heard (Heard 2013). Asexual reproduction simply replicates chromosomes through mitosis and is as such similar to what happens when cells divide.

11.1.1. *Don't be too rigid – plasticity is also important*

Here we must make it a little bit more complicated. Apologies. We have seen in previous chapters the importance of phenotypic plasticity. For those who like gardening, we have all learned that plants not only need to be seeded but they require care. Too few or too much watering will kill even the best varieties. Phenotypic variation is observed along an environmental gradient (e.g. humidity). The offspring of a beautiful rose will bear little phenotypic resemblance to their ancestor, if not irrigated sufficiently or too much. What is transmitted to the offspring is not the capacity to reproduce faithfully the phenotypic variation of the parents, but the capacity to reproduce phenotypes depending on the environmental gradients to which the individual is exposed to. Phenotypic resemblance between generations relies thus on the way in which they react upon environmental changes. It is this capacity to produce phenotypic variants in response to an environment that is heritable, not the phenotype itself. This capacity was defined as "norm of reaction" (see also Chapter 7) by Richard Woltereck in 1909.

Environmental events differ in intensity and duration (Figure 11.2). There are sudden highly intense events such as ice storms or heat waves, longer-lasting mild events such as climate change, and very long-lasting events such as the ice age. Individuals must cope with these diverse environments to be able to survive and reproduce (Penuelas et al. 2013; Kilvitis et al. 2014; Soen et al. 2015). On the molecular level, these environmental events can be and are often thought to be regulated by different mechanisms. They do so by using different molecular mechanisms and types of adaptation. Sudden changes in the environment (minutes, hours, days) induce short-term physiological adaptations, i.e. a stress response. Such physiological adaptations are reversible but non-heritable. These short-term responses are regulated by changes in gene expression that are quickly reversed when the stimulus disappears. Longer-lasting environmental exposures (week(s), months, years) within an individual's lifespan are called acclimation processes. Here, several examples were explained in Chapters 8 and 9, showing epigenetic modifications to take place. These epigenetic modifications are transient, but can be transmitted to subsequent generation(s). Recent studies indicate that acclimation reflects a transition from transient regulation by the central nervous system (CNS) to more stable adaptation at the level of peripheral tissues and organs (Horowitz 2014). This rapid adaptation is maintained over cell generations in the affected organism, but is mostly not inherited over many generations. In long-lasting environmental changes, which span over very many generations, adaptations are mediated by selection of genetic mutations. Genetic adaptations are transgenerationally heritable over generations and revisions of a change are rare (mostly irreversible).

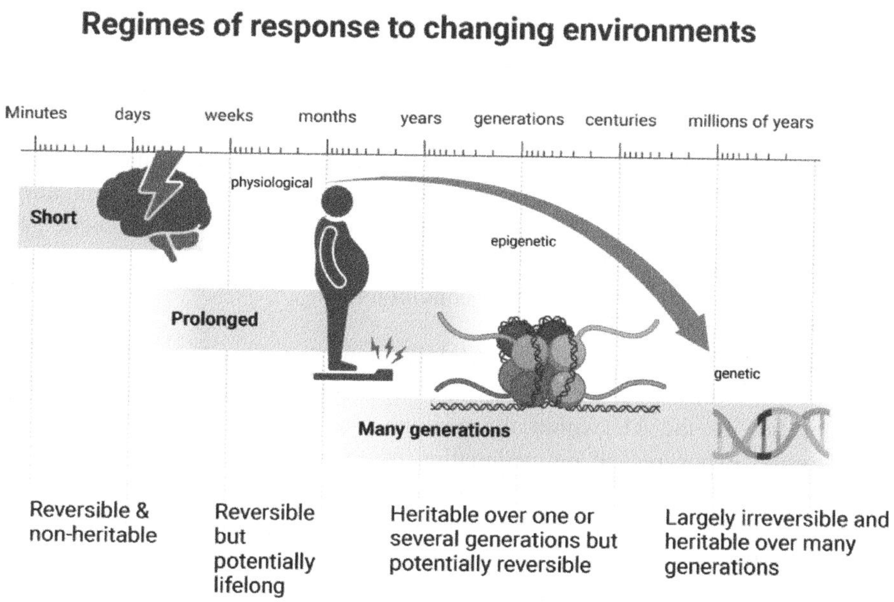

Figure 11.2. *Relations between the duration of environmental change, response type and its stability. For a color version of this figure, see www.iste.co.uk/grunau/epigenetics.zip*

All of these types of adaptation mechanisms can lead to different phenotypic variants that can be selected for. Researchers are still aiming to understand how these adaptation mechanisms function and how they intertwine. It has been suggested that less persisting epigenetic changes may provide phenotypic variation, allowing adaptation prior to the appearance of mutations (Stajic et al. 2019).

Phylogenetic trees are mostly constructed based on genetic mutations, for example, in the Hyaenidae family (Westbury et al. 2021). But genomes among species are often similar, especially when closely related. Sympatric speciation is an example of speciation in which one species diverge into two subspecies within an overlapping home range. A study in the blind mole rat, *Spalax galili*, conducted in Israel, investigated an early speciation event where two populations live on different soils, basalt and chalk. The animals rarely intermate. Researchers identified low genetic variability, but epigenetic differences between both subspecies (Li et al. 2020). These results may hint that epimutations occur before genetic changes.

The design of lab experimental setups mimicking evolution processes is tricky because lab experiments cannot reflect natural conditions, while natural conditions are multifactorial and as such complex (often impossible) to disentangle. One of the first laboratory experiments was done in a simple model-species, the budding yeast (*Saccharomyces cerevisiae)*, showing that also the degree of an epigenetic regulation (gene silencing) influences the evolution outcome. The study showed that especially intermediate gene silencing facilitates adaptation, due to an increase in the population size, which accelerates the acquisition of mutations (genetic assimilation) (Stajic et al. 2019). These results promote an intertwined process of epigenetic gene regulation with genetic mutations, and as such support the results in sympatric speciation in *Spalax galili* under natural conditions.

The interplay between genome and epigenome has also been shown in the epigenetic marks on transposable elements (TE) in which methylation marks silence the re-location of the TE within the genome (see Chapter 3). With a stronger environmental exposure and de-methylation of TE region, it can start relocating. These findings also show that the intensity of environmental change affects the epigenetic response. Thus, if the environmental change is of low (minor change), moderate (within the organism reaction norm) or high (stressful) intensity, results show different responses in the epigenome and physiology. We can find examples for high-intensity changes in Chapter 6. Moderate changes, i.e. within the reaction norm, have been conducted for instance on male wild guinea pigs, which have been exposed to 10°C increase in temperature for a duration of 2 months. DNA methylation changes were detected in genes in the liver which are important for thermoregulation, spermatogenesis and gene regulation in fathers and sons (Weyrich et al. 2016, 2019). Low-intensity changes are hard to monitor, because they generate a minor epigenetic and physiological response.

In conclusion, the reaction norm (the range in which an individual copes), the time duration and intensity of the environmental alteration strongly act together in bringing about the diversity of living beings using different regimes of response to changing environments.

11.1.2. *Bringing everything together*

Now let us try to put all this together in the light of evolution: we have learned earlier that epigenetic information is heritable. The process is not yet fully understood but we have seen in Chapter 2 the mechanisms by which, for example, DNA methylation patterns are reproduced. The stability of epimutations is a matter of a lively scientific debate and has been studied under carefully controlled laboratory conditions. Adam Klosin and colleagues (Klosin et al. 2017) have shown that epigenetically coded phenotypes can be induced during 5 generations and then transmitted over 14 subsequent generations in the roundworm *Caenorhabditis elegans*, a popular model for multicellular organisms. Filippo Ciabrelli in Giacomo Cavallis group in Montpellier (Ciabrelli et al. 2017) could even demonstrate epigenetic inheritance over more than 50 generations in the fruit fly *Drosophila melanogaster*. Interestingly, the stability of transmission depends on environmental conditions, in this case the temperature in which the flies were raised.

Probably the longest documented epigenetic silencing effect was described in the year 2021 in the laboratory of Antony M. Jose at the University of Maryland again in *Caenorhabditis elegans*. They introduced genes of fluorescent proteins into the worms` genomes that made them glow in the dark. Like this, these worms were easily distinguishable from non-glowing worms. *C. elegans* exits in two sexes: one is a hermaphrodite which can produce both sperm and eggs, and one is a male which produces only sperm. Hermaphrodites and their offspring were always nicely glowing. In contrast, when male worms which carried the genes for fluorescence proteins were crossed with wild-type hermaphrodites, the offspring inherited the paternal genes, but despite this, they usually weakly glowed or did not glow at all. The genes were epigenetically silenced. This inactivate state was then heritable for more than 300 generations (Devanapally et al. 2021).

In natural conditions, there are few studies that investigate the stability of epigenetic inheritance, but it was for instance shown for 10 asexual generations in alligator weed (Shi et al. 2019). Thus, epigenetic marks can clearly be heritable, but they can also revert to the initial state occasionally. Theoretical considerations put the reversion threshold where epimutations should contribute to evolution at about 1 in 1,000 reversions of epimutations to their original state (Kronholm and Collins 2016). In the above-mentioned *Arabidopsis* Clark Kent epimutant reversal of the phenotype was observed in 1–3% of plants. Other studies have reported reversal in a wide range from 60% of individuals or only 1 in every 10,000 individual (Kronholm and Collins 2016; Roquis et al. 2016). Sometimes the epigenotype might be too "floppy" and unstable to allow for evolution under moderate selective pressure, but sometimes its heritability is largely sufficient to contribute to evolution.

In summary, epigenetic inheritance can reach over several ten to hundreds of generation but can also be ephemeral, a characteristic that is often called "metastable", "transient" or in the literature of the 1930s an "enduring modification" ("Dauermodifikation") (Jollos 1939) (see also Chapter 1).

We have also seen that epigenetic information influences the function of genes and as such generates phenotypic variants. It has therefore a very similar function to genetic information and can contribute to evolution processes. Wouldn't it be useful to combine both? Indeed, in nature, neither a genome nor an epigenome exists alone. Pure DNA can only be produced in a test tube, while in the living cell, it is always present as part of the chromatin and as such associated with epigenetic information. And naturally, an epigenome can only exist when a genome is present. One can think of both ink (epigenetic marker) and paper (DNA) that makes the book we are reading. Both are necessary to produce this text and both are important. In biology, such obligatory interplay is called "a system" and scientists have proposed to call the genotype–epigenotype combination an "inheritance system" (Cosseau et al. 2016; Huang 2022).

It is this inheritance system that is the origin of heritable phenotypic variations on which selection can act and evolution can occur. How much of the heritable phenotypic variations are caused by epigenetic or genetic means depends on the type of organism, its type of reproduction and its environment. A rough estimate could be that phenotypic traits that are metastable for some 4 to 10 generations are probably epigenetic. Or in Shakespeare's words:

> *A ring the county wears,*
> *That downward hath succeeded in his house*
> *From son to son,* ***some four or five descents***
> *Since the first father wore it: this ring he holds*
> *In most rich choice.*
>
> (Shakespeare 1778)

11.2. Conclusions and further readings

Dobzhansky's often paraphrased words "Nothing in biology makes sense except in the light of evolution" are certainly also true when it comes to epigenetics. Whether its role is to accelerate evolution, to "buy time" for the accumulation of genetic mutation that produce phenotypes with higher fitness or because it is simply *ying* of the genetic *yang* remains to be established and these questions will occupy interested minds for a long time to come. For further reading, we recommend

Eva Jablonka and Marion J. Lamb "Inheritance Systems and the Extended Evolutionary Synthesis", Cambridge University Press (2020)

Etienne Danchin "L'hérédité comme on ne vous l'a jamais racontée", HUMENSCIENCES (2021)

11.3 References

Chen, L. and Wiens, J.J. (2021). Multicellularity and sex helped shape the Tree of Life. *Proc Biol Sci*, 288, 20211265.

Ciabrelli, F., Comoglio, F., Fellous, S., Bonev, B., Ninova, M., Szabo, Q., Xuéreb, A., Klopp, C., Aravin, A., Paro, R. et al. (2017). Stable Polycomb-dependent transgenerational inheritance of chromatin states in *Drosophila*. *Nat Genet*, 49, 876–886.

Cosseau, C., Wolkenhauer, O., Padalino, G., Geyer, K.K., Hoffmann, K.F., Grunau, C. (2016). (Epi)genetic inheritance in *Schistosoma mansoni*: A systems approach. *Trends Parasitol*, 33, 285–294.

Devanapally, S., Raman, P., Chey, M., Allgood, S., Ettefa, F., Diop, M., Lin, Y., Cho, Y.E., Jose, A.M. (2021). Mating can initiate stable RNA silencing that overcomes epigenetic recovery. *Nat Commun*, 12, 4239.

Heard, E. (2013). *Epigenetics and Development*. Academic Press, San Diego.

Horowitz, M. (2014). Heat acclimation, epigenetics, and cytoprotection memory. *Compr Physiol.*, 4, 199–230.

Huang, S. (2022). Towards a unification of the 2 meanings of "epigenetics". *PLOS Biol*, 20, e3001944.

Jollos, V. (1939). *Grundbegriffe der vererbungslehre, insbesondere Mutation, Dauermodifikation, Modifikation*. Borntraeger, Berlin.

Kilvitis, H.J., Alvarez, M., Foust, C.M., Schrey, A.W., Robertson, M., Richards, C.L. (2014). Ecological epigenetics. *Ecol Genom Ecol Evol Genes Genomes*, 781, 191–210.

Klosin, A., Casas, E., Hidalgo-Carcedo, C., Vavouri, T., Lehner, B. (2017). Transgenerational transmission of environmental information in *C. elegans*. *Science*, 356, 320–323.

Kronholm, I. and Collins, S. (2016). Epigenetic mutations can both help and hinder adaptive evolution. *Molecular Ecology*, 25, 1856–1868.

Li, K., Zhang, S., Song, X., Weyrich, A., Wang, Y., Liu, X., Wan, N., Liu, J., Lövy, M., Cui, H. et al. (2020). Genome evolution of blind subterranean mole rats: Adaptive peripatric versus sympatric speciation. *Proc Natl Acad Sci USA*, 117, 32499–32508.

Penuelas, J., Sardans, J., Estiarte, M., Ogaya, R., Carnicer, J., Coll, M., Barbeta, A., Rivas-Ulbach, A., Llusià, J., Garbulsky, M. et al. (2013). Evidence of current impact of climate change on life: A walk from genes to the biosphere. *Global Change Biology*, 19, 2303–2338.

Rohde, A., Grunau, C., De Beck, L., Van Montagu, M., Rosenthal, A., Boerjan, W. (1999). Carpel, a new Arabidopsis epi-mutant of the SUPERMAN gene: Phenotypic analysis and DNA methylation status. *Plant Cell Physiol*, 40, 961–972.

Roquis, D., Rognon, A., Chapparo, C., Boissier, J., Arancibia, N., Cosseau, C., Parrinello, H., Grunau, C. (2016). Frequency and mitotic heritability of epimutations in *Schistosoma mansoni*. *Mol Ecol*, 25, 1741–1758.

Shakespeare, W. (1778). *All's Well, That Ends Well. A Comedy. As it is Acted at the Theatres-Royal in Drury-Lane and Covent-Garden*. Gale ECCO, London.

Shi, W., Chen, X., Gao, L., Xu, C.-Y., Ou, X., Bossdorf, O., Yang, J., Geng, Y. (2019). Transient stability of epigenetic population differentiation in a clonal invader. *Front Plant Sci*, 9(1851).

Soen Y., Knafo M. & Elgart M. (2015). A principle of organization which facilitates broad Lamarckian-like adaptations by improvisation, *Biology Direct*, 10, 68.

Stajic, D., Perfeito, L. Jansen, L.E.T. (2019). Epigenetic gene silencing alters the mechanisms and rate of evolutionary adaptation. *Nat Ecol Evol*, 3, 491–498.

Thomas, F., Lefévre, T., Raymond, T. (2016). *Biologie évolutive*. De Boeck, Brussels.

Westbury, M.V., Le Duc, D., Duchêne, D.A., Krishnan, A., Prost, S., Rutschmann, S., Gaur, J.H., Dalén, L., Weyrich, A., Norén, K. et al. (2021). Ecological specialization and evolutionary reticulation in extant hyaenidae. *Mol Biol Evol*, 38, 3884–3897.

Weyrich, A., Lenz, D., Jeschek, M., Chung, T.H., Rübensam, K., Göritz, F., Jewgenow, K., Fickel, J. (2016). Paternal intergenerational epigenetic response to heat exposure in male Wild guinea pigs. *Mol Ecol*, 25, 1729–1740.

Weyrich, A., Lenz, D., Fickel, J. (2018). Environmental change-dependent inherited epigenetic response. *Genes*, 10(1), 4. doi: 10.3390/genes10010004.

Weyrich, A., Lenz, D., Fickel, J. (2019). Environmental change-dependent transgenerational epigenetic response. *Genes*, 10(1), 4. doi: 10.3390/genes10010004.

12

Epigenetics and Society: Epigenetics in the French Press

Michel DUBOIS[1], **Catherine GUASPARE**[1] **and Séverine LOUVEL**[2]
[1] *GEMASS, CNRS, Sorbonne Université, Paris, France*
[2] *Pacte, CNRS, Sciences Po Grenoble, Université Grenoble Alpes, France*

12.1. Introduction

Since the 2000s, advances in epigenetic research and their possible implications for society have been widely reported in the mainstream media. There is constant media coverage of epigenetic discoveries in newspapers (such as Zimmer (2018)), magazines (*The New York Times*) (Cloud 2010; Lewino 2018), mainstream books (Carey 2013; De Rosnay 2018; Giacobino 2018), radio programs (Feinberg 2004; Martin 2017; Rebeihi 2018) and television (such as Holt and Paterson (2007); Serfaty (2015)). In the English-speaking world, this extensive media coverage of epigenetics has not gone unnoticed. A number of social science studies have looked at the presence of epigenetics in radio broadcasts (Waggoner and Uller 2015), in the national and international general press (Stelmach and Nerlich 2015; Lappé 2016; Robison 2016; Seitz and Schuol 2017; Dyke et al. 2020), as well as in the general and scientific news press (Dubois and Guaspare 2020). This research is complemented by studies on the circulation of epigenetics in other types of media: digital social networks, in particular Twitter/X (Nerlich et al. 2020); websites referenced on Google, compared with the English-speaking national and international press (Dubois et al. 2019).

In contrast, the presence of epigenetics in the French-speaking media in general, and particularly in France, has not been studied. In this chapter, we analyze the

Epigenetics in Ecology and Evolution,
coordinated by Christoph GRUNAU and Stéphane MAURY. © ISTE Ltd 2025.

recent dynamics of media coverage of epigenetics in the French-speaking press, focusing on three main aspects: the modes of circulation of epigenetics in the press; the content of the interpretative repertoire offered by the media and the reasons for the appeal of epigenetics in generalist and specialized press.

First, we look at how epigenetics is circulated and how it is framed by the media. Are scientific discoveries more likely to be relayed by the news or popular science press, or by the national or local press? Which research areas or potential applications attract the most attention? Do the articles mainly aim to inform readers, or rather to influence their consumption behavior (e.g. by promoting products and services that could come from epigenetic discoveries)?

Next, we study the interpretative repertoire relating to epigenetics in the French-speaking press. In other words, are there metaphors or narratives that guide the dominant representations of epigenetics in this media space? Research carried out in the English-speaking public sphere has highlighted the emergence of a repertoire specific to epigenetics, and compared it with the repertoire that accompanied the rise of genetics. In particular, the great metaphors that referred to the genome as the "book of life" or the "code" have given way to a series of partial metaphors. Epigenetic modifications are described as marks or switches that turn genes on or off (Nerlich et al. 2020); metaphors borrowed from cooking or music portray an opposition between a fixed genetic foundation (likened to a recipe or a musical score) and the epigenome that drives the biological machinery (the cook or the performer) (Dubois et al. 2019). In the 1990s, the presentation of the genome as a "book of life" and of genes as a series of instructions to be followed was decisive in the advent of the "DNA mystique" in the cultural and media space (Nelkin and Lindee 1995) and, more broadly, in the multiplication of social uses of genetics (Nerlich et al. 2002). Thirty years later, does the press offer an interpretative repertoire specific to epigenetics that provides the general public with keys to understanding and acting?

Lastly, we explore the reasons behind the press' enthusiasm for epigenetics. What messages do newspapers consider relevant, and how are they likely to appeal to readers? Previous research has shown that the international media send out contrasting messages. Some media content emphasizes on the individual "empowerment" that epigenetics gives access to, and the promise of (re)taking control of our health by adopting a lifestyle capable of favorably reprogramming our epigenome. This scope of expectations regarding epigenetics is also very present on the Internet (Dubois et al. 2019) and digital social networks (Nerlich et al. 2020). However, other media content emphasizes the "epigenetic damage" resulting from a poor diet, smoking or stress, and which is passed down from one generation to the next (Dyke et al. 2020). Press narratives often combine fatality (for those who suffer from such damage), with the duty of prevention and a sense of responsibility towards our descendants, more often formulated with mothers in mind (Lappé 2016). Are these ambivalent messages also found in the French-speaking press?

To answer these questions, we have built up and analyzed an original corpus of French-speaking press articles published between 2017 and 2021. After presenting our data and methodology, we provide an overview of the corpus: the press media; the words and categories used. We then detail four axes of analysis: the relationship between genetics and epigenetics; epigenetics and health; consumerist-type epigenetics; and lastly, epigenetics and environmental exposure. We conclude by discussing the reasons for the appeal of epigenetics in the French-speaking press.

12.2. Data and methodology

For the purposes of analysis, an original database dedicated to the public circulation of epigenetics has been produced. This database focuses on what has become known as everyday epigenetics (Dubois et al. 2019). Data extraction was produced from the Europresse© database, a full-text database of national, regional and international press articles, covering a five-year period: from January 2017 to December 2021. As our previous study focused on the English-speaking world, we chose to concentrate on the French-speaking press, to determine the extent to which it was possible to differentiate these areas of public circulation.

The extraction was performed using the following query:

> TEXT = epigen* & (meditat* | coach* | beauty | recipe | cosmetics | diet | relaxation | yoga | “wellness” | fertility | nutrition | shampoo | food | fitness | optimis* | sport* | lifestyle).

The raw extraction of press articles from this query resulted in an initial set of 548 distinct references. This initial raw corpus was then filtered and coded. The purpose of the filtering was to remove duplicates, as well as to ensure that the articles identified were not simply repetitions of academic press releases, or articles of an essentially academic nature. The aim was to retain only extra-academic production. This filtering process led to the removal of around 150 references and the consolidation of a base of around 390 references, in other words, an average of six to seven articles per month for the period under consideration, with an overall volume of around half a million words (456,836). This database was then standardized, coded and imported into the Iramuteq© text analysis tool. Each reference was identified on the basis of a set of six variables: source (press title), target audience (general or professional), category of epigenetic use, type of medium (press, blog, website), mode of intervention in the public space (informational or commercial) and year. While it seems relatively easy to qualify the 390 references in the database in terms of variables such as the media, modes of intervention or target audience, the qualification of the category of use required collective standardization work. This work resulted in the identification of a limited number of dominant categories linked to the epigenetics of everyday life. Only keeping the categories that represent more than 5% of our corpus, we observe six main categories in descending order: health (22% of our corpus), nutrition (15%), procreation (9%), age and aging (8%), well-being (7%) and beauty (7%). These six categories alone account for almost 70% of the references in our corpus, while the remaining 30%, below the 5% reference threshold, are unevenly distributed across 10 secondary categories.

12.3. Epigenetics press

A scan of the French-speaking press titles in the database (French, Quebec, Swiss and Belgian) reveals that the circulation of epigenetics is ensured by a wide variety of journalistic formats. Indeed, epigenetics appears in numerous general-interest press titles (magazines, dailies, weeklies), with national coverage (*Le Monde*, *Le Figaro*, *Le Nouvel Observateur*) as well as regional or local coverage (*Ouest France*, *24 heures*), and in the specialist press, the latter mainly, and unsurprisingly, in science news magazines (*Sciences et Avenir*, *La Recherche*, *Québec Science*) or those focusing on health issues (*Dr Good!*, *Santé+*).

Women's magazines (*Madame Figaro*, *Biba*) also relay epigenetic information, and more unexpectedly, business and financial newspapers (*Les Echos*, *La Tribune*) or consumer information (*Le Particulier*, *Protégez-vous*). Lastly, the trade press (industrial, financial, insurance, sports) is also taking part, albeit to a lesser extent, with the exception of the medical trade press (*Le Quotidien du médecin*), in the dissemination of public discourse on epigenetics.

While in 2017, interest in epigenetics was mainly driven by science magazines (*La Recherche*, *Sciences et avenir*) and a few generalist press titles with national coverage (*Le Figaro*, *Le Matin*, etc.) from 2018 onwards, new press outlets, both national (*Le Monde*, *Les Echos*, etc.) and local (*Ouest France*, *La Provence*, etc.), began to address the topic, driving up occurrences (55 articles tackling epigenetics in 2017, 99 in 2018). As for magazines targeting specific audiences (women's magazines, economic and financial news, health), it was from 2019 onwards that they multiplied the number of articles referring to epigenetics, with a more marked presence of women's press in 2021.

While epigenetic discoveries are covered by all media, it is to be expected that, depending on their thematic focus or even political inclination, as well as their target audience, the media will differ in their treatment of the information. Indeed, science news magazines and major general-interest daily newspapers with national coverage echo advancements in the field, through articles focused on work carried out in research laboratories, and frequently giving scientists a voice. However, daily newspapers tend to link these scientific advances to social issues such as obesity ("Obesity. Poverty, an aggravating factor", *Le Monde*, 2017), physical health prevention ("Young French people move less than their elders, a major future health problem", *Le Figaro*, 2020), as well as mental health issues such as suicide prevention ("When science opens the door to suicide prevention", *Le Devoir*, 2017) or depression ("We are not all equal in this respect", *Le Figaro*, 2017). While the local general press also relays epigenetic information, the framing is sometimes more educational, pointing out the "right" behaviors to adopt ("Pregnancy: a healthy lifestyle", *La République des Pyrénées*, 2021) to prevent the health disorders that can be induced by adopting or maintaining harmful lifestyle habits (diet, smoking, stress, lack of activity, etc.) ("Do you know the real benefits of olive oil?", *Ouest France*, 2021; "Sperm quality can be rapidly improved by changing one's diet", *Corse Matin*, 2020).

As for the opinion or political press, it tends to use epigenetic research to reinforce pre-existing positions or orientations. For example, the conservative magazine *Valeurs actuelles* uses epigenetics to warn about the risks to the health of children born through medically assisted reproduction or surrogate motherhood, arising from the use and development of these medical procreation techniques ("When PMA becomes a public health issue", *Valeurs actuelles*, 2019). In contrast, the Christian newspaper *La Vie*, which claims a "humanist philosophy", relies on epigenetics to highlight the transformative and restorative potential ("The children of our enemies. It is possible to repair children's brains!", *La Vie*, 2020) opened up by this "revolutionary medicine" (*La Vie*, 2019).

More generally, magazines dedicated to a targeted readership approach advances in epigenetics through the themes they promote. Women's magazines, for example, are developing an approach to epigenetics that focuses on maintaining the beauty and youthfulness of the skin ("Under the mask, ten years younger" *Madame Figaro*, 2021), as well as of the body and its vitality ("Scientists' recipe for looking 3 years younger in 8 weeks" *Pleine Vie*, 2021), through the adoption of the "right" gestures, which often involves purchasing the "right" products ("Can a serum change your beauty routine?" *Gaël*, 2020). In the same vein, health magazines emphasize risk prevention and good habits to adopt in daily life in order to preserve health and stay in shape (through sport, meditation, yoga and, of course, a healthy, well-considered diet). We can see that these articles have multiplied from 2020 onwards, with a notable diversification in 2021 towards beauty or anti-aging products, or even "wellness" services, revealing the rise of a cosmetics and self-care market backed by advances in epigenetics.

12.4. Words and categories

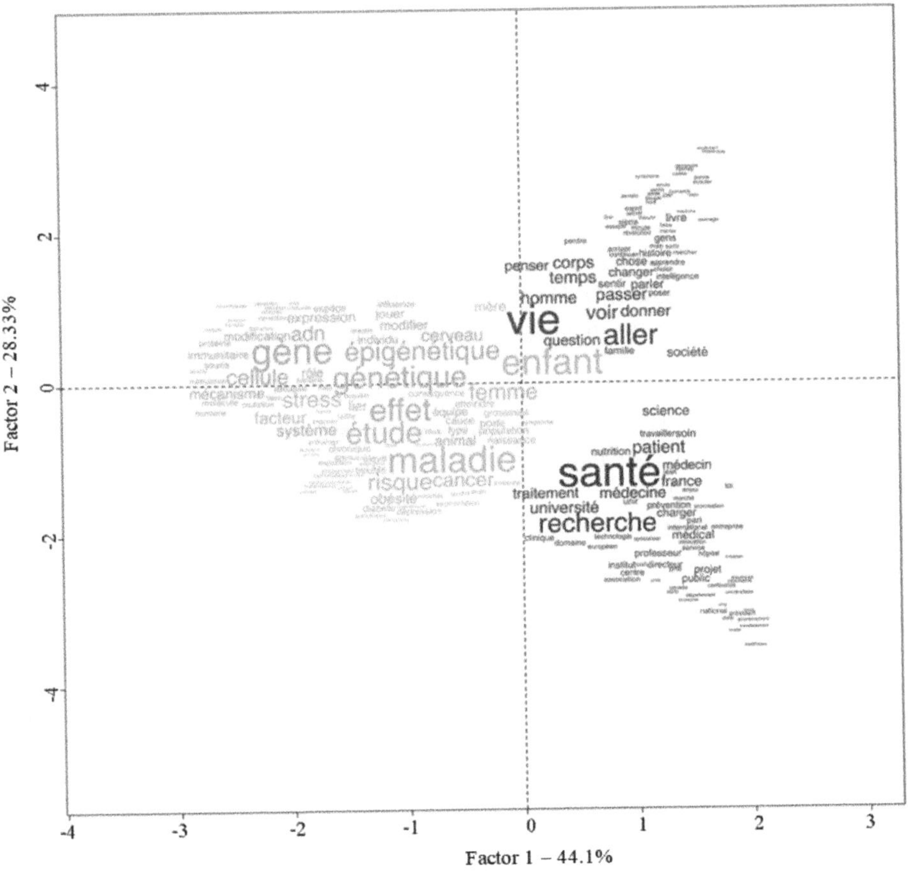

Figure 12.1. *Correspondence factor analysis (CFA) of classes produced by descending hierarchical classification (DHC). Character size is proportional to frequency per class – horizontal axis, factor 1: 44.1% – vertical axis, factor 2: 28.33%. For a color version of this figure, see www.iste.co.uk/grunau/epigenetics.zip*

The Iramuteq© text analysis tool (www.iramuteq.org) is used to produce an initial overall description of our corpus. Computer-assisted analysis of textual data generally involves multidimensional statistical analysis of texts, with frequency and specificity measures, factorial correspondence analysis and automatic classification. Here, we focus on the main results of this process.

As a first general observation, the correspondence analysis table in Figure 12.1 shows that our corpus groups together four distinct lexical classes, in unequal proportions. For the sake of readability, we have only included the first 50 terms in each class, with a visualization scale proportional to the frequency of use.

The first class (Figure 12.1b) represents over a third (33.79%) of our corpus. It highlights the strong link between the public circulation of epigenetics and the personal development sector and its various markets. The aim here is to define the conditions for a "healthy lifestyle", so as to identify the conditions for "well-being", for "harmony between body and mind", for our ability to act to "change our daily lives", as well as to identify the variety of services (advice, coaching, etc.) and resources (books) available to help us achieve these goals. Here is just one excerpt from this first class: "[...] epigeneva draws its DNA from the epigenetic lifestyle, the marriage between beauty, well-being and wild nature. The virtues of the planet have been studied and extracted to offer the best to the bodies and minds of women around the world" (*Le Figaro*, 2020). The second class (Figure 12.1c) represents 17.51% of our corpus. Here, the dominant repertoire is primarily medical: the aim is to associate advances in epigenetics with the understanding, prevention and sometimes even treatment of a certain number of "risks" of "diseases" or "disorders". Obesity, diabetes, cancer, cardiovascular pathologies, stress and hypertension, depression and anxiety are frequently cited in articles in the French-speaking press. An excerpt among others from this second class: "[...] our way of living can change the expression of 85 of our genes, as in the case of the gene responsible for breast cancer, which a woman can pass on to her daughter" (*TV Mag*, 2018). The third class (Figure 12.1a) represents 26.39% of our corpus. This class concentrates the bulk of the scientific and technological terminological repertoire. Press articles use this lexical class to identify concepts specific to epigenetics – "methylation", "histone", "chromatin", etc. – and the way in which they more or less profoundly renew the understanding and use of more established concepts, notably the gene, DNA, the link between gene and environment and, more broadly, genetics and genomics. An extract among others from this third class: "[...] both at the level of DNA methylation and chromatin modifications throughout our body, epigenetic marks reflect our interaction with the environment [...]" (*Science et Vie*, 2020). Lastly, the fourth and last class, representing 22.31% of our corpus (Figure 12.1d), corresponds to the organizational and professional repertoire coupled with the scientific and medical advances of everyday epigenetics. These include "centers", "associations", "institutes" or "departments", "conferences" or "commissions". This professional and organizational aspect is developing both in the public sector, with the activities of doctors and scientists in universities and hospitals, for example, and in the private sector, with a "range" of training courses designed to enhance employee "performance". We have just one excerpt from this fourth class: "[...] we realize how far we still have to go [...] we need to develop occupational health medicine by demonstrating the interdependence between physiological balance, employee work capacity and company profitability" (*Les Echos*, 2018).

Beyond this first general characterization of the lexical classes, Table 12.1 shows the terminological content of the six main categories coded in the corpus: health, nutrition, procreation, aging, well-being and beauty. A specificity score per term has been calculated for all categories, and for each of the 6 main categories, the 10 most specific words were retained. Table 12.1 presents these 60 terms in alphabetical order, with their specificity scores by category. The higher the score, the more specific the term is to the category in question. To make the table easier to read, a color gradient draws attention to the most contrasting cells. The darker the cell, the more specific the word is to the observed category.

Without needing to go into a detailed reading of the table, these specificity measurements provide a useful addition to the description of the four main lexical classes. Each column can be read as a list of keywords whose relative importance evolves according to the thematic focus of our press corpus. Articles focusing on the link between epigenetics and health, for example, will differ from others by placing emphasis on the diagnosis, treatment and prevention of cancer, particularly breast cancer. Articles focusing on the link between epigenetics and well-being will differ from others through the emphasis they place on meditation, the search for happiness and confidence, and the identification of stressful situations. As a final example, the link between epigenetics and beauty can be seen in the various terms used by the pharmaceutical industry in its communications. In recent years, epigenetics has become an additional marketing argument used by certain brands to sell an ever-expanding range of skincare products and creams designed to "turn back the hands of time". It is interesting to note that there are some rare overlaps between categories, in particular between "aging" and "beauty", which come together to varying degrees around certain terms, such as "anti-ageing".

	Santé [Health]	Nutrition [Nutrition]	Procréation [Procreation]	Vieillissement [Aging]	Bien-être [Well-being]	Beauté [Beauty]
aliment [food]	-3.18	55.65	-4.53	-1.16	-2.83	-3.57
anti âge [anti-aging]	-4.86	-3.28	-2.18	14.16	-1.25	14.83
bonheur [happiness]	-3.33	-1.38	-0.42	-1.04	23.52	0.39
calmer [calming]	-2.19	-2.58	-0.58	29.40	-0.98	-0.91
cancer [cancer]	58.34	1.21	-10.86	-5.70	-10.86	-11.47
centenaire [centenarian]	-1.37	0.24	-1.55	20.80	-0.89	-0.82
cheveu [hair]	-3.91	-2.41	-1.32	-0.30	2.53	29.79
chronologique [chronological]	-2.46	-1.32	-0.88	20.18	-0.50	-0.14
confiance [confidence]	-0.74	-1.18	-0.88	-1.23	16.49	0.31
conscience [consciousness]	-3.67	0.28	-1.65	1.09	18.33	-0.45
cosmétique [cosmetics]	-5.32	-4.66	-3.10	2.47	-1.06	47.95
couple [couple]	-1.63	-2.20	33.40	-1.58	-1.51	-1.40
crème [cream]	-6.84	-0.42	-2.02	0.87	-1.68	48.29
diagnostic [diagnosis]	26.28	-5.85	1.53	-3.28	-2.71	-0.48
enfant [child]	-3.07	-0.42	56.72	-22.33	-1.21	-19.77
espérance [hope]	-0.56	-1.78	-1.88	21.01	-0.91	-1.47
femme [female]	-6.91	-3.98	98.27	-4.53	-5.70	0.55
fertilité [fertility]	-2.04	-1.70	31.88	-1.98	-1.80	-1.67
gamme [range]	-2.51	0.66	-3.18	0.51	-1.10	37.94
gpa [surrogacy]	-4.20	-2.90	42.70	-1.76	-1.10	-1.03
gras [fat]	-2.52	59.71	-2.01	-1.65	-4.44	-0.38
huile [oil]	-8.82	37.19	-4.52	1.15	-0.32	-0.26
immortalité [immortality]	-1.76	-0.95	-0.63	16.10	-0.36	-0.33
immunitaire [immune]	15.02	5.72	-5.92	-3.91	-0.45	-2.40
longévité [longevity]	0.41	-0.32	-3.97	26.82	-0.22	-2.12
maladie [illness]	26.00	-0.81	-2.54	1.02	-3.59	-8.68
marque [brand]	-4.66	-4.48	0.66	-1.16	-0.43	39.81
méditation [meditation]	-5.78	-5.37	-5.73	1.39	69.42	-2.14
mère [mother]	-5.50	-2.92	40.47	-3.41	-0.98	-5.80
nutrition [nutrition]	-3.88	41.37	-2.13	-1.45	-1.88	-2.13
obésité [obesity]	-0.47	44.22	0.72	-10.65	-6.31	-7.05
ocytocine [oxytocin]	-1.68	-1.39	-0.41	-0.84	15.58	-0.49
olive [olive]	-5.74	34.05	-2.05	-1.88	-1.18	-0.26
patient [patient]	46.11	-1.31	-3.37	-5.14	0.31	-2.26
peau [skin]	-18.55	-12.61	-5.88	8.69	1.26	136.73
pma [assisted reproduction]	-4.25	-4.92	64.52	-2.99	-1.87	-1.74
positif [positive]	-0.40	-3.46	-1.66	-0.81	20.66	-0.25
prévention [prevention]	14.87	1.24	-5.18	0.27	-0.78	-0.94
psychanalyste [psychoanalyst]	-4.75	-3.21	43.52	-1.96	-1.22	-1.14
retraite [retirement]	-1.98	-1.24	-1.34	12.32	0.61	-0.71
sein [breast]	12.53	-0.87	-1.02	-4.23	0.47	-1.57
sérum [serum]	-7.26	-2.88	-2.59	2.28	-1.49	51.12
soi [self]	-0.59	-0.58	-2.91	-0.51	14.16	1.34
soin [care]	-0.44	-5.22	-3.85	1.46	3.44	53.59
sperme [sperm]	-2.51	-4.41	44.38	-2.69	-1.68	-1.56
stress [stress]	-0.74	-1.77	-3.25	-4.51	22.28	0.79
sucre [sugar]	-0.56	37.07	-2.35	-1.00	-1.96	-0.64
surpoids [overweight]	-0.66	34.53	-0.39	-3.87	-1.59	-2.25
traitement [treatment]	23.85	-2.38	-0.28	1.33	-0.73	-1.45
tumeur [tumor]	15.32	-2.99	-1.23	-2.46	-1.54	-1.43
tumoral [tumorous]	12.17	0.36	-1.34	-1.23	-0.77	-0.71
vieillir [aging]	-4.09	-4.35	-2.89	33.50	-0.31	-0.87
visage [face]	-7.26	-3.91	-2.59	6.31	-0.83	30.08
vitamine [vitamin]	-9.78	47.18	-2.98	0.72	-1.92	0.88

Table 12.1. *Lexical terms (on a color scale from white to green, the lowest negative values are the whitest and the highest positive values are dark green) by main categories (health, nutrition, reproduction, aging, well-being, beauty) and specificity scores*

The lexicometric approach adopted in this section enables us to broadly characterize our corpus. In-depth analysis, however, requires a more qualitative approach. The next section of this chapter develops this analysis around four main axes: the relationship between genetics and epigenetics; epigenetics and health; consumerist epigenetics; and epigenetics and environmental exposure.

12.5. A look at epigenetics in the French press

12.5.1. *The relationship between epigenetics and genetics*

Like the English-speaking press (Dubois et al. 2019), the French-speaking press tends to present a homogeneous vision of epigenetics, and to organize it in a rather strong opposition to genetics.

First of all, it presents epigenetics as having revolutionized our knowledge of living organisms. A "new, cutting-edge scientific discipline", epigenetics is said to challenge the belief that genetic sequences and their mutations entirely determine phenotypes. It would bring about a "real revolution in biology" that would challenge the power of genetics: "Genes do not do everything" (*Le Monde*, 2021); "We are not determined solely by our genes" (*Le Journal de Saône et Loire*, 2018); "We are not just genetic beings, each of us is the sum of what we have experienced, our epigenetics" (*Télérama*, 2020). In particular, genetic heredity is no longer inevitable (*La Presse*, 2018).

However, the press does not portray a shift from "all genetics" to "all epigenetics". It defines epigenetics as "a new layer added on top of the mechanisms of traditional genetics" (*Le Monde*, 2021) and as the study of mechanisms "that enable genes to be influenced without being modified" (*Le Progrès*, 2018). While the notion of genetic heritage remains important (the expression recurs 60 times in the corpus), this heritage is presented as being "modulated" by epigenetics and expressed differently due to epigenetic factors. Some articles give a digital estimate of the role of genetic inheritance, attributing three times less importance to it than to epigenetics: 30% for gene activity in general (*La Vie*, 2019) and life expectancy (*Sciences et Avenir*, 2020); 20 to 30% for health (*Le Figaro*, 2020). Headlines in the French-speaking press use the same metaphors as those in the English-speaking press to describe how epigenetic mechanisms "activate" or "express" the genome. The musical metaphor has been particularly popular following the publication of J. De Rosnay's book (*De Rosnay* 2018): "If the genome is a musical instrument, epigenetics will make it possible to add personal nuances" (*Le Figaro*, 2018). However, the epigenome is a director (*Le Courrier international*, 2021) or a cook (*Québec Science*, 2018). It is also compared to the switch that turns certain genes on or off (*L'Obs*, 2017), or to the reading code used to read the DNA book (*Le Devoir*, 2021). Metaphors of the genetic program (14 occurrences) or the genetic code (23 mentions) are still present in the press. Epigenetics is also compared to software (*Le Temps*, 2018) or to applications that provide the necessary instructions for its execution: "On the one hand, you have a fixed program that makes [the computer] work, and on top of that, apps that allow you to modulate it" (*Sciences et Avenir*, 2018).

Lastly, epigenetics is seen as "a science that studies the influence of the environment on the expression of our genes" (*Sciences et Avenir*, 2021) or "the science of interactions between genes and the environment" (*Le Figaro*, 2020). While epigenetics "puts the environment in the spotlight" (*La Recherche*, 2017) – where genetics tends to forget its role – this environment is often poorly defined (as including lifestyle habits, stress factors, diet, etc.) or not defined at all, with a simple mention of the environment "in the broadest sense". What matters is the extent to which these environments are harmful or beneficial via gene activation.

12.5.2. *Epigenetics and health*

The themes of health and disease are very present. Epigenetics is put forward as a determining factor in the good health of adults and the development of children, and as a key to understanding the onset and progression of a long list of chronic non-transmissible diseases, including cancers, cardiovascular diseases, obesity, type 2 diabetes, depression, anorexia, Parkinson's disease and Alzheimer's disease (cited in order of importance in the corpus). Thus, the press popularizes the idea of "epigenetic diseases": "Today we know that a large proportion of diseases are caused by epigenetics" (*Science et Vie*, 2020), sometimes contrasting them with diseases of purely genetic origin: "Just because you have the 'wrong' gene (except in certain cases) does not mean that a pathology will necessarily occur" (*Science et Vie*, 2021). The idea of epigenetic causes or factors has two meanings, which press articles use separately or together (Auroy and Louvel 2022). On the one hand, it refers to the epigenetic alterations involved in the cellular dysfunctions that accompany disease, such as the transformation of cancer cells or the formation of adipose tissue in obese individuals (*Le Figaro*, 2019). On the other hand, diseases are said to be of epigenetic origin when environmental factors disrupt the functioning of biological systems via epigenetic mechanisms, which disrupt, for example, the immune system of children born by caesarean section (*Courrier international*, 2020) or adults subjected to intense stress (*Le Quotidien du Médecin*, 2021) or the metabolic system of obese people (*Dr Good!*, 2021).

The press also presents epigenetics as contributing to medical progress in terms of prevention, screening, diagnosis and treatment. Here, the possibility of "erasing" epigenetic marks is highlighted: "We know that epigenetic marks, unlike genetic alterations of DNA, are potentially reversible. This makes them interesting therapeutic targets" (*Le Temps*, 2018). Epigenetics is contributing to the rise of "personalized" medicine, a polysemous term that refers, in this context, to two definitions (Auroy and Louvel 2022). On the one hand, epigenetics extends the effort that began with genomics to develop so-called precision medicine, which relies on molecular information to establish a diagnosis, predict disease progression or patient response to treatment. The potential contribution of epigenetics to precision oncology is particularly emphasized: the development of epigenetic tests to detect the presence of tumor cells earlier and more reliably (*L'Avenir*, 2017; *Le Monde*, 2018) or to identify subcategories of patients whose epigenomic profile predicts better treatment efficacy; the development of individualized treatments described as epidrugs (*Le Temps*, 2018): "[epigenetics] will help clinicians to quickly prescribe the most optimal treatment and spare many patients the heavy side effects of unnecessary chemo" (*Sciences et Avenir*, 2018). On the other hand, while the massive benefits of epigenetics for precision medicine are not expected for another decade (*Québec Science*, 2019), epigenetics could more immediately contribute to the implementation of "person-centered" preventive medicine and "improved use of our personal resources to defend ourselves" (*Les Echos*, 2019). This "epigenetic medicine" involves "pampering our epigenome" (*Sciences et Avenir*, 2019) by following simple, science-based lifestyle rules – such as exercising, reducing meat consumption or limiting stress: "Thanks to our knowledge of epigenetics and the microbiota, we will know what type of disease we are at risk of having, and how to modify our lifestyle to avoid it" (*Sciences et Avenir*, 2021).

12.5.3. *Consumer epigenetics*

The development of everyday epigenetics coincides with the rise of new commercial sectors. This consumerist epigenetics is applied to an ever-increasing variety of products and services. Here are a few examples. Researchers in their laboratories are not always aware of it, but everyday epigenetics sometimes takes the form of spa cures:

> A treatment specifically built around the concept of epigenetics: that is what the Thalazur group, France's leading thalassotherapy company, is going to offer from the 2022 season in eight of its nine Thalasso establishments [...]. Epigenetics, a new cutting-edge scientific discipline, offers a new understanding of life and sheds light on the

> diseases of civilization by studying the impact of the environment and our lifestyles – diet, stress, physical activity, social relationships, pleasure of living, etc. – on the way our genetic heritage is expressed [...] (*Les Echos*, 2021).

Everyday epigenetics is also embodied in a variety of coaches, coaching methods and institutes that are dedicated to well-being, which claim to be aligned with the latest advances in research. This is the case, for example, of the creator of the Résilience Institute, who, in an interview, emphasized that "neuroscience and epigenetics have highlighted the importance of the four dimensions: mental, physical, emotional and spiritual. A person at the top of their game, who is physically fit, connected to their emotions, anchored in positive values and fully aware of themselves and their relationship to the world becomes a better version of themselves" (*Le Temps*, 2018). The cosmetics industry represents another branch of commercial development for epigenetics that is particularly present in the press. The proliferation of so-called "epigenetic" creams and serums seems to make epigenetics a new branch of cosmetics:

> [...] the new Sisley serum [...] acts at the 'epigenetic' level against spots. In other words, by preventing the external factors that cause pigmentation disorders. [...] Aging is not only linked to our genetic heritage. It is mainly induced by our habits and living conditions, which, according to some scientists, account for nearly 80% of the process. This field of study is called epigenetics (*Les Echos*, 2021).

Last but not least, epigenetics sometimes also refers to kitchen therapy, i.e. the idea that our quality of life is closely linked to our choice of food and its preparation: "[...] cooking should be about doing good. Intuitively, [we know] what epigenetics is rediscovering today: we are co-creators of our health" (*Madame Figaro*, 2019).

Spa cures, coaching methods, beauty products, cooking recipes, etc. – though this catalog of everyday epigenetics seems extremely heterogeneous, a few common guidelines do emerge. Two simple ideas in particular: on the one hand, epigenetics would be capable of "changing our lives" to a greater or lesser extent, and on the other, this change ultimately depends on our behavior alone. The press, for example, devotes considerable space to testimonials from wellness professionals wishing to share the transformation they themselves have experienced: "[...] seeing how meditation had changed my life, it not only opened my mind but also my heart, I wanted to share this with my patients and healthcare professionals" (*La Recherche*, 2019). These changes can be quite different in scale. In some cases, they represent a true "paradigm shift that gives citizens control over their physiological data" (*Dr Good!*, 2020). In other cases, the changes may be more limited and "a matter of common sense, but not so simple to maintain over time" (*Le Point*, 2021). The press also stresses the importance of a few high-profile popularizers who invite their readers to become aware of their ability to act on themselves: "The new revolution in biology, known as epigenetics, shows that your daily behavior – what you eat, the exercise you do, your resistance to stress, the lifestyle you adopt, etc. – will inhibit or activate some of your genes. You are like the conductor of a symphony, co-author of your life, your health, your balance" (De Rosnay 2018). Here, we touch on the central idea that any change ultimately rests on our ability to make the "right" choices. To illustrate, this discourse, which reduces everyday epigenetics to consumer choices, is very present among epigenetic nutrition experts interviewed in the press: "[...] to get the most out of olive oil, you need to know how to choose it properly" (*Acadie Nouvelle*, 2018); "[...] you do not go hunting for fat, you choose it better" (*Valeurs actuelles*, 2019); "[...] keeping you healthy by helping you choose the right foods, practice a sport that suits your morphology" (*La Tribune*, 2017).

12.5.4. *Epigenetics and environmental exposure*

One feature of epigenetics that is widely present in our press corpus is its ability to induce health disorders, not just to individuals living in harmful environments (pollution, food, tobacco, etc.), but also in their descendants, even though they are not directly exposed to these contaminating environments. More remarkably, the traumas experienced by "ancestors" exposed to particularly adverse situations (war, conflict, abuse, etc.) could also leave "traces in the genes" (*Le Particulier*, 2021) of their descendants and thus be passed on, like "invisible baggage" (*Courrier international*, 2021), from one generation to the next. The press thus conveys an understanding of this "transgenerational transmission" as a set of "marks" (*Le Monde*, 2018), "traces" (*Le Figaro*, 2018) or "scars" (*La Vie*, 2019) "inscribed in the genetic heritage" and "inherited" (*Les Echos*, 2018) by the individual at a prenatal stage through biological mechanisms that activate an "epigenetic memory" (*Le Temps*, 2020) of the cells.

While it relays "increasingly established" scientific advances, drawing a link between lifestyles (smoking, nutrition, lack of physical activity or sleep, etc.) or the harmfulness of certain conditions of existence (polluted, stressful or violent environments, etc.) (*Santé+*, 2018), and the development of a range of pathologies (cancers, obesity, hypertension, psychological disorders, etc.) over the course of a lifetime, it carries a vision of this "fetal programming" (*Le courrier de l'Est*, 2017) as an inevitable and automatic inscription in the body of a developing being of the environmental experience of its parents, especially the mother. Contaminants "enter" the body (*Le Devoir*, 2021), "cross the placenta barrier" (*20 minutes*, 2019), to leave their biological "imprint" (*Le Figaro*, 2018), with consequences that are deferred in time: "If I am pregnant, endocrine disruptors will enter me, through what I eat, what I touch, maybe what I breathe, and this will directly affect the environment of the fetus" (*Le Devoir*, 2021).

The mother's womb, often referred to by the biological term "uterus", and "what goes on there" (*Le Droit*, 2019), is the focus of attention because: "What a pregnant woman puts in her mouth, the air she breathes and even her emotions influence the development of her future baby, and the health of the adult he or she will become. So much so that obesity, diabetes or cardiovascular diseases can take root... in the womb" (*Le Droit*, 2019).

However, the role of fathers in transmitting future health problems also exists, albeit much less so when compared to the treatment given to the womb. The press thus relays the "latest" scientific findings, which point, for example, to the role of overweight (*Le Droit*, 2019) or paternal stress (*Québec Science*, 2018) in the onset of subsequent disorders in their offspring.

However, a second feature of epigenetics enables the press to develop a very different, positive message. Epigenetic marks can be erased thanks to one of the epigenome's remarkable properties: its flexibility. By adopting beneficial lifestyles, individuals are able to act on their health and "restore gene expression in the right direction" (*Métro*, 2019).

The main message conveyed by the press focuses on the theme of prevention, in which information plays a central role. If future mothers are aware of the harmful nature of certain lifestyles or environments for their child's future (*Le Devoir*, 2021), they will be able to act on these environments, for example by adopting new behaviors (diet, stopping smoking, stress management, etc.). The same applies to fathers, for whom we need to step up our educational efforts: "While women are generally informed and avoid exposure to various substances during pregnancy, this is not always the case for men, who can also transmit 'environmental' information during reproduction. The spermatozoon not only transmits its DNA but also its epigenome, which can be affected by various factors" (*La Recherche*, 2019).

Beyond prevention, restoring balance by acting on the epigenome means interrupting the cycle of transmission between generations (*Courrier international*, 2021), which is a theme particularly relevant for inherited trauma (Dubois and Guaspare 2020).

While the press emphasizes individual responsibility in health management, it also highlights the powerlessness of individuals in the face of certain exposures, particularly to all kinds of toxic products or pollutants, for which answers can only be provided on a societal scale (*La correspondance économique*, 2018). In a similar vein, social and economic adversity translates into health inequalities, which should also lead to the adoption of health policies that are directed primarily at women: "The transmission of a genetic inheritance is in the very nature of things. However, this metabolic inheritance is just as heavy to bear, and unfair to these adults in the making. A massive challenge awaits health systems" (*Le Quotidien du Médecin*, 2019).

12.6. Discussion: the appeal and visibility of epigenetics

How can we explain the social success of everyday epigenetics? While there are no surveys to provide a definitive answer to this question, it can be hypothesized that a significant part of this success stems from the way in which its public circulation is frequently associated with a discourse of empowerment. As illustrated by the numerous press excerpts quoted above, and this is true in both the French- and English-speaking worlds, epigenetics is publicly constructed as a research field offering the general public not only the opportunity to become aware that DNA is not an inescapable inevitability, but also to "take control" of their bodies and their health through a myriad of simple everyday consumer choices. It is hardly surprising that this promise of control and change should find a favorable following or attention. Who among us has not wanted, at one time or another, to believe in such a promise? The previous sections have shown how this promise of control is accompanied by the development of a wide range of products and services. New markets are developing in various fields – skincare, wellness, nutrition, fertility coaching, etc. – with a particular focus on pregnant women.

However, the importance of this consumerist, often naively optimistic conception of epigenetics, should not lead us to adopt an overly simplistic view of its public visibility. A recent survey, *Les Français et la science 2021* (Bauer et al. 2021), highlighted a paradox: while the public's interest in life sciences remains high and sustained, a majority of the respondents remain unaware of the latest advances in research. Compared to scientific and technological advances such as artificial intelligence, or biotechnologies such as GMOs, the public visibility of epigenetics remains low. The survey shows that just under 5 in 10 French people (46.4%) have "often", "a little" or "rarely" heard about epigenetics. By comparison, 8 out of 10 French people have heard of artificial intelligence and GMOs. The survey also shows that the visibility of epigenetics varies significantly according to classic socio-demographic factors, such as educational level or age. Unsurprisingly, the population with a level of education higher than bac + 3 is over-represented by almost 15 percentage points compared to the general average of the population claiming to be informed. Similarly, the youngest segment of the population, the 18–24 year-olds, is over-represented by four points compared to the general average. In short, it would be difficult to avoid reflecting on the relation between the intrinsic quality of public discourse on epigenetics and the general conditions of its exposure and reception.

12.7. Conclusion

This chapter presents, for the first time, a comprehensive study of the public circulation of epigenetics in the French-speaking world. It usefully complements the few works available in the English-speaking world, showing that the French-speaking press has been able, on its own scale, to appropriate the latest advances in the life sciences. In addition to describing the data collected and the press media present in our corpus, the analysis highlighted the way in which the press more or less artificially constructs the opposition between epigenetics and genetics. It also emphasizes the fact that consumerist form of epigenetics, which is very present in the public space, all too often promotes a naively reductionist vision of health disparities. It is as if the overexposure of individual consumer choices – epigenetics as a principle of empowerment – renders the more structural aspects over which individuals often have no control invisible. In the section devoted to environmental exposure, it was particularly stressed that while the discourse on collective responsibility and the importance of public policies is not totally absent, it still remains marginal at best.

This study has a number of limitations that should be highlighted. On the one hand, the period considered (2017–2021) is too short to identify any major changing trends. It will be useful to repeat this analysis periodically, so as to determine the extent to which the uses and terminological repertoires identified are changing over time, or not. On the other hand, our corpus has a unique source, the Europresse© database. This database is undeniably useful, but it shadows most of the public circulation of epigenetics that are online, such as in notebooks, blogs, forums, social networks or websites. In short, it is an extremely rich space of purely digital circulation that deserves a study in its own right. It should be noted, however, that the press framings identified are not unlike those highlighted by an earlier study, based on an English corpus mixing press and digital media (Dubois et al. 2019).

Despite these limitations, our analysis, like any future exploitation of our database, can be useful: it provides insights to the scientific community about the state of collective representations, and can, to some extent, help it to develop and target its awareness-raising and communication activities. After all, there is no reason why we cannot learn from the past, and in particular from the history of genetics and genomics, and thus avoid fueling exaggerated social expectations that are often destined to turn into future frustrations and disappointments. Such work would undeniably provide an opportunity to develop original multidisciplinary collaborations between social sciences and life sciences.

12.8. References

Auroy, L. and Louvel, S. (2022). Épigénétique et cancérologie. Deux visages de la personnalisation de la médecine. *Médecine/Sciences*, 38(3), 296–302.

Bauer, M., Dubois, M., Hervois, P. (2021). Les Français et la science 2021 : représentations sociales de la science 1972–2020. Research Report, Université de Lorraine, Nancy.

Carey, N. (2012). *The Epigenetics Revolution: How Modern Biology is Rewriting our Understanding of Genetics, Disease, and Inheritance*. Columbia University Press, New York.

Cloud, J. (2010). Why your DNA isn't your destiny. *Time Magazine*.

De Rosnay, J. (2018). *La symphonie du vivant. Comment l'épigénétique va changer votre vie*. Éditions Les Liens Qui Libèrent, Paris.

Dubois, M. and Guaspare, C. (2020). From cellular memory to the memory of trauma: Social epigenetics and its public circulation. *Social Science Information*, 59(1), 144–183.

Dubois, M., Louvel, S., Le Goff, A., Guaspare, C., Allard, P. (2019). Epigenetics in the public sphere: Interdisciplinary perspectives. *Environmental Epigenetics*, 5(4), dvz019.

Dyke, S.O., Ennis, C.A., Joly, Y., Walter, J., Siebert, R., Pastinen, T. (2020). Communicating science: Epigenetics in the spotlight. *Environmental Epigenetics*, 6(1), dvaa015.

Feinberg, A. (2004). Interview with feinberg andrew. Examining epigenetics. *NPR* [Online]. Available at: https://www.npr.org/templates/story/story.php?storyId=3613288 [Accessed 29 March 2022].

Giacobino, A. (2018). *Peut-on se libérer de ses gènes ? L'épigénétique*. Stock, Paris.

Holt, S. and Paterson, N. (2007). Ghost in your genes. *Nova PBS* [Online]. Available at: pbs.org.

Lappé, M. (2016). Epigenetics, media coverage, and parent responsibilities in the post-genomic era. *Current Genetic Medicine Reports*, 4(3), 92–97.

Lewino, F. (2018). Médecine : et si l'épigénétique changeait votre existence. *Le Point*.

Martin, N. (2017). Epigénétique : quand le pouvoir des gènes atteint ses limites. *France Culture* [Online]. Available at: https://www.radiofrance.fr/franceculture/podcasts/la-methode-scientifique/epigenetique-quand-le-pouvoir-des-genes-atteint-ses-limites-5711225.

Nelkin, D. and Lindee, M.S. (1995). *The DNA Mystique: The Gene as a Cultural Icon*. Freeman, New York.

Nerlich, B., Dingwall, R., Clarke, D.D. (2002). The book of life: How the completion of the Human genome project was revealed to the public. *Health*, 6(4), 445–469.

Nerlich, B., Stelmach, A., Ennis, C. (2020). How to do things with epigenetics: An investigation into the use of metaphors to promote alternative approaches to health and social science, and their implications for interdisciplinary collaboration. *Social Science Information*, 59(1), 59–92.

Rebeihi, A. (2018). Qu'est-ce que l'épigénétique ? Grand bien vous fasse ! *France Inter* [Online]. Available at: https://www.radiofrance.fr/franceinter/podcasts/grand-bien-vous-fasse.

Robison, S.K. (2016). The political implications of epigenetics: Emerging narratives and ideologies. *Politics and the Life Sciences*, 35(2), 30–53.

Seitz, S. and Schuol, S. (2017). State of the public discourse on epigenetics. In *Epigenetics, Ethical, Legal and Social Aspects*, Heil, R., Seitz, S.B., Koënig, H., Robienski, J. (eds). Springer, Berlin, 109–124.

Serfaty, L. (2015). Les nouveaux secrets de notre hérédité. *Arte France*.

Stelmach, A. and Nerlich, B. (2015). Metaphors in search of a target: The curious case of epigenetics. *New Genetics and Society*, 34(2), 196–218.

Waggoner, M. and Uller, T. (2015). Epigenetic determinism in science and society. *New Genetics and Society*, 34(2), 177–195.

Zimmer, C. (2018). The famine ended 70 years ago, but dutch genes still bear scars. *The New York Times*.

List of Authors

Frédéric BANTIGNIES
Institut de Génétique Humaine, CNRS
Université de Montpellier
France

Natacha BIES ETHEVE
LGDP
Université de Perpignan Via Domitia
France

Frédéric BRUNET
IGFL
ENS Lyon
France

Francesco CALZAFERRI
IBMM, CNRS
Montpellier
France

Raphaëlle CHAIX
Éco-anthropologie (UMR 7206)
Muséum National d'Histoire Naturelle
CNRS
Université Paris Cité
France

Séverine CHAMBEYRON
Institut de Génétique Humaine, CNRS
Université de Montpellier
France

Vincent COUSTHAM
INRAE, NuMeA
Université de Pau et des Pays de l'Adour
Saint-Pée-sur-Nivelle
and
IJPB, INRAE
Université Paris-Saclay
AgroParisTech
Versailles
France

Michel DUBOIS
GEMASS, CNRS
Sorbonne Université
Paris
France

Patricia GIBERT
LBBE, CNRS
Université Claude Bernard Lyon 1
Villeurbanne
France

Christoph GRUNAU
IHPE
Université de Perpignan Via Domitia
France

Catherine GUASPARE
GEMASS, CNRS
Sorbonne Université
Paris
France

Julie LECLERCQ
AGAP Institut
CIRAD
INRAE
Institut Agro Montpellier
Université de Montpellier
France

Marie LOPEZ
IBMM, CNRS
Montpellier
France

Séverine LOUVEL
Pacte, CNRS
Sciences Po Grenoble
Université Grenoble Alpes
France

Nelia LUVIANO
IHPE
Université de Perpignan Via Domitia
France

Stéphane MAURY
P2E
Université d'Orléans
France

Nicolas NÈGRE
DGIMI
Université de Montpellier
France

Frédérique PITEL
GenPhySE
INRAE
Castanet-Tolosan
France

Christophe PLOMION
INRAE, Biogeco
Université de Bordeaux
Cestas
France

Arnaud POCHEVILLE
CRBE
Université de Toulouse
France

Dominique THIS
AGAP Institut
CIRAD
INRAE
Institut Agro Montpellier
Université de Montpellier
France

Patrice THIS
AGAP Institut
CIRAD
INRAE
Institut Agro Montpellier
Université de Montpellier
and
UMT Géno-Vigne
IFV
Montpellier
France

Cristina VIEIRA
LBBE, CNRS
Université Claude Bernard Lyon 1
Villeurbanne
France

Clémentine VITTE
GQE – Le Moulon
Gif-sur-Yvette
France

Alexandra WEYRICH
German Centre for Integrative
Biodiversity Research
University of Leipzig
and
Leibniz Institute for Zoo and Wildlife
Research
Berlin
Germany

Index

For some entries that appear frequently in the book, only each chapter's first occurrence has been listed.

E, F

M, N

P, Q

Printed and bound by CPI Group (UK) Ltd, Croydon, CR0 4YY
13/09/2026
14961073-0004